Oxford Handbook of
Clinical Surgery

Oxford Handbook of Clinical Surgery

G. R. McLATCHIE FRCS

Consultant Surgeon
The General Hospital
Hartlepool

with the assistance of

S. PARAMESWARAN FRCS

Surgical Registrar

Oxford New York Tokyo
OXFORD UNIVERSITY PRESS
1990

Oxford University Press, Walton Street, Oxford OX2 6DP

Oxford New York Toronto
Delhi Bombay Calcutta Madras Karachi
Petaling Jaya Singapore Hong Kong Tokyo
Nairobi Dar es Salaam Cape Town
Melbourne Auckland

and associated companies in
Berlin Ibadan

Oxford is a trade mark of Oxford University Press

Published in the United States
by Oxford University Press, New York

British Library Cataloguing in Publication Data
McLatchie, Greg R.
Oxford handbook of clinical surgery.
1. Medicine. Surgery
I. Title
617
ISBN 0-19-261710-9

Library of Congress Cataloging in Publication Data
McLatchie, Greg R.
Oxford handbook of clinical surgery/G. R. McLatchie
with the assistance of S. Parameswaran.
1. Surgery—Handbooks, manuals, etc. 2. Surgery, Operative—
Handbooks, manuals, etc. I. Parameswaran, S. (Sandrasekeram)
II. Title.
[DNLM: 1. Surgery—handbooks. 2. Surgery, Operative—handbooks.
WO 39 M47850]
RD37.M45   1990   617-dc20   90-7051
ISBN 0-19-261710-9

Typeset by Joshua Associates Ltd, Oxford
Printed in Great Britain by
Courier International, Tiptree, Essex

Acknowledgements

We are grateful to colleagues and friends who have helped produce this book, the idea of which was first suggested by Mr Gordon MacBain, consultant surgeon at the Southern General Hospital, Glasgow. We have received considerable support from the staff of Oxford University Press, and are also indebted to Mr J. Rhind and Dr J. Daniel for their contributions and our surgical teachers especially Mr J. S. F. Hutchison, Mr M. K. Browne, Mr J. Neilson, Mr D. Young, Mr A. Young, and the late Mr I. McLennan whose practical advice and anecdotes pepper the pages; to Mr K. Davies and Mr P. Gardiner for illustrations and to our secretaries, Mrs Kate Wright and Mrs Vera Spaldin who prepared the manuscript.

1990

G. R. M.
S. P.

For Ross, Cameron, Ailidh,
Claire, Calum, and Lakshmi

Abbreviations

A&E	accident and emergency department
ACTH	adrenocorticotrophic hormone
APUD	amine precursor uptake decarboxylase (cells)
ATC	anterior tibial compartment
BCC	basal cell carcinoma
BSA	body surface area
BSE	breast self-examination
CBD	common bile duct
CE	clinical examination
CoHb	carboxyhaemoglobin
CPR	cardiopulmonary resuscitation
CSF	cerebrospinal fluid
CVP	central venous pressure
CVS	cardiovascular system
DSA	digital subtraction angiography
DVT	deep venous thrombosis
ECF	extracellular fluid
ECG	electrocardiogram
EMG	electromyography
ERCP	endoscopic retrograde cholangiopancreatography
ESR	erythrocyte sedimentation rate
FBC	full blood count
FNAB	fine needle aspiration biopsy
FOB	fecal occult blood
FSH	follicle stimulating hormone
GA	general anaesthetic
GFR	glomerular filtration rate
GI	gastrointestinal
GP	general practitioner
HCG	human chorionic gonadotrophin
HIV	human immunodeficiency virus
HLA	human leucocyte-associated (antigen)
HSV	highly selective vagotomy
ICF	intracellular fluid
IM	intramuscular
IV	intravenous
IVC	intravenous cholangiography
IVI	intravenous infusion
IVP	intravenous pyelogram
LA	local anaesthetic
LH	luteinizing hormone
LMM	lentigo maligna melanoma
MSSU	midstream specimen of urine
MTBE	methyl tertiary butyl ether
NM	nodular melanoma
OP	outpatient
PIF	prolactin release inhibiting factor
PTC	percutaneous transhepatic cholangiography
PTE	pulmonary thrombo-embolism
PTH	parathyroid hormone
PTT	partial thromboplastin time

PUO	pyrexia of unknown origin
RA	regional anaesthetic
RBC	red blood cell
SCC	squamous cell carcinoma
SOPD	surgical outpatients department
TB	tuberculosis
TENS	transcutaneous electrical stimulation
TPHA	*Treponema pallidum* haemagglutination assay
TPIT	*Treponema pallidum* immobilization test
TPN	total parenteral nutrition
TSH	thyroid stimulating hormone
U&E	urea and electrolytes
US	ultrasound
VIP	vasoactive inhibitory peptide
VMA	vallinylmandelic acid

Contents

1 The surgical patient

The 'no-lose' philosophy in surgery

Pascal's 'no-lose' philosophy relates to belief in God. If God does exist and we live our lives believing in Him then everything is gained when we die; if God does not exist then at least nothing is lost.

Clinical implications

The 'no-lose' philosophy is occasionally applied in practice but in a different context. When faced with a problem the outcome of which is doubtful, a doctor may err on the side of 'no-lose' in giving the prognosis to relatives. If the patient dies then the prediction was correct, if the patient lives then everything is won. In either event the doctor does not lose.

Dangers of 'no-lose'

Application of the philosophy to all areas of medical practice carries considerable risks especially when investigating patients (in order not to lose, the diagnosis must be made). This may be illustrated by a case report.

A patient recovers well after oversewing of a perforated duodenal ulcer, but on checking his blood an elevated serum calcium is noted. After further investigation for a possible parathyroid adenoma, selective venous cannulation of the neck veins was carried out. No tumour was demonstrated but the patient developed infective complications from which he died. At post-mortem no parathyroid tumour was found.

Advice for surgeons

- Prevent iatrogenic disease. Beware of invasive investigations which may be expensive, dangerous, or unnecessary. Is the risk of the procedure worth taking? Would you carry out the same investigation on yourself?
- How important is it to make the diagnosis? If it is academic do not cause your patients anxiety or discomfort, or expose them to risk.
- Avoid the attitude: 'At least this won't cause any harm even if it does no good' when prescribing drugs, treatment, or investigation. If it does no good, why do it at all?
- When discussing the prognosis with the patients or relatives be as honest as possible. Base your statements on your own experience, that of your seniors, and the relevant literature.

The 'no-lose' philosophy has considerable potential for loss. Further, it may adversely influence decision-making and prevent the solution of clinical and ethical problems.

Reference

Galbraith, S. (1978). The 'no-lose' philosophy in medicine. *Journal of Medical Ethics*, **4**, 61–3.

4 Reassurance

Reassurance is commonly referred to in medical and surgical practice. It is believed to relieve anxiety and to improve the patient's physical condition. It is good practice to give reassurance.

Implications of indiscriminate reassurance

'Normal' people as well as obsessional patients have intrusive thoughts from time to time which are disturbing. For example, the patient who presents with nausea may interpret the symptom as being due to gastric cancer instead of alcohol-induced gastritis due to his regular drinking. He needs reassurance that cancer is not the cause of his problem, and that the symptoms are due to excessive alcohol intake. The necessary investigations are carried out and are negative. The doctor then reassures the patient that all is well, that cancer plays no part. At first the patient is relieved but then sets about seeking further reassurance, 'Why did you do the investigations if you did not suspect cancer?' 'What if you have missed something?' 'I never thought I had cancer. Why did you mention it?' This response can lead to recurrent out-patient attendances, further investigations, further reassurance, and further anxiety. The vicious circle so set up has been described as the 'fear syndrome'.

Guidelines for dealing with 'fear syndrome'

- Identify why the patient has come for help. Listen and decide the role anxiety plays in the presentation.
- Remember that 'malingerers' get malignancies too. Carry out an unbiased thorough assessment. Further investigations may not be needed.
- Anxious patients make anxious doctors. Make your judgements independently, irrespective of how anxious the patient appears to be.
- Do not feed doubts by providing irrelevant information.
- Speak clearly so that the patient understands.
- Show awareness that anxiety and symptoms are part of any clinical problem.
- Give the patient a chance to ask questions after imparting information.
- Be sure that the patient has understood. The presence of a less anxious relative may be helpful in the SOPD.

Reference

Warwick, H. M. C. and Salkovskis, P. M. (1985). Reassurance. *British Medical Journal*, **290**, 1028.

Ethics and the surgeon

Duties of surgeons to patients

A surgeon must maintain the highest professional standards and practise his profession without a profit motive.
He must:
- be obligated to preserve human life
- be loyal to his patients
- summon a second opinion if a certain type of treatment is outwith his ability
- maintain confidentiality on his knowledge of patients
- give emergency care where indicated as a duty unless others are able and willing to administer such care

Duties of surgeons to one another

Surgeons must:
- behave respectfully and professionally towards colleagues
- not attempt to or succeed in enticing patients from surgical colleagues

Unethical practices

- Self-advertisement
- Collaboration in medical practice where clinical independence is not maintained
- Receiving monies, other than proper professional fees
- Acts or advice which could weaken the mental or physical status of a person and which could result in profit of some kind for the surgeon

Caveats

- Beware of new discoveries and techniques unless they are properly tried and tested.
- Give certification or testify only to that which you can personally verify.

Reference

General Medical Council (1989). Advice on standards of professional conduct and on medical ethics. In *Professional Conduct and Discipline: Fitness to Practice*, pp. 18–28. GMC, London.

Insure yourself against professional negligence and ensure that your subscription is up to date.

Common surgical reasons for allegations of negligence

Amputation of the wrong digit or limb/operating on the wrong side This is virtually indefensible and is due to carelessness in the patient/doctor relationship. If you are operating, speak to the patient on the pre-operative ward round and identify the side. Record it in the notes. Mark the side/digit with a waterproof marker pen yourself. Speak to the patient again in the anaesthetic room and repeat the process. *Do not permit the induction of anaesthesia until you are certain.*

Leaving swabs or instruments in the patient It is the total responsibility of the operating surgeon to ensure that nothing is left in the patient. Satisfy yourself that the swab count, etc. is correct. If in doubt X-ray the patient on the operating table.

Removing the wrong organ/removing a solitary organ (when there should be two!) When one of paired organs is diseased, ensure that it is the diseased organ which is removed. When there is a diseased or damaged organ of a pair ensure that its mirror image is *present* and *functional* (e.g. kidney). If the operation is vital for the well-being of the patient then his consent should be obtained. Under exceptional circumstances (e.g. road traffic accidents) it may not be possible to obtain consent.

Ligating ducts/ureters/arteries, etc. Be aware of the peri-operative risks of each procedure you carry out, e.g biliary surgery is fraught with the danger of damage to the bile ducts; colonic (right and left) surgery may lead to damage to the ureters or duodenum; laparoscopic sterilization may lead to small intestinal injury with subsequent pelvic abscess or peritonitis.

If your patient is not recovering as predicted, carry out further investigations to ascertain why. Laparotomy may be necessary.

It may be *accidental* to cause damage to a structure during surgery, but it is *negligent* not to act if the patient has signs which suggest such damage.

Operating on the wrong patient This results from breakdown of patient identification. Check that each patient corresponds with the list both numerically and for the surgical procedure. Risks are reduced if you recognize your own indelible marker. Check the notes and patient identification in the anaesthetic room *before anaesthesia* is induced (see above).

Failing to X-ray fractures/applying splints too tightly/applying plaster casts too tightly Look for the clinical signs of fractures. X-ray if suspicious. Beware the quiet fractures—scaphoid (may not show radiologically for 10–14 days), C_7/C_8 fractures (X-ray in 'swimming' position. Ask for $C_7/C_8/T_1$ views). Check splints and plasters after 24 hours. Give the patient a warning card to return if there is pain, discomfort, or numbness.

Wrong transfusion/wrong drugs/wrong dose Transfusion mistakes can be avoided by checking carefully that the name, hospital number, and date of birth correspond to the label on the blood bag. Blood samples for cross-matching should be accurately labelled immediately the blood is taken. Drugs and dosages should be clearly legible on the Kardex. If in doubt, check with your senior or consultant.

Communication with patients

Communication is the act of imparting (knowledge) or exchanging (thoughts, feelings, or ideas) by speech, writing, or gestures. Doctors must be able to communicate successfully with patients, colleagues, nursing staff, and administrators.

Five areas of communication

What to tell The truth if at all possible. Establish the diagnosis by histology (e.g. malignant disease) or overwhelming radiological or biochemical evidence. Use clear non-esoteric language. Tell the truth calmly. Sit at the same level as the person to whom you are speaking. Discuss treatment options.

When to tell When all relevant results are available, a full diagnosis with implications of treatment and prognosis can be given. It may be easier to give the diagnosis in stages: a clinical impression in the SOPD, the results of relevant investigations or histology in SOPD on the ward, and the operative findings once the patient has recovered sufficiently to understand (usually the first postoperative day). Try to tell the patient and relatives as soon as possible.

Whom to tell Tell the patient. Use discretion when the prognosis is very poor. Permit the patient to ask questions. He has a right to know what is happening to him. Discuss the clinical implications of the diagnosis with the closest relatives. Reassure them that a truthful approach will permit maximum co-operation from the patient and also justify future admissions, treatments, or continued follow-up at hospital, etc.

Where to tell Speak to the patient or his relatives in privacy—not in the corridor. If in the open ward, draw the screens and ask the nurse allocated to the patient's care to accompany you.

Who tells? Junior hospital doctors (housemen–senior registrar), consultants, staff nurses, or sisters. Establish the ward and consultant's policy. Nurses are often asked the diagnosis or result of an operation during the delivery of care. They are frequently better at speaking to patients than doctors are, and should be involved. After telling the patient be prepared to talk to him again. When the initial shock has passed there may be many questions. Others may be relieved to have a diagnosis for their troublesome symptoms. Some may accept the situation without further discussion.

References

Maguire, P. (1985). Helping patients cope with cancer. Tape cassette. Hoechst UK Ltd, Pharmaceutical Division, Middlesex.

Maguire, P. (1985). Helping the doctor cope with the cancer patient. Tape cassette, Hoechst UK Ltd, Pharmaceutical Division, Middlesex. lt

Maguire, P. (1985). Consequences of poor communication between nurses and patients. *Nursing*, **2**, 1115–18.

Communication with colleagues

Personal anxiety and frustrations are rarely caused by patients but they can be caused by outside sources such as colleagues, nursing, administrative, and laboratory staff.

Communication with hospital doctors

When making requests for clinical consultations, write a letter to the consultant concerned. Be brief and clear. Check that this is the person that your chief wants to see the patient. Ask him for his opinion or advice on management. Do not refer at registrar or houseman level alone without informing the consultants involved.

When asked to see a patient, go the same day if possible. Write your opinion in the case notes stating clearly what you recommend. If in doubt, discuss it with your colleagues on your own firm.

Communication with general practitioners (GPs)

Telephone the GP in the case of an OP admission or death of a patient. When discharging, give the discharge letter to the patient to deliver to the GP by hand. Mark it 'This must be delivered to your GP's surgery as soon as possible'. Even so, almost 25 per cent of letters remain undelivered and there may be a 4-week delay before the GP receives any details at all. The handwritten discharge should be delivered within 4 days, and the typed discharge summary with all results within one week.

Be polite in telephone calls and letters. State the diagnosis, treatment, and prognosis clearly with dates of follow-up visits. Tell the GP what you have told the patient and his family.

X-ray and laboratory colleagues

Is the investigation really necessary? If there is doubt about the correct investigation, telephone for advice. Complete request forms correctly and include clinical data where necessary, e.g. symptoms and sigmoidoscopy findings in a barium enema request.

Anaesthetists

Where there are specific medical problems or drug treatments, make sure that the anaesthetist is informed about them. It may pay to find out which laboratory investigations and X-rays the anaesthetist concerned routinely likes done before being given an operating list for the following day.

Administrators

Complete all official forms regarding employment and contracts as quickly as possible. Produce the certificate of your Defence Union. Complete holiday and study leave forms clearly, and ask your chief to agree to your request. When making enquiries be polite and calm. You are not the only one who has hassle in their life.

Nurses

Co-operate with the nursing staff. Introduce yourself on arrival to sister or staff nurse. They will help you learn the ropes. Do your ward work efficiently, dovetailing with the delivery of nursing care. Let sister or staff nurse know when you are going to lunch. It might save you being buzzed. Do an evening ward round to check on problem patients and drug requirements. It usually lets you have a good night's sleep.

References

Marston, A. (1978). Mistakes in communication, In *Current surgical practice* Vol. 2, (eds J. Hadfield and M. Hobsley), pp. 1–10. Edward Arnold, London.

Penney, T. M. (1988). Delayed communication between hospitals and general practitioners: where does the problem lie? *British Medical Journal*, **297**, 28–9.

Beware the patient with a label

Once a patient has been labelled with a diagnosis it becomes increasingly difficult to review his situation. New symptoms tend to be attributed to the diagnostic label, especially where the disease is chronic.

Example

A 53-year-old woman who was a 'known' Crohn's disease sufferer has had multiple right groin and right iliac fossa sinuses for the past 10 years. Her treatment includes Salazopyrin and low-dose steroids. Her case notes are encyclopaedic. When a new registrar arrives at the hospital he is quite firmly informed by the patient that 'Nothing further can be done. I just have to live with it'. He spends some considerable time going through her notes, only to discover that the patient last had any barium studies done more than 8 years ago. He initiates new studies, including sinography, with the patronizing support of his consultant.

When the results come through there is no evidence of active Crohn's disease, but sinography confirms several sinuses all confluent on the caecal region.

The patient undergoes laparotomy and resection of the right colon with ileocolic anastomosis. Postoperatively the sinuses heal and all medication is stopped. At review a year later the patient remains well.

Moral

All medical practice involves detective work which is not always rewarding. When presented with the statement: 'This patient is a "known" case of . . .', adopt an attitude of healthy cynicism and check that there is documentary proof of the diagnosis—usually ultimately in the form of a histology report. When the patient has encyclopaedic notes, make a summary of the main features. This will permit an appraisal of the situation and also makes life easier for your colleagues who may see the patient later.

Beware the patient you do not like

When excessive bias is introduced, logical decision-making becomes almost impossible. One human weakness is to moralize on the actions of others or to patronize them because of their station in life, speech, physical characteristics, etc. Lack of sleep (a common junior surgeon's problem) may also introduce serious bias or misjudgement.

Examples

- It is 3 a.m. and your buzzer goes off. You are the surgical registrar on call. It has been a busy day and tomorrow you have a full day's list of horrendoplasties assisting Mr Grump. The call has come from the A & E department. Your houseman wishes you to see a young man with a suspected perforated duodenal ulcer. If his diagnosis is right you may well be up for the rest of the night. As usual your room is miles from the A & E department and, what's more, it's raining. Even before you have got out of bed you have decided that you don't want to admit this patient.

- It is the New Year's day post-receiving ward round. You are the registrar on call over the festive period. Your consultant has 'popped out' to visit friends. Although you've been dry, the ward is full of young men and women suffering from hangovers and alcoholic gastritis. You feel cheated, and casually dismiss most of the patients without properly examining them. You are astonished when one is readmitted in a severely toxic state. At laparotomy you have to resect a large segment of ischaemic bowel as the result of a mesenteric infarction. The houseman reminds you that his haemoglobin was 18.3 g/dl (polycythaemia).

Moral for surgeons

When on call, train yourself to become resigned to the time commitment, excessive though it often is. Establish a checklist for clinical examination, review all relevant results, and discuss them with your colleagues if possible. Resist at all times the temptation to self-pity and to moralizing. They will detract from your clinical competence. 'Professionalism' is a learned skill which combines good surgical practice with unbiased decision-making.

If you genuinely cannot cope because of fatigue, report it to your colleagues and your seniors. The system will not collapse without you, and most consultants are sympathetic.

The elderly surgical patient

A large proportion of general and orthopaedic surgical procedures are carried out on patients older than 65 years.

Problems associated with ageing

- Multi-system disease and multiple medications
- Impaired mental acuity
- Difficulties with history-taking and in obtaining informed consent
- Late presentation. Patients often associate symptoms with 'plain old age'.

Aims of surgery

- To comfort if not to cure
- Palliation or cure to allow return to home circumstances and family (the ideal)
- If surgery is going to be of no benefit, do not consider it

Specific measures

- Consider each patient's problems individually.
- Involve the anaesthetist, geriatrician, rehabilitation services, the GP, the family as and when indicated.
- Assess the home circumstances. Who looks after the invalid husband, the dog, the cat, etc. Arrange for support for those at home.
- Assess renal, cardiovascular and metabolic status. Do ECG, chest X-ray, U&E, FBC, MSSU, blood sugar (blood gases may be relevant).
- Assess nutritional status.
- Prevent deep venous thrombosis if surgery is contemplated.
- Give prophylactic antibiotics for gastrointestinal, biliary tract, vascular, and prosthetic surgery.

Postoperative care

- Provide high-dependency nursing areas.
- Maintain CVS, respiratory status, fluid and electrolyte balance.
- Identify bowel or bladder problems.
- Avoid pressure sores.
- Mobilize early.
- Speak to the relatives. Try to get the patient home, if appropriate. This prevents depression and accelerates recovery.

Prognosis

The mortality rate is higher due to CVS disease. This can be reduced by good management.

References

Crosby, D. L. (1987). Management of the elderly surgical patient. *British Journal of Hospital Medicine*, **38**, 135–8.

Gray, J. A. M. (1982). Practising prevention in old age. *British Medical Journal*, **285**, 545–7.

Relevant points in the history of patients with acute abdominal pain

- Ask the patient what the complaint is.
- Determine the onset, constancy, relieving factors, and exacerbating factors.
- *Age* is important in predicting the pathology.
- *Pain* is a most important symptom. Determine its onset (i.e. sudden or gradual), character (dull, sharp, colicky), site, severity, movement (e.g. periumbilical to right iliac fossa in acute appendicitis). Ask if anything relieves or exacerbates the pain, e.g. movement or coughing in peritonitis, food in peptic ulcer disease.
- *Vomiting*. Note onset and time, type (projectile), persistence, content. Did pain precede vomiting? If so the cause of vomiting is usually surgical.
- *Diarrhoea/constipation*. These may be relevant in cases of intermittent or complete intestinal obstruction.
- *Temperature*. Is the patient pyrexial? Are there rigors? These suggest bacterial infection.
- *Past medical history*. Ask about previous operations, disease, family history, recent injury, menstrual periods.
- *Drugs*. Ask about current or recent medication. Specifically ask about hypoglycaemics, steroids, sedatives, hypotensives, non-steroidal anti-inflammatory drugs, anticoagulants, antibiotics, and cardiac drugs.
- In patients who have sustained trauma, get details of what happened from the patient or other witnesses. Start resuscitation if required at the same time.

History-taking and annotation of notes

Fortunately in general surgery the patient's problem is often obvious, e.g. 'I have a lump in my groin', 'I have terrible pain in my stomach'. Nevertheless it is important to establish clearly:

- The presenting complaint, its time of onset and progression, e.g. better or worse
- A previous history of similar problems
- The past medical and surgical history including any operations
- The name and nature of any drugs taken by the patient
- The family and social history
- There is usually enough time even in emergencies to establish a clear history of the functioning of the body systems, picking up any gross abnormalities

Annotation of notes

- Make your notes brief and to the point. Write legibly. Increasingly, many departments have an admission protocol with questions to which boxes are appended. As each question is answered the box is ticked. The data allows the records to be computerized and facilitates surgical audit.
- When describing significant features found on examination use simple line drawings to illustrate the point, e.g.

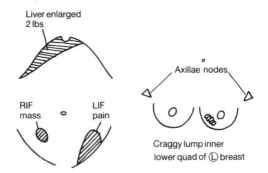

Liver enlarged 2 lbs

RIF mass   LIF pain

Axillae nodes

Craggy lump inner lower quad of Ⓛ breast

- Use illustrations to annotate your operation notes, e.g.

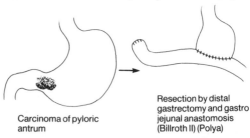

Carcinoma of pyloric antrum

Resection by distal gastrectomy and gastro jejunal anastomosis (Billroth II) (Polya)

- When you go on your ward round write a brief update of your patient's condition.
- Record any changes in your patient's condition in the notes.
- Ask visiting consultants/registrars to write their instructions and opinions in the notes. They usually do.
- File results when they arrive. If the results are telephoned as an emergency, always write them in the clinical continuation sheet.
- Prepare the brief discharge summary letter to give to the patient on the day of leaving hospital.
- Ensure that all relevant results and the full discharge summary are sent to the GP as soon as possible (7–10 days is ideal).

The regions of the abdomen

For descriptive purposes the abdomen is divided by an imaginary grid into nine regions. Many surgeons use four quadrants and an umbilical region.

Horizontal planes

- *Subcostal* passes through the lower edge of the tenth costal cartilage and corresponds to the lower border of L_2.
- *Transtubercular* passes through the tubercles of the iliac crest and corresponds to the lower border of L_2.
- *Transpyloric* is midway between the jugular notch and symphysis pubis and is said to pass through the pylorus of the stomach which is in fact a mobile structure. The plane corresponds to the lower border of L_1. It may be used in addition to the grid.

Vertical planes

The subcostal and transtubercular planes divide the abdominal cavity into three zones which are further divided into regions by a pair of vertical planes which pass through the mid inguinal point and ninth costal cartilage.

Thus in the upper zone the three regions are right and left hypochondrium and epigastrium; in the middle zone right and left lumbar and umbilical; in the lower zone right and left iliac (fossa) and suprapubic.

Relationships of abdominal organs to regions

The regions present a guide to the position of the intra-abdominal organs. This may vary in the same patient depending on whether a mobile organ is loaded or not and in different patients according to body habitus.

- *Right hypochondrium* (RH): Liver, gallbladder, hepatic flexure of colon
- *Left hypochondrium* (LH): Most of stomach, splenic flexure, spleen, tail of pancreas
- *Epigastrium*: Stomach, first, second, and fourth parts of duodenum, upper half of transverse colon, head and body of pancreas
- *Right lumbar* (RL): Ascending colon, part of right kidney
- *Left lumbar* (LL): Part of descending colon, part of left kidney
- *Umbilical*: Lower half of transverse colon, and part of duodenum, small bowel loops, bifurcation of aorta and inferior vena cava. Part of both ureters.
- *Right iliac fossa* (RIF): Caecum, appendix
- *Left iliac fossa* (LIF): Part of descending colon
- *Suprapubic*: Bladder, lower parts of ureters, sigmoid colon, rectum

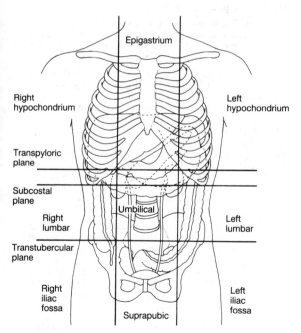

Regions of the abdomen

Examination of the abdomen

Follow the order: inspection, palpation, percussion, auscultation.

Inspection

- *General contours*: Look for the 5Fs as possible causes of protuberance or distension—fat, faeces, fluid (ascites), fetus, flatus
- *Eversion of the umbilicus*: ?Due to malignant disease (pregnancy also causes this)
- *Localized enlargement*: ?Due to organomegaly, e.g. spleen, liver, etc.
- *Retraction*: Scaphoid abdomen may be seen in dehydration; emaciation
- *Appearance of the skin*: Wrinkled, dry, tense, oedema, etc. Note site of scars and striae
- *Pigmentation*: Distribution of abdominal hair
- *Tortuous veins*: ?Portal or inferior vena caval obstruction.

Movement Does the abdomen move freely with respiration? Diminished movement may suggest inflammation. Excessive movement is seen when the abdominal muscles are used as accessory respiratory muscles.

Pulsation The abdominal aorta can be seen to pulsate in thin patients and in enlargement due to aneurysm. NB. pulsating liver in tricuspid regurgitation. Transmitted pulsation from the right ventricle causes epigastric pulsation.

Peristalsis Is it visible? Note the site. In pyloric stenosis peristaltic waves pass from left to right. In transverse colonic obstruction peristalsis crosses from right to left. Small-bowel obstruction causes a ladder pattern in the centre.

Palpation

Make sure your hands are warm. Ask the patient to lie flat. Palpate gently the four quadrants. Always start in the right iliac fossa (unless it is where there is pain). Examine the most painful quadrant last. Is there tenderness, guarding, enlarged organs? Then palpate each quadrant more deeply. Determine the size, shape, consistency and mobility of any swellings.

Specifically feel:

- *Liver*: Start in the right iliac fossa. Dip the fingers to detect the edge as the patient breathes. The liver moves with respiration. Note its size (in finger breadths below the costal margin), consistency, the presence of irregularities on its surface and edge (?metastases), tenderness and pulsation. Draw a diagram.
- *Spleen*: The spleen moves with respiration. Begin as for the liver and palpate towards the left upper quadrant. Use the other hand placed in the left loin for counter pressure. Note its consistency, the notch, tenderness, direction and degree of enlargement. Note that if you can 'get above the swelling' it is not a spleen—more probably a kidney. 'Rotation of the patient 45° to the right increases the chance of feeling the spleen'.

- *Kidneys*: Feel bimanually. The kidneys move slightly with respiration. 'Ballott' the kidney between the two hands.
- *Uterus and bladder*: Can be felt in the hypogastrium or beyond when gravid or full respectively.

When there is marked ascites, a fluid 'thrill' can be elicited. Have a colleague place the side of his hand in the patient's midline. By tapping the finger against the patient's side a 'thrill' can be felt by the other hand placed on the opposite side. Intestinal or omental swellings can sometimes be ballotted in ascites.

Percussion

Confirm enlargement of liver, spleen, bladder. Detect shifting dullness by percussion (ascites). Start from the midline to the flanks. If they are dull move the patient so one flank is uppermost. If it is now resonant to percussion there is free intra-abdominal fluid.

Auscultation

Note normal borborygmi. Sounds are diminished or absent in peritonitis, increased in obstruction. In pyloric obstruction splashing may be heard without the stethoscope by rocking the abdomen (succussion splash).

Groins

Look for herniae, glands, abnormal vessels.

Genitals

Look for urethral discharge, vaginal discharge, warts, abnormalities (endocrine disease).

Rectal examination

Note haemorrhoids, anal tags, anal fissure, perianal warts. Insert the gloved and lubricated index (pad first and directed posteriorly). Feel the normal prostate anteriorly or the normal cervix. Feel for masses. Is there blood, mucus, pus on finger? Test for FOB. Are the faeces normal?

Carry out proctoscopy/sigmoidoscopy if indicated.

Dysphagia and heartburn

These are the commonest symptoms of oesophageal disorders and may occur separately or together. Dysphagia (difficulty with swallowing) may be oropharyngeal or oesophageal. When dysphagia and heartburn occur together this suggests that there is lower oesophageal spasm in association with reflux oesophagitis.

Symptoms

The main symptoms are difficulty with swallowing, retrosternal pain, and discomfort. Remember that oesophageal pain may mimic the pain of myocardial ischaemia. When there is doubt do an ECG.

Causes

- *Pharynx*: Pharyngeal pouch, pharyngeal webs
- *Oesophageal*: Carcinoma, benign tumours, disorders of peristalsis (achalasia of the cardia, neurological incoordination), strictures, e.g. due to reflux oesophagitis in hiatus hernias or stricture due to corrosives, Barrett's ulcer
- *Lesions of the lumen*: Impacted foreign body
- *Extrinsic pressure*: Malignant mediastinal lymph nodes, para-oesophageal hernia, dysphagia lusoria caused by abnormal blood vessels
- *Autoimmune disease*: Scleroderma
- *Plummer-Vinson syndrome*: Hypochromic anaemia, cheilosis, and koilonychia associated with upper oesophageal webs. Occurs in middle-aged women.
- *Neurological disorders*: Myasthenia gravis, disorders of brain stem, functional dysphagia

Common investigations

Radiology The investigation of choice is a barium swallow. When the barium reaches the stomach the patient should be screened in the Trendelenberg position (tilted) if hiatus hernia is suspected.

Oesophagoscopy and biopsy This requires considerable experience and there is a well-documented risk, albeit small, of perforation. Either the flexible or the rigid (Negus) endoscope may be used. The lesion is directly visualized and a biopsy taken. Rigid oesophagoscopy is less frequently used except for specific purposes, e.g. the injection of varices may be easier.

Biopsy This may be taken by biopsy forceps and the specimen sent for histology. Increasingly, brush cytology of the oesophagus is giving accurate results especially in patients with early oesophageal cancer. A fine brush is passed down the endoscope to abrade the mucosa. The material obtained is then smeared directly onto slides and examined after staining. This lessens the risk of perforation and bleeding.

Intraluminal pH measurements and manometry Intraoesophageal pH can be measured by indwelling glass electrodes swallowed by the patient and maintained at a measured distance from the incisors. Acid regurgitation is present if the pH is less than 4 over a measured length of oesophagus. 24-hour telemetry studies can also be used.

Manometry This method studies changes in the oesophagus. It is most useful in functional disorders when there are minimal radiographic changes.

Dyspepsia

Dyspepsia is defined as upper abdominal discomfort related to eating or drinking, which may often be described as pain by the patient.

Common causes

Hiatus hernia, peptic ulceration, cholelithiasis, carcinoma of the stomach, pancreatic disease, ?myocardial ischaemia.

Specific questions in the history

Ask about abdominal pain and discomfort—its site and character, e.g. constant, colicky, radiation relationship to food and bowel action. Is there flatulence, i.e. eructation of air or passage of flatus? Is there nausea or vomiting in association with the pain? Does this relieve it?

Investigations

Upper GI endoscopy This can often be performed without sedation or by giving the patient an intravenous injection of Diazemuls 10 mg. With forward-viewing endoscopes most lesions of the upper GI tract can be visualized directly. All gastric ulcers should be biopsied at four edges to exclude gastric cancer; brushings for cytology are equally rewarding. Endoscopy, in expert hands, is accurate and causes a low risk of perforation.

Radiology Straight abdominal films may demonstrate gallstones in 20 per cent of patients. They may also demonstrate gas in the biliary tree suggestive of choledocho- or cholecystoduodenal fistula, cholangitis, or following the recent passage of a stone through the ampulla of Vater.

Barium meal This will demonstrate most peptic ulcers—both gastric and duodenal. When double contrast (air and barium) radiology is used, many lesions distal to the duodenal cap may be identified. Most duodenal ulcers occur in the first part of the duodenum in the duodenal 'cap' or 'bulb'. Contrast techniques are also being increasingly used for stomach lesions, especially in Japan. Most benign ulcers occur on the lesser curve where they appear as a niche. Most gastric cancers appear at the antrum or in sites not normally affected by benign ulceration, e.g. the greater curve. Hiatus hernia can be demonstrated by placing the patient in the Trendelenberg position.

Gall bladdder disease

The two common methods of investigation of the patient with suspected gallstones are oral cholecystography and grey-scale ultrasonography. Oral cholecystography will not outline the gall bladder in the presence of a serum bilirubin level greater than 30–40 mmol/l. Grey-scale ultrasonography is the investigation of choice in jaundiced and non-jaundiced patients. It images the whole biliary tree and liver.

Endoscopic retrograde cholangiopancreatography (ERCP)
This has widespread use in both biliary and pancreatic disease. It
can be used diagnostically to visualize radiologically the pancreatic and biliary ducts. Cytology specimens can also be taken.
Therapeutically, procedures like endoscopic papillotomy or
retrieval of bile duct stones can be carried out.

Intravenous cholangiography (IVC) is now rarely performed
but may be indicated occasionally (see diagram). It involves the
injection of a water-soluble compound of iodine which is
excreted by the liver to display the duct system. Absence of
opacification of the gall bladder when there is normal liver
function suggests obstruction of the cystic duct.

Percutaneous transhepatic cholangiography (PTC) Direct
injection of contrast medium into dilated intrahepatic ducts
using a Chiba needle is of value in the investigation of the jaundiced patient. PTC outlines the hepatic ductal system indicating
the site of obstruction.

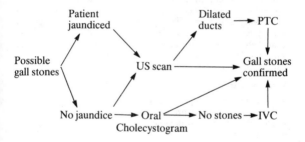

Constipation

Always ask the patient what he means by constipation. It refers to defecation twice a week, or less. The stools are hard and pain is experienced in 25 per cent of cases. Each year in the UK 450 000 people will consult their GP with the symptom. The incidence is highest in women and becomes more common with increasing age.

Causes

- *Dietary*, e.g. low fibre intake
- *Colorectal disease*, e.g. fissure, prolapse, diverticular disease, neoplasm
- *Metabolic disease*, e.g. hypothyroidism, diabetes mellitus
- *Neurological disease or injury*
- *Psychiatric illness*, e.g. depression
- *Unknown*, but the colon is physiologically and anatomically normal (the largest group)

Clinical features

Many patients are asymptomatic, do not seek advice and may self-treat with laxatives. In others there may be associated abdominal pain, bloating, and alternating diarrhoea (irritable bowel syndrome). In the elderly, faecal impaction is a feature associated with reduced rectal sensation, multiple medications, immobility, and poor toilet facilities. More than 40 per cent of elderly acute admissions have incidental faecal impaction (Read *et al.* 1985). In some young women constipation may be severe, with painful defecation occurring every 4–6 weeks associated with abdominal pain. Soiling is a feature in many cases.

Investigations

1. Exclude anorectal and colonic pathology
- Rectal examination/proctosigmoidoscopy
- Double-contrast barium enema
- Colonoscopy

2. Exclude metabolic and drug causes
- Careful history, blood sugar, thyroid function tests, serum calcium, parathormone, serum porphyrins

3. Colonic transit times In specific patients (those with normal-diameter colons and severe symptoms)
- Plain abdominal X-ray after swallowing 20 radio-opaque PVC shapes—80 per cent are excreted by normal people after 5 days (Hinton *et al.* 1969)

4. Anorectal physiology
- In normal individuals the internal sphincter relaxes when distended by a balloon (rectoanal reflex). If absent this suggests Hirschsprung's disease.
- Assessment of the function of puborectalis and external anal sphincter by electromyography (EMG).

5. Psychological assessment
- Many patients are disturbed, depressed or anxious (Preston *et al.* 1984). Make sure that pathology has been confidently excluded before applying the label.

Treatment

Medical
- Increase dietary fibre, use bran or bulking agents
- Avoid long-term drugs like senna, use oral lactulose
- Suppositories like bisacodyl may be valuable
- Phosphate enemas
- Drugs which act specifically on the myenteric plexus, e.g. cisapride, may be valuable but the long-term effects are unknown

Surgery is rarely indicated in the management of idiopathic constipation unless it is intractable and associated with abdominal pain.

References

Hinton, J. M. *et al.* (1969). A new method for studying gut transit times using radio-opaque markers. *Gut*, **10**, 842–7.

Preston, D. M. *et al.* (1984). Psychiatric assessment of patients with severe constipation. *Gut*, **25**, 582–3.

Read, N. W. *et al.* (1985). Anorectal function in elderly patients with faecal impaction. *Gastroenterology*, **89**, 959–66.

Risks of treatment and informed consent

'Doctors must tell patients any facts necessary for them to decide whether they want the operation. Exactly how much the doctor should tell is a matter for discretion, but real and forseeable risks should be disclosed.' (National Consumer Council 1983).

Criteria for the validity of a patient's consent

- *Capacity*: The patient must be over 16 years old and have the mental capacity to give consent. Children, therefore, need their parents' informed consent. The mentally confused and elderly need the informed consent of nearest relatives. If there are none, the informed consent of the Health Authority acting for the patient may be obtained.
- *Information*: As full and adequate an explanation as possible of all investigations and procedures should be given. In the UK at present this is a matter for clinical judgement.
- *Voluntariness*: Consent should be obtained without force or deceit and no pressure should be brought to bear on the patient.

Informed consent may be

- *Implied by circumstances*: The patient implies that he is prepared to undergo preliminary procedures e.g. ECG, X-rays by attending hospital at a given time and date.
- *Verbal*: e.g. for blood sampling and other invasive procedures.
- *Written*: When an investigation or operation is to be carried out which requires an anaesthetic written consent must be obtained on the standard form supplied by the Health Authority for the purpose.

How much do patients wish to know?

In surveys of drug prescribing, 93 per cent of patients wished to know why they are given drugs, 90 per cent the common side effects, and 80 per cent the rare side effects; yet doctors and pharmacists preferred not to disclose these. It is probable that the patient attending hospital for surgery has similar expectations (Jombert and Lasagna 1975).

Problems facing the doctor

The doctor's decision-making is based on the 'probable' outcome for certain procedures and surgical conditions. How much of this probabilistic information should be offered to allow the patient to give 'informed' consent? What the patient is told is also coloured by the doctor's attitude and not the perception of the patient's attitude to risks and gains. It has therefore been suggested that the responsibility for risk-taking should be shifted more towards the patient and away from the doctor. Where there are treatment options with differential risks and benefits, the choice of treatment A as opposed to B will reflect the patient's attitude to risk versus benefit (McNeil 1982).

Guidelines for 'informed' consent

- Assess your patients carefully.
- Ask patients how much they want to know about their condition and the proposed procedure.
- Give adequate and reasonable explanations for all procedures.
- Answer their questions as completely as possible.
- Where there are treatment choices, the quality of decision-making might be improved by eliciting patient preferences.
- Ask the nurse who has been delegated to care for the patient to accompany you. She may be better able than you at clarifying areas of misunderstanding and can also act as an advocate on behalf of the patient.

References

Jombert, P. and Lasagna, L. (1975). Patient package inserts: nature, notions and needs. *Clinical Pharmacology and Therapeutics*, **18**, 507–13.

McNeil, B. J. (1982). On the elicitation of preferences for alternative therapies. *New England Journal of Medicine*, **306**, 1259–62.

National Consumer Council (1983). *Patients' Rights*. HMSO, London.

Diagnosis of death

There is no legal definition of death in the UK. The criteria, left to clinical judgement, are cessation of circulation and respiration.

Somatic death

Diagnosis Apnoea, pulselessness, absent heart sounds, fixed dilated pupils. Auscultate chest and praecordium. Fundic ophthalmoscopy unreliable. If in doubt do an ECG.

Brain death (death of the brainstem)

Diagnosis
- Comatose patient. Coma not due to depressant drugs, primary hypothermia ($< 35\,°C$), metabolic or endocrine disturbances.
- Ventilated patient due to inadequate or absent spontaneous respiration. Exclude neuromuscular blocking drugs and curare-like drugs.
- Diagnosis of a disorder which can lead to brain stem death must be established, e.g. severe head injury, intracranial haemorrhage. If primary pathology is in doubt continue clinical investigations.

Diagnostic tests: all brainstem reflexes must be absent
- Fixed dilated pupils with absent light reflexes
- Absent corneal reflexes
- Absent eye movements on caloric testing (20 ml ice water in ears in turn). Visualize tympanic membranes first to exclude false negatives.
- Absent motor responses in the cranial nerve distribution in response to stimulation of face, limbs, and trunk
- Absent gag reflex to catheter in trachea
- Absent cough reflex
- Absent respiratory movements when patient disconnected from ventilator ($P_{Co_2} \geqslant 50$ mmHg/6.7 kPa)

Death must be confirmed by the consultant in charge of the patient or his deputy (registered more than 5 years). Experience in making the diagnosis is necessary. The opinion of another suitably experienced doctor is also required. The tests must be done by both doctors, separately or together, and repeated to ensure permanent absence of response within a suitable period, usually 24 hours.

Criteria in the USA

If the diagnosis has to be established within 6 hours of apparent cessation of cerebral function, then EEG confirmation is required. Diagnosis is also permitted in intoxication with drugs (self-poisoning, narcotics overdose not responding to antidotes in a patient with apparent cessation of cerebral activity).

After death

Inform relatives, GP, consultant. Provide a death certificate. If brain dead obtain permission for organ harvesting (heart, lungs, kidney, liver, cornea, etc.) from the relatives. Point out that they are being asked to act as agents in expressing what they believe to be the wishes of the patient. The best person to ask may be the consultant but a senior nurse, a chaplain (or other religious figure), or the family doctor may be more appropriate. If there is doubt about the cause or circumstances of death inform the Coroner or Procurator Fiscal.

References

Cadaveric organs for transplantation (1983). (A code of practice including the diagnosis of brain death.) HMSO, London.

Jennett, B., Gleave, J., and Wilson, P. (1981). Brain death in three neurosurgical units. *British Medical Journal*, **282**, 533.

Medical Consultants (1981). *Journal of the American Medical Association*, **246**, 2184.

Suicide

The suicide rate in the UK is currently 12.5 per 100 000. The society EXIT, founded in 1935, concentrates on giving advice to people on how to commit suicide painlessly and effectively. It also aims to achieve legislation to permit voluntary euthanasia. However, to date this has not been passed in the UK.

Patients at risk

- The recently bereaved (p. 42)
- Cancer patients have a five times increased risk
- Men over 55 years with oral cancer and a history of alcohol abuse
- Women of any age often suffering from gynaecological or breast cancer. (In both these latter groups the treatment of the disease involves disfigurement and a change of body image. Additionally, women may suffer a sense of loss of femininity.)

Action

Patients about to undergo disfiguring surgery for any reason should be carefully counselled in the period after confirmation of the diagnosis and before surgery. Doctors should discuss all treatment options and implications clearly. The support of a 'mastectomy counsellor' or 'stomatherapist' is invaluable.

References

Morris, C. A. (1985). Self concept as altered by the diagnosis of cancer. *Nursing Clinics of North America*, **20**, 611–30.

Patt, N. (1983). Identification of risk factors for suicide amongst cancer patients. *Oncology*, **2**, 246.

Euthanasia

Euthanasia is the painless termination of life at the request of the patient concerned. In the UK it is illegal to administer any drug to accelerate death, irrespective of how compassionate the motive may be. The law holds that the intention to kill is malicious, and such action would be classified as murder.

Withholding treatment from handicapped babies may also be seen as a form of euthanasia, but this is not simply a medical decision. The parents must be involved and understand the alternatives fully. Many handicapped children (e.g. Down's syndrome) enjoy life. This positive aspect must be considered when speaking to parents about the implications of a child's handicap.

Many doctors in the Netherlands have purposely courted litigation by indicating that a patient's death was due not to natural causes but euthanasia. Where court cases ensued the doctors involved were acquitted, but there has been subsequent pressure on the Dutch government to review the penalties relating to doctors and their dying patients.

In the UK in 1985 the British Voluntary Euthanasia Society (EXIT) reported that 75 per cent of a sample of the general public surveyed felt that 'The Law should allow adults to receive medical help to an immediate peaceful death if they suffer from an incurable illness that is intolerable to them', but only 15 per cent of doctors agreed with this.

Why do patients request euthanasia?

Terminally ill people may have several reasons:
- Fear of awaiting death
- Intractable pain
- Disfigurement
- Fear of becoming a burden
- To establish whether euthanasia is in fact an option in British medical care

Why have parents requested non-treatment for newborn babies?

- Down's syndrome
- Severe spina bifida
- Severe chromosomal abnormalities
- Jehovah's Witnesses and other religious groups. (Doctors have sought protection by making the child a ward of court.)

Guidelines

- There are no social or legal criteria for withholding treatment. The decision is difficult and is not purely medical.
- Most patients' fears can be eliminated by skilled care and support.
- Most patients' symptoms can be controlled by appropriate medication.

- In a few patients, increasingly large doses of drugs must be given to relieve symptoms. In them the risk of unconsciousness and hypostatic pneumonia is increased.
- Learn to recognize when attempts to cure have been exhausted, and direct your energies into caring for the patient. There may be a need for the absolute control of discomfort in those who are dying.
- Always discuss the care of terminally ill patients with your senior colleagues.

References

British Medical Association (1971). *The problem of euthanasia*. British Medical Association, London.

Dawson, J. (1986). Easeful death. *British Medical Journal*, **293**, 1187–8.

Death after bereavement

In the UK 2 per cent of men and 8 per cent of women between 45 and 60 years suffer conjugal bereavement. In those people over 75 years, 30 per cent of men and 65 per cent of women are widowed. The estimated population suffering bereavement is about 4 million.

Risks for the bereaved

Risk of death for the survivor of conjugal bereavement is greatest in the first 6 months for men, and the second year after for women. Men and those bereaved when younger tolerate bereavement badly, and have a poorer prognosis.

The risks seem to be related to loss of care experienced by the survivor which can lead to self-neglect, often the case in widowers. Reduced resistance to infection may also be a factor brought on by an unhealthy existence due to excessive drinking or smoking. Grief itself may suppress lymphocyte function, which could increase the risk of cancer. Those who remarry early after bereavement live relatively normal lives afterwards.

Causes of death

Cardiovascular disease, cancers, suicides, and accidents.

Bereavement counselling

Professional counselling by nurses, doctors, and self-help groups reduces the mortality of the bereaved population. The aim is to provide 'therapeutic communication'—in other words, sympathy.

Moral

It is possible to die of a broken heart. Although there is rarely enough time to spend lengthy periods with bereaved families, the effort should be made. Even a one-off interview can have a settling effect. This should be followed up by the district nurse or general practitioner and is one good reason why the GP should be informed of the death of a patient promptly.

References

Kraus, A. S. and Lilienfield, A. M. (1959). Some epidemiological aspects of the high mortality rate in the young widowed group. *Journal of Chronic Diseases*, **10**, 207–17.

McAvoy, B. R. (1986). Death after bereavement. *British Medical Journal*, **293**, 835.

Parks, C. M. (1980). Bereavement counselling: does it work? *British Medical Journal*, **281**, 3–6.

2 Inflammation, infection, and burns

Inflammation

Inflammation is the response of the body to injury. It can be acute or chronic.

Causes

- *Mechanical trauma*: Possibly the commonest cause. This was recognized, by John Hunter who differentiated it from suppurative inflammation (associated with infection and digestive softening of the tissues).
- *Infection* due to bacteria, viruses, parasites, fungi, protozoa
- *Chemical and physical agents* (heat, cold, radiation)
- *Ischaemia*
- *Hypersensitivity*

Acute inflammation

e.g. cellulitis (infective). A surgical incision is an example of mechanical trauma which produces an acute inflammatory response followed by rapid resolution and healing (unless infected).

Cardinal signs Calor, rubor, rumor, dolor, and funtio laesa. (Heat, redness, swelling, pain, and impaired function.)

Development

- *Tissue injury or the injection of organisms through a break in the epithelium*: Vascular response and exudate
- *Vascular response*: Capillary vasoconstriction then vasodilatation—increased capillary blood flow then stasis. Increased vascular permeability leads to active emigration of polymorphonuclear leukocytes and passive carriage of RBCs to the interstitial space. Phagocytic neutrophil polymorphs (macrophages) derived from monocytes engulf the bacteria.
- *Exudate*: composed of fluid and cells from the capillaries, protein rich. Main cells are polymorphs with some RBCs. Other cells are eosinophils (?antiparasitic, type I hypersensitivity reaction), mast cells (release chemicals). Lymphocytes, plasma cells, and macrophages.

Outcome Depends on host (age, blood supply, immune system) and organisms (dose, route of introduction, virulence, exotoxins (hyaluronidase coagulase) or endotoxins (due to breakdown of bacterial cell walls).
Possibilities include:
- Resolution with return to normal
- Suppression: pus with or without ulcer formation
- Spread: through the tissues (cellulitis), tracking (sinus or fistula formation) empyema, lymphangitis, bacteraemia, septicaemia
- Chronic inflammation
- Fibrosis with scarring
- Death (due to multiple organ failure)

Chronic inflammation

This may follow acute inflammation but may begin as a chronic process, e.g. TB. Cellular damage and inflammation occur at the same time as healing.

Types
- Failure of resolution (of acute inflammation)
- Chronic inflammatory infections and disease

Cell types Lymphocytes, plasma cells, macrophages.

Examples
- Failure of resolution of acute infection leads to the formation of granulation tissue accompanied by fibrosis and regeneration. Chronic abscess formation may result from failure of resolution of acute suppuration with chronic pus formation or pyogenic membrane if localized. Pus should be drained. If the patient is treated with antibiotics alone the pattern of inflammation may be altered with the development of a hard lump—an 'antibioma'. In this the infection continues at a low grade. Antibiomas should be excised, especially in the breast where they may mimic cancer.
- The chronic infections, e.g. tuberculosis, leprosy, syphilis characterized by granuloma formation and chronic inflammatory disease, like cirrhosis of the liver which lacks granuloma formation but is characterized by disordered nodular regeneration and fibrosis

Granuloma A circumscribed collection of macrophages (Syn. histiocytes, monocytes in the blood stream) often multinucleate with central necrosis.

Types
- *Foreign body granuloma*: e.g. retained suture. The macrophages are unable to phagocytose all of the stitch.
- *Immune granuloma*: e.g. TB, Crohn's disease

Wounds and wound healing

Classification: non-surgical and surgical

In both types healing will occur, but because of optimum conditions a surgical incision will generally heal by *first intention*.

First intention healing The sutures approximate the edges of the wound, allowing only a narrow zone of granulation tissue. Union occurs rapidly with minimal scar formation. There are three phases:

1. *Early lag phase* (0–3 days). This is the phase of inflammation. There is fibrinous adhesion only of the edges. Wound strength depends on the sutures. Wound breakdown is the result of poor technique during this phase.
2. *Proliferative phase* (3 days–3 weeks). Wound strength progresses with the formation of granulation tissue and collagen. The tensile strength of the wound increases rapidly—fibroplasia (American terminology).
3. *Remodelling phase* (3 weeks–1 year). There is continuous reorientation and maturation of collagen fibres. At one year 70 per cent or more of the original tissue may be regained, i.e. scar tissue is never as strong as unwounded tissue.

Second intention healing The wound gapes because of destruction of tissue by trauma or infection and the elastic pull of the dermis on each side. This defect initially fills with blood clot which dries to form a scab. Under the scab, in small uninfected wounds re-epithelialization begins from the wound edges.

The scab is gradually lifted at its edges until it falls off. In larger wounds the growth of epithelium is more easily seen. De-epithelialization is enhanced in a moist environment where scab formation is reduced.

Wounds heal from below upwards. Capillary loops bud and fibroblasts proliferate to form collagen. This gives a velvet appearance to the wound and is called *granulation tissue*. Fusion of this wound begins at its base. As the myofibroblasts contract the wound gets smaller.

These changes are accompanied by the migration of squamous epithelial cells—*re-epithelialization*. Once skin cover is complete the stimulus from further granulation ceases and the final result is *scar tissue*.

80 per cent of closure of an epithelial defect, however, is due to wound contraction and not re-epithelialization.

The difference between first intention and second intention healing is quantitative only.

Local factors affecting healing

- Bacterial contamination and infection
- The presence of a foreign body within the wound, e.g. organic material, glass
- The presence of necrotic tissue
- Poor tissue vascularity
- Tension and oedema

General factors delaying healing

- Advanced age
- Nutritional deficiencies, low vitamin C and K, hypoprotein-aemia, zinc deficiency
- Chronic steroid therapy
- Poorly controlled diabetes mellitus, malignant disease associated with hypoproteinaemia, renal disease.
- Patients receiving chemotherapy or radiotherapy

Traumatic wounds

Traumatic wounds are usually contaminated or contain devitalized tissue.

Classification

Tidy:
- Lacerations
- Injuries with minimal skin loss

Untidy:
- Lacerations with marginal necrosis
- Avulsion injuries with skin loss
- Crush injuries

Management

Primary or delayed closure? Tidy wounds can be closed primarily. Untidy wounds need wound excision with removal of debris and dead tissue. Delayed closure is indicated in heavily contaminated wounds and those greater than 6 hours old. Following wound excision the wound is dressed with Gelanet gauze or lightly packed. An occlusive dressing is then applied, e.g. Opsite, and the wound inspected daily under sterile conditions. If there is no evidence of further necrosis or inflammation the would can be closed at 48–72 hours.

Irrigation There are agents in soil which potentiate infection. Contaminated wounds should be irrigated with a high pressure system at greater than 65 kPa. Commercial systems are available (Water-Pik irrigation system). A large syringe (30–50 ml) with a 10 gauge needle is also satisfactory.

Manual scrubbing causes local oedema and decreases host defences. If necessary use a porous sponge with a mild aqueous antiseptic like Savlon or cetrimide. They are effective detergents in removing grease and dirt. This will prevent traumatic tattooing.

Wound excision All non-viable tissue should be excised with a knife or scissors. There are high concentrations of bacteria in necrotic tissue and around foreign bodies. Their removal improves wound healing.

When viability is in doubt give IV fluorescein (15 mg/kg). If there is no fluorescein, excise. If patchy, observe but excision may still be needed. In muscle, colour, bleeding and contraction decide viability.

Antibiotics Early administration of prophylactic systemic antibiotics (within three hours of injury) improves wound healing. Cephazolin 0.5–1 g every 6–12 hours IV reaches high concentrations in the wound and is effective against sensitive Gram-positive and Gram-negative organisms. Aminoglycosides may also be administered, e.g. gentamicin 80 mg IV every 8 hours. It is important to measure serum levels because of their low toxic/therapeutic ratios and adjust the dose preoperatively so that effective levels are present when the wound is closed.

The use of topical antibiotics is well established in burns and contaminated wounds. Effective agents include aminoglycosides, cephalosporins, and pencillins.

References

Haury, B., Rodeheaver, G., Venski, J., Edgerton, M. T., and Edlich, R. F. (1978). Debridement: an essential component of traumatic wound care. *American Journal of Surgery*, **135**, 238–42.

Polk, H. C., Trachtenberg, L., and Finn, M. P. (1980). Antibiotic activity in surgical incisions: The basis for prophylaxis in selected operations. *Journal of the American Medical Association*, **244**, 1353.

Soft-tissue infection

Cellulitis

This can occur in all soft connective tissue. It is commonly caused by B-haemolytic streptococci (*Streptococcus pyogenes*). Enzyme production leads to a reddened appearance with bleb formation.

Management
- Therapeutic trial IM or slow IV (0.6–1.2 g) of aqueous penicillin at 4-hourly intervals for 8–12 hours to distinguish from erysipelas
- Early incision and fasciotomy, if necessary
- Gram staining. Continue penicillin if streptococci are present.

Polymicrobial infections can extend deeper. These are a danger in diabetics and patients with vascular occlusive problems or traumatic injuries. There are oedema and crepitations. Differentiate from clostridial infections by Gram staining. Give an aminoglycoside combined with an anti-anaerobic agents (chloramphenicol or clindamycin). Incise and remove dead tissue.

Fasciitis

Suspect when pain, oedema, and skin necrosis appear within 24–48 hours of injury or operation.

Signs The skin may be normal, oedematous, or mottled. There may be spiking fever, hypotension, mental confusion.

Treatment
- IV fluids. Extensive excision. Ladder incision in abdominal wall parallel to the ribs.
- Antibiotics on the basis of Gram stain.
- Dress incisions with moist gauze impregnated with 0.5 per cent silver nitrate. Change these frequently under general anaesthetic.

Myositis

The bacterial infection has penetrated and destroyed muscle bundles.

Cause Usually *Clostridium perfringens*, a Gram-positive anaerobic spore-forming rod.

Management
- Recognize the condition—there is oedema and serosanguinous exudate; exposed muscle is swollen and ranges from salmon pink to deep green/black appearance. Crepitation may be present. There is a 'sickly-sweet' smell.
- Excise dead tissue—until healthy contactile bleeding muscle is encountered.
- Carry out Gram staining immediately. Streptococcal myositis rarely needs amputation, clostridial may.
- Replace extracellular fluid (ECF) deficit. Transfuse if needed.

- Broad-spectrum antibiotics (tetracycline or chloramphenicol) plus high-dose penicillin (IM or IV 0.6–1.2 g every 2 hours).
- Excision of all infected tissue may necessitate full thickness excision of abdominal wall with prosthetic mesh replacement.

Amputations are performed by guillotine technique and the wounds packed open.

There is no proven advantage to hydrogen peroxide dressings or the use of hyperbaric oxygen, but passive immunization in proven clostridial gangrene has been suggested using a variety of bacterially derived products.

Reference

Polk, H. C. and Galland, R. B. (1982). Enhancement of non-specific host defense mechanisms In *Infection and the surgical patient* (ed. J. R. Polk), pp. 101–10. Churchill Livingstone, Edinburgh.

Tetanus

This is a rare infection in the UK but is common in many parts of the world with a mortality rate of about 60 per cent. The causal organism, *Clostridium tetani*, produces a powerful exotoxin which has a high molecular weight and is specifically neurotoxic. It is poorly eliminated from the blood by the kidneys and enters the spinal cord, by way of peripheral nerves, where it blocks inhibitory spinal reflexes.

Clinical features

There may be a history of injury in 60 per cent of patients, but this may be so trivial as to go unnoticed. The circumstances of contamination must be considered, e.g. injuries during sport or gardening (dung), application of dung to the umbilical cord of neonates.

Trismus, dysphagia, and stiffness of the neck, the back, and the abdomen are common. Generalized spasms with the danger of asphyxiation may develop. Characteristic signs are the risus sardonicus and opisthotonus (severe hyperextension of the back and neck due to spasm). By then the diagnosis is obvious. If the incubation period is short the prognosis is poor.

Differential diagnosis

Dental or jaw infection, drug reactions, hysteria.

Treatment

- Benzyl penicillin 1 g every 6 hours IV
- Human antitetanus immunoglobulin 30 units/kg/IMI
- If a wound is present it should be excised
- Culture and microscopy of all material excised
- Some patients will need to be admitted to an intensive care unit for ventilatory support.

Subsequent support

Depends on the severity of the attack. Muscle spasm can be relieved by diazepam orally or parenterally. Morphine is given for pain. If there is marked salivation or respiratory difficulties tracheostomy, paralysis with muscle relaxants, and positive pressure ventilation are indicated. Fluid balance and nutritional needs must be carefully monitored. Regular physiotherapy is required during the recovery period.

Complications

Hypertension, cardiac arrhythmias, sweating, and salivation. Control with benzodiazepines, opiates, and occasionally beta-blockers. Death is due to cardiac arrest, cerebral haemorrhage, stress haematemesis and melaena, bronchopneumonia, and pulmonary embolus.

Prevention

- Immunization of children—adsorbed tetanus toxoid 0.5 ml IM repeated after 6–8 weeks and again after 4–6 months. Booster doses of 0.5 ml are given 5 years after primary immunization.
- Patients with new wounds attending A&E—0.5 ml IM unless a booster has been given in the previous year.

Skin grafts

Definition

A skin graft is a segment of dermis and epidermis separated from its blood supply and transferred to another area of the body.

Types

Partial thickness (Thiersch) consists of epidermis and upper capillary dermis but not epidermal appendages. The graft is cut with a dermatome to about 0.25 mm thick. If kept moist it can be stored at 40 °C for 3–4 weeks. The donor area heals quickly, and further skin can be harvested after 14 days. It should be covered with an Opsite dressing.

- *Main uses*: Skin cover for burned patients. Tissue culture can increase harvest for widespread burns. Minor skin surgery to close defects.
- *Donor sites*: Arms, thighs, buttocks
- *Successful 'take'*: Depends on the vascular bed of the recipient area, prevention of seroma, haematoma, infection, and movement of the graft. Apply a bolus pressure dressing with foam or wool.

Full thickness (Wolfe) contains dermis and epidermis. Donor area may need closure with split graft if suturing is not possible.

- *Main uses*: Facial reconstruction, resurfacing the hand
- *Donor sites*: Pre- and postauricular areas, nasolabial fold, supraclavicular fossa, antecubital fossa, groin, instep of foot
- *Successful 'take'*: Depends on good vascular bed, immobility of graft. Apply with a tie-over bolus dressing.

Skin flaps

Skin and its subcutaneous tissue can be transferred from one area to another provided its vascular pedicle is maintained.

Types

Random flaps survive on blood vessels from the subdermal skin plexus. They are rotated or transposed to areas of adjacent skin loss. Larger flaps may be created by joining two random flaps, e.g. *the tube pedicle flap*. Such flaps can be raised as an abdominal, acromiopectoral, clavicular, or scapular tube. After 6 weeks one end of the tube is transferred to its transfer area, e.g. the wrist. A further 3 weeks then elapses before the tube is divided. The abdominal origin is repaired and the donor area left for 7 days before trimming the flap to make it site neatly on the defect and closing the skin.

Axial flaps contain a recognized arteriovenous system and may be at least 4 times as long as their base, e.g. deltopectoral flap based on perforating branches of the internal mammary plus veins, groin flap on superficial circumflex iliac plus veins, forehead flap on anterior branch of superficial temporal artery.

Muscle and myocutaneous flaps

Muscle or muscle and skin may be transposed axially to cover bare bone or large defects.

Example: Latissimus dorsi flaps for breast reconstruction after mastectomy.

Fasciocutaneous flaps

Include deep fascia and the vascular plexus superficial to it. Blood vessels in the subcutaneous tissue are not disrupted.

Example: Shin flaps raised on the medial or lateral aspects of the leg may be transposed over the tibia.

Microvascular free flaps

In this axial flap its neurovascular bundle is anastomosed to that in a donor area.

Example: Free groin flap transfer for jaw reconstruction.

Burns

Causes

Most are due to fire, steam, or scalding liquids. Some are caused by UV light (sunburn), irradiation, electricity, chemicals, and friction.

Classification

- *Partial thickness*: epidermis only (superficial). Minimal tissue damage. *Pain* is characteristic. Heals in 5–10 days. Can extend to depths of dermis (deep). Regeneration is from sweat gland and hair follicle remnants. Heals in 15–30 days depending on extent and presence of infection.
- *Full thickness*: dry, brown, white, or black appearance. Firm to touch. *Pain* not characteristic, due to destruction of pain receptors.

Severity and prognosis

- Depth
- Age—old and young have worse prognosis
- Body surface area (BSA)—more than 10 per cent (children), more than 15 per cent adults—IV fluids
- Associated disease—diabetes, congestive, cardiac failure, pulmonary disease—prognosis worse
- Site—respiratory tract—worse prognosis, hands—morbidity

Rule of nines (Wallace)

Each arm = 9 per cent, head = 9 per cent, anterior trunk = 18 per cent, posterior trunk = 18 per cent, each leg = 18 per cent, perineum = 1 per cent (adults), head = 18 per cent (child under 5 years), head = 13 per cent (child over 5 years) (see diagrams)

First aid

1. Stop the burning process—douse burning clothes with water, remove clothing in scalds. Immerse affected limbs in cold water.
2. Ensure a patent airway—remove from smoke, CPR if necessary.
3. Avoid contamination—cover the burn with a clean sheet or cloth soaked in cold tap water.
4. Transfer the patient to hospital.

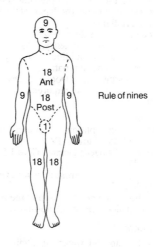

Rule of nines

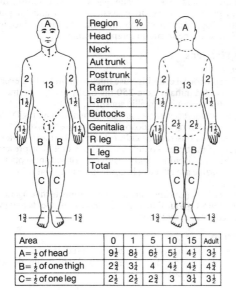

Region	%
Head	
Neck	
Aut trunk	
Post trunk	
R arm	
L arm	
Buttocks	
Genitalia	
R leg	
L leg	
Total	

Area	0	1	5	10	15	Adult
A = ½ of head	$9\frac{1}{2}$	$8\frac{1}{2}$	$6\frac{1}{2}$	$5\frac{1}{2}$	$4\frac{1}{2}$	$3\frac{1}{2}$
B = ½ of one thigh	$2\frac{3}{4}$	$3\frac{1}{4}$	4	$4\frac{1}{2}$	$4\frac{1}{2}$	$4\frac{3}{4}$
C = ½ of one leg	$2\frac{1}{2}$	$2\frac{1}{2}$	$2\frac{3}{4}$	3	$3\frac{1}{4}$	$3\frac{1}{2}$

Burns

Burns (*cont.*)

Hospital care

- *Outpatient*: if less than 15 per cent BSA in adults, less than 10 per cent in children (both partial thickness), less than 2 per cent BSA (full thickness)
- *Inpatient*: if greater than above
  - *airway*—maintain patency, do blood gases, carboxy haemoglobin. Humidified O_2 by mask. Respiratory stridor—consider intubation.
  - *fluid therapy*. The Brooke Army Hospital formula is commonly used but requirements should be tailored to each patient.

First 24 hours

- *Colloid* (plasma, plasma substitutes, dextron) Give 0.5 ml/kg/per cent BSA burned. IV
- *Electrolytes* (Ringer's lactate). Give 1.5 ml/kg/per cent BSA burned. IV
- *Water* (5 per cent dextrose in water). Give 2000 ml IVI (less for children)

Half the above volumes are given in the second 24 hours, after which most patients can drink normally.

Wound care

Clean with soap and water on admission (under analgesia). Puncture blisters. Aseptic technique. Use dry occlusive dressings in partial thickness to prevent infection and for patient transfer. Topical antibacterials (mafenide acetate) reduce infection. Excise or incise circumferential eschar.

Complications

- *Sepsis*
  - Streptococcal cellulitis—use flucloxacillin 500 mg every six hours IM initially then 250 mg orally every 6 hours. Give erythromycin if penicillin sensitive.
  - *Pseudomonas aeruginosa*—use gentamicin 80 mg IM or IV every 8 hours. Cephalosporins are also effective.
  In all situations regular cultures should be taken (twice per week) of the burns wounds. Antibiotics should be prescribed on the basis of culture and sensitivity. Topical antiseptics such as flamazine are effective in controlling bacterial colonization required
- *Stress ulcers* (Curling's ulcer): give cytoprotective agents like sucralfate 1 g four times daily by mouth. Antacids and IV H_2 antagonists are less effective.

Nutrition

Metabolic requirements can be increased fourfold. A high-protein/high-calorie diet is required. Involve the dietician/nutritional team early.

AIDS and the surgeon

Infection with the human immunodeficiency virus (HIV) results from the passage of infected body fluids (usually blood) from one person to another by several methods: anal and vaginal sexual intercourse, shared needles amongst drug abusers, and infected blood products. Infection with the virus is permanent, and all carriers will eventually manifest the disease. Surgeons are at risk because they come into contact with patients' body fluids in the most uncontrollable of circumstances and may injure themselves by glove puncture or from their needles and knives.

Categories of patients with HIV who may need surgical care

- Patients with AIDS who need surgery
- Patients with anorectal disease related to homosexuality who may have AIDS or HIV infection
- Patients with injuries or other surgical problems

High-risk procedures

- Lymph node, lung, or brain biopsy for the diagnosis of opportunistic infection or HIV
- Procedures to deal with unusual malignancies, e.g. Kaposi's sarcoma, B-cell and non-Hodgkin's lymphoma, squamous carcinoma of the mouth
- Anorectal disease, abscesses, and sepsis in homosexuals and drug addicts
- Abdominal operations and accident surgery in HIV-infected people
- Venous access procedures to central veins for blood sampling treatments or parenteral nutrition

Patients at risk

- Homosexual and bisexual males
- Intravenous drug abusers
- Haemophiliacs and other recipients of blood products before routine testing was introduced
- Indigenous population of sub-Saharan African countries
- Sexual partners of the above groups and their children

The surgeon's responsibility to the AIDS- or HIV-infected patient

The UK General Medical Council has made it clear that surgeons are obliged, if required, to operate on patients with AIDS or HIV infection.

Precautions for surgeons

- Know about HIV infection and how to manage such patients.
- Identify groups of patients at risk. Routine HIV testing without consent may be illegal.
- Protect nursing staff from contamination by body fluids, and ensure correct disposal of contaminated material.

- Theatre technique
  —Remove unnecessary equipment or personnel. Surgeon and staff only should operate.
  —Staff with open wounds should be excluded from theatre.
  —Plastic aprons, gloves (two pairs), glasses or visors should be worn.
  —Aprons (plastic), boots, and disposable drapes are recommended.
  —After surgery all disposable equipment should be discarded in specific containers.
  —The theatre should be cleaned with dilute hypochlorite solution.
  —All recovery staff must be aware of the protocol for the care of the HIV patient.
  —Blood-stained dressings should be replaced before the patient returns to the ward.
- Surgical technique:
  Use a no-touch technique if possible. Use stapling devices for GI anastomosis and skin closure. Avoid sharp instruments. Ensure haemostasis.

Accidental injury

If there is an injury, wash the wound with soap and water and let it bleed. Report the incident to infection control. Take blood for future testing.

References

Sim, A. (1988). Aids and the surgeon. *Surgery*, **59**, 1396–401.
Surgeons and HIV (1988). *British Medical Journal*, **296**, 80.

Viral hepatitis

This is the commonest liver disease in the world. It can cause acute liver failure with death resulting from extensive liver cell necrosis. It may also lead to chronic active hepatitis and in the case of hepatitis B, to the development of primary liver cancer.

Types

Hepatitis A Is caused by an enteric virus and was known formerly as infectious hepatitis. The antibody to the virus is anti-HAV. This is the commonest form of jaundice in children and young adults. Spread is by the faecal–oral route. The incubation period is about 1 month. IgM is present in the serum during the acute infection. IgG indicates past infection, and is detected in the serum of 40–50 per cent of the urban population of the UK.

Hepatitis B (formerly serum hepatitis) Is carried by 5 per cent of the world's population. Infection is largely blood-borne and can be transmitted by blood transfusion, inoculation, sharing syringes (drug addicts), sexual intercourse during menstruation (with an infected partner), homosexuality. The incubation period is 12 weeks. Antigens appear in the serum—HBsAg, the surface antigen; HBcAg, the hepatitis core antigen; HBeAg, the 'e' antigen; the Dane particle; double standard DNA and DNA polymerase activity. Antibodies formed against these antigens (anti-HBs, anti-HBc, anti-HBe) can be detected in the peripheral blood.

The antigen is a risk for hospital staff, especially in theatres, renal dialysis units, and transplant units.

Hepatitis C Is a variant which can cause a range of problems from a mild illness to cirrhosis of the liver. Its incubation period is around 4 weeks. It is detected in 1 in 150 screened blood donations.

Hepatitis vaccine

Should be offered to 'at risk' members of staff. Staff at risk include those who work with residential mentally handicapped patients, those in haemophilia centres, laboratory workers, pathologists, dentists, theatre staff, accident and emergency staff, staff in liver and renal units, staff in genitourinary clinics. Others may also qualify for immunization. Suspected contacts may also be immunized if the regime is started within the incubation period in addition to the administration of hyperimmune serum.

Reference

Finch, R. G. (1987). Time for action on hepatitis B immunisation. *British Medical Journal*, **294**, 197–8.

Measures to prevent spread of viral infections amongst medical and nursing staff

Modes of transmission of viral disease:

- Direct percutaneous inoculation of infected blood, e.g. needle prick, scalpel wound
- Entry of infection through minute skin abrasions after contact with spilled blood
- Entry of infection through mucosal surfaces after contamination with blood, e.g. accidental splashing of the eyes
- Transfer of infection by formites, e.g. contaminated equipment

Procedures designed to minimize transmission

- Identify infected patients by serology.
- Specific procedures for the care of infected patients, e.g. barrier nursing in a single room. Inform all staff involved with the patient of the diagnosis.
- Disposal of disposable items related to the patient's care
- Specific treatment and sterilization of used non-disposable equipment
- Additional precautions for surgeons and theatre staff (see page 64)
  - Wear a plastic apron beneath a disposable gown
  - Wear two pairs of gloves
  - Wear goggles to protect the eyes
  - Prevent accidental injuries
  - Minimize blood spillage and contamination
  - Wash all equipment with dilute hypochlorite solution or glutaraldehyde at the end of the procedure
- Use specific measures for disposal of the dead.

Actinomycosis and syphilis

Both are rare diseases.

Actinomycosis

A chronic granulomatous suppurative disease caused by a gram-positive anaerobe, *Actinomyces israeli*, which is commonly harmless and present in the pharynx. The development of the disease usually follows entry through an infected tooth socket.

Clinical features The head and neck are the commonest sites. 40 per cent of patients have either thoracic or abdominal involvement. The liver may become involved by blood-borne spread leading to multiple abscesses and so-called 'honeycomb liver'.

Characteristically there is induration with non-tender nodule formation and abscesses which discharge to develop into sinuses. Secondary infection is common.

Diagnosis
- Yellowish club-shaped 'sulphur granules' can be seen in the pus.
- Gram-positive mycelia on microscopy

Treatment Surgical excision. Long-term antibiotics, usually penicillin (up to 2–4 g daily IM or slow IV in 2–4 divided doses).

Syphilis

Now very rare indeed. It should, however, be considered and excluded in the differential diagnosis of ulcers (especially genital), aortic valve disease, thoraco-abdominal aneurysms and unusual bone infection (p 457). The causal organism is the spirochaete *Treponema pallidum*.

Clinical features
Types:
- Hunterian chancre (primary)
- Perineal lesions, coppery rash, PUO, swollen lymph nodes (secondary) } curable with treatment
- Vascular disease (syphilitic aortitis) osteitis and gumma formation (tertiary)
- Tabes dorsalis, psychiatric disease (quaternary) } lesions remain despite treatment
- Congenital syphilis

Tests
- Dark-ground microscopy of ulcer fluid
- VDRL (venereal diseases research laboratory)—non-specific
- TPHA (*Treponema pallidum*) haemagglutination assay—specific

If you suspect the diagnosis, ask advice from the laboratory.

Treatment Usually Procaine penicillin 0.6–1.2 g daily IM for 2–4 weeks

Tuberculosis

Primary pulmonary tuberculosis

(Hope, R. A. and Longmore, J. M. (1985). *Oxford handbook of clinical medicine*, p. 180).

This is usually treated with a combination of drugs, e.g. streptomycin, isoniazid, rifamipacin, ethambutol. Surgical measures are almost entirely employed for dealing with residual lesions which persist after adequate conservative treatment. When surgical therapy is indicated, segmental or lobar resection is the treatment of choice in nearly 90 per cent of cases. Thoracoplasty is necessary only when the risks of resection are unduly high or the disease too widespread for safe removal.

Intestinal tuberculosis

This can present as a confusing acute illness in recent immigrants.

The most commonly affected areas are: the small intestine (lower ileum) and the ileocaecal region.

Small intestinal region Small intestinal tuberculous lesions consist of circular ulcerative areas involving localized segments which end up as multiple strictures.

Signs and symptoms Diarrhoea. Stools have a particularly fetid smell and contain pus and occult blood. There is weight loss, and the patient is often already receiving treatment for pulmonary tuberculosis.

Diagnosis Barium meal often discloses complete absence of filling in the lower ileum, caecum, and most of ascending colon.

Treatment Prolonged chemotherapy. Operation is required in the rare event of perforation, and occasionally cicatrization causes intestinal obstruction and calls for surgical intervention.

Ileocaecal region Ileocaecal and colonic lesions are of the hypertrophic type.

Signs and symptoms There is a mass in the right iliac fossa. Thickening of subserous and submucous coats produce narrowing of the lumen and there is simultaneous nodal enlargement, peritoneal nodular deposits, and adhesions.

Weight loss, fever and night sweats, abdominal pain, gastro-intestinal upsets and changes in bowel habits are all common symptoms.

Diagnosis Family history, sputum for microscopy and culture, chest X-ray, ESR, barium meal and enema.

Treatment Early laparotomy and conservative resection or stricturoplasty. Avoid blind loops and short-circuiting operations.

Genito-urinary tuberculosis

Genito-urinary tuberculosis is always secondary to a tuberculous focus elsewhere.

Signs and symptoms The disease may present with:

- Symptoms related to the urinary tract, including repeated attacks of urinary tract infection despite adequate therapy, frequency of micturition, dysuria, incontinence, haematuria, vague loin pain. Chyluria is common in some parts of the world.
- Symptoms related to renal destruction including disturbances of uraemia, fatigue, weight loss, and anaemia
- Recurrent inflammatory lesions in the genital tract, e.g. chronic epididymo-orchitis
- Constitutional symptoms alone including weight loss, loss of appetite, night sweats, fever

Diagnosis
- Urine examination
- Radiological examination
- Cystourethroscopy

Treatment
Mainly medical, but if complications arise partial resection of the kidney may be required and nephrectomy in advanced lesions. Other surgical procedures that might be necesary are surgery for ureteric structure and cystoplasty.

Bone and joint tuberculosis

New disease is seen less and less in the UK. However, you may encounter features of past disease, e.g. Pott's disease, in patients. Learn to recognize them (page 456). Bone and joint tuberculosis is haematogenous in origin. The disease starts either in the synovial membrane or in intra-articular bone.

Signs and symptoms Locally the patient complains of ache in the joint which is worse on exertion or at night. The joint becomes increasingly stiff, partly because movement is painful and partly because movement is limited by adhesion formation, muscle spasm and bone destruction. As the bone is destroyed the joint may dislocate until deformity is obvious. In the spine swelling is not visible until a considerable quantity of tuberculous pus has collected and stiffness may be too slight to be noticed; therefore a mild ache may be the only symptom of a potentially crippling disease.

Systemically the patient feels unwell, listless, febrile and has night sweats.

Diagnosis Haematology and immunology—ESR and white cell count are raised, the latter with lymphyocytosis. The Mantoux test is positive.

Treatment Early in the disease the pathological progress can be arrested and cured with antituberculous drugs. If treatment is deferred until pus has formed, surgery will be required to evacuate the tuberculous abscess.

3 Common surgical problems

Tumours

The S mnemonic may be useful during examination: Swelling, Site, Size, Shape, Surface, Sensitivity, Skin, and other Structures (attachment to). Tumours are the result of uncontrolled and disordered cell proliferation which serves no useful function (syn: neoplasm; new growth).

Classification

- *Behavioural*: benign or malignant
- *Structural*: tissue of organ

Benign tumours are slow growing, usually encapsulated, do not spread, do not recur after excision, often multiple, do not usually endanger life. Effects are due to size and site. Histology: well differentiated, low mitotic rate, resemble tissue of origin.

Malignant tumours expand and infiltrate locally, encapsulation is rare, metastasize to other organs via blood, lymphatics or body spaces, endanger life if untreated. Histology: varying degrees of differentiation from tissue of origin, pleomorphic (variable cell shapes), high mitotic rate in comparison with benign tumours.

Structural classification

Tissue of origin	Tumour types
Epithelium	Benign: papilloma, adenoma (glandular epithelium) Malignant: carcinoma (adenocarcinoma, squamous cell carcinoma indicate cell types)
Connective tissue	Benign: fibroma (fibrous tissue), lipoma (fat), chondroma (cartilage), osteoma (bone), leiomyoma (smooth muscle), rhabdomyoma (striated muscle), etc. Malignant: sarcoma, e.g. fibrosarcoma, osteosarcoma, etc. (if well differentiated). Spindle cell sarcoma, etc. (if poorly differentiated). The benign to malignant differentiation can be difficult histologically in sarcomas
Neural tissue	These arise from nerve cells, nerve sheaths and supporting tissues, e.g. astrocytoma, medulloblastoma, neurilemmoma, etc.
Haemopoietic	The leukaemias, Hodgkin's disease, multiple myeloma, lymphosarcoma, reticulosarcoma

Melanocytes	Melanoma
Mixed origins	e.g. Fibro-adenoma, nephroblastoma, teratoma (all three germ layers), chorio-carcinoma
Developmental blastomas	e.g. Neuroblastoma (adrenal medulla), nephroblastoma (kidney), retinoblastoma (eye). These are malignant tumours of childhood. Spread is by the bloodstream except CNS tumours, e.g. medulloblastoma (brain) which does not metastasize outwith the skull. The tumours are usually poorly differentiated

Cysts

A cyst is a collection of fluid in a sac lined by endo- or epithelium which usually secretes the fluid.

Types

True cysts are lined by endo- or epithelium. False cysts are the result of exudation or degeneration, e.g. pseudocyst of pancreas, cystic degeneration in a tumour, etc.

Classification

Congenital

Sequestration dermoid: due to displacement of epithelium along embryonic fissures during closure, e.g. skin. Sites include outer and inner borders of orbit, midline of the body, anterior triangle of neck (brachial cyst), (cf. implantation dermoid due to skin implantation from injury).

Tubulo-dermoid/tubulo-embryonic: abnormal budding of tubular structures, e.g. enteric cysts, postanal dermoid, thyroglossal cyst.

Dilatation of vestigial remnants: e.g. urachal cysts, vitello-intestinal cysts, paradental cysts, hydatid of Morgagni, Rathke's pouch, branchial cleft cysts.

Acquired

Retention cysts: due to the blocking of a glandular or excretory duct, e.g. sebaceous cyst (sweat gland); ranula (salivary gland) and cysts of the pancreas, gall bladder, parotid, breast, epididymis, Bartholin's glands, hydronephrosis, hydrosalpinx.

Distension cysts: due to the distension of closed cavities as a result of exudation or secretion, e.g. thyroid or ovarian cysts; hygroma (lymphatic cysts), hydrocoele, ganglia, bursas (latter three are false cysts).

Cystic tumours: e.g. cystadenoma, cystadenocarcinoma of ovary

Parasitic cysts: e.g. hydatid cysts (*Taenia echinococcus*)

Pseudocysts: due to necrosis or haemorrhage—liquefaction and encapsulation, e.g. necrotic tumours, cerebral softening, pseudocyst of pancreas, etc.

Clinical features

Smooth, spherical swelling which may be soft and fluctuant when palpated in two planes with the fingers at right angles to each other. This depends on the tenseness of its contents, which if extreme may produce pain in the cyst or the surrounding tissue. If the fluid is clear the swelling will transilluminate. Ultrasound and aspiration of contents are methods of determining whether a given swelling is cystic and may differentiate a cyst from a lipoma.

Clinical effects

May compress surrounding tissues. May produce pain if complications supervene.

Complications

Cysts may burst or discharge spontaneously. They are also subject to infection, torsion if on a pedicle, haemorrhage, and calcification.

Treatment

Excision, marsupialization (deroofing and suture of the lining to skin), or drainage according to site.

Ulcers

An ulcer is a breach in an epithelial surface.

Classification

- *Specific* (e.g. TB)
- *Non-specific* (traumatic, pyogenic, vascular, neurogenic)
- *Malignant* (basal or squamous cell carcinoma, carcinoma of stomach, colon, breast etc., sarcomas)

Features to note on examination

- *Site*: Neck, groin, and axilla (TB); legs and feet (vascular); anywhere (malignant)
- *Surface*: Usually depressed. Elevated in malignancy, vascular granulations.
- *Size*: Measure the ulcer. Is it large by comparison to the length of history?
- *Shape*: Oval, circular, serpiginous, straight edges
- *Edge*: Eroded (actively spreading), shelved (healing), punched out (syphilitic), rolled or everted (malignant)
- *Base*: Fixed to underlying structures? Mobile? Indurated? Penetrating?
- *Discharge*: Purulent (infection), watery (TB), bleeding (granulation or malignancy)
- *Pain*: Usually occurs during the extension phase of non-specific ulcers. In diabetic patients ulcers are relatively painless.
- *Number*: Widespread locally (?local infection like cellulitis), widespread generally (constitutional upset)
- *Progress*: Short history (pyogenic), chronic (vascular or trophic, e.g. phlebitic syndrome, decubitus ulceration of paraplegia)
- *Lymph nodes* in the region of an ulcer may indicate secondary infection or malignant change.

Natural history

- *Extension*: There is discharge, thickened base, inflamed margin. Slough and exudates cover the surface.
- *Transition*: Slough separates and the base becomes clean. The discharge becomes scanty, the margins less inflamed.
- *Repair*: Granulation becomes fibrous tissue and forms a scar after re-epithelialization.

Investigations

History, biopsy and histology, serology as indicated by presentation.

Treatment

Treat the underlying cause in specific and malignant ulcers. Use desloughing agents, curettage, skin grafting for non-specific ulcers. Chronic ulcers may respond to ultraviolet light.

Sinuses and fistulas

Definitions

A *sinus* is a blind epithelial track, lined by granulation tissue which extends from a free surface into the tissues.

Examples: pilonidal sinus, perianal sinus.

A *fistula* is an abnormal communication between two epithelial surfaces. It is also lined by granulation tissue and colonized by bacteria.

Examples: Fistula in ano, pancreaticocutaneous fistula, colovesical, vesicovaginal fistual, etc.

Causes
- Specific disease, e.g. Crohn's
- Abscess formation and inadequate drainage, e.g. perianal abscess may lead to fistula in ano
- Penetrating wounds
- Iatrogenic
- Neoplastic

Non-closure is due to:
- Necrotic material in a sinus, e.g. bone
- The presence of foreign material, e.g. a suture
- Inadequate drainage of an abscess
- Distal obstruction (intraluminal, intra- or extramural)
- Continuing sepsis, e.g. infection or discharge of faeces—chronic infection e.g. TB, actinomycosis
- Ramification of a sinus or fistula may lead to complex architecture and epithelialization. This can occur in persisting disease like Crohn's or cancer.

Investigation

Establish the extent by sinography or a fistulogram.

Treatment

Provide adequate drainage by: laying it open, removing septic material, and excising the track completely. Scrape all granulations away from the walls and pack the cavity left to allow it to heal by granulation. The ends of a fistula should be disconnected. Recurrence may be prevented by tissue interposition, e.g. interposition of omentum between bladder and bowel in colorectal fistula.

Where a sinus or fistula leads to a cavity it may be necessary to close the cavity completely or disconnect a fistula and repair the organs involved.

Most fistulas will close spontaneously if there is adequate distal drainage, accurate measurement of output with adequate fluid and nutritional replacement. High-output GI fistulas (more than 200 ml/day) should be treated by giving the patient nil orally, ensuring drainage by nasogastric suction and adequate bowel evacuation and giving the patient IV feeding. Elemental diets are safer than IV feeding. They can be instituted by either tube feeding or jejunostomy below a high GI fistula. If the fistula fails to close after 7–10 days, definitive surgery should be considered.

Reference

Soeters, P. B., Ebeid, A. M., and Fischer, J. E. (1979). Review of 404 patients with gastrointestinal fistulas. Impact of parenteral nutrition. *Annals of Surgery*, **190**, 189–202.

Gangrene

Definitions

Gangrene is necrosis with putrefaction, usually by saprophytic organisms in the dead tissue (*wet gangrene*). The term is also applied to necrosis of tissue with desiccation or mummification (*dry gangrene*) following infarction (usually of the extremities).

Slough is dead tissue. It becomes separated from living tissue during the process of ulceration or necrosis.

Causes

Gangrene most commonly results from failure of the blood supply to tissue. It can also be the result of physical or chemical trauma, e.g. frostbite (cold), trench foot (immersion in water), intra-arterial injection of thiopentone, ergot preparations, carbolic acid.

Types

- *Vascular*: due to arterial or venous disease. The commonest cause is atherosclerosis of senility.
- *Traumatic*: e.g. strangulated hernia, intimal flap leading to arterial occlusion.
- *Thermal or physical*: burns and scalds, frost bite, trench foot, irradiation.
- *Infective*: Due to sepsis, e.g. gangrenous appendicitis, gas gangrene (*clostridium perfringens*).

Clinical appearances

Dry gangrene The affected limb, digit, or organ is black because of breakdown of haemoglobin, dry, and shrivelled. Dry gangrene shows little or no tendency to spread.

Moist gangrene Veins as well as arteries are blocked. Pain is initially severe but lessens as the patient becomes more toxic. There is always infection. The skin and superficial tissues become blistered and gases elaborated from putrefaction lead to emphysema and crepitus. There is a broad zone of ulceration which separates it from normal tissue. Proximal spread is a feature leading to septicaemia and death.

Separation of gangrene

In dry gangrene a zone of demarcation appears between the dead and viable tissue and separation begins to take place by aseptic ulceration in a few days.

In moist gangrene the zone is vague and separation is by septic ulceration usually resulting in greater loss of tissue or septicaemia.

Treatment

- *Limb salvage procedures*: Conservative excision may be successful if the proximal blood supply is good or can be improved
- *Amputation*: This is life-saving in traumatic, spreading, or gas gangrene

General measures

Correct cardiac arrhythmias or failure. Control diabetes, pain. Correct anaemia. Give antibiotics, IV broad-spectrum, e.g. cefuroxime 750 mg IV + metronidazole 500 mg IV 8 hourly.

Local measures

- Expose the affected limb to the air
- Release pus by desloughing to remove dead skin to assist demarcation

N.B. Make every effort to keep dry gangrene dry and to convert moist gangrene to the dry variety. Laparotomy is mandatory for the removal of gangrenous bowel, but unfortunately often too late to save the patient.

4 Pre- and postoperative care

Preparation for theatre

Operations should be carried out under optimal conditions to ensure the best possible result for the patient. Therefore certain basic rules must be observed.

Blood

- Ensure that the patient is not anaemic. Do a blood haemo-globin estimation and full blood count (FBC) before even minor operations.
- If the patient is anaemic (rule of thumb, haemoglobin > 10 g/dl) postpone an elective procedure until it is investigated and corrected. If an emergency operation is necessary, correct the anaemia by infusion of blood or packed cells.
- Remember that black patients may have sickle-cell problems. Order a sickle-cell test. Exclude blood disease and bleeding tendencies, especially if the patient is jaundiced or has liver disease. Vitamin K analogue injections 10 mg 8 hourly may be required to correct abnormal prothrombin time.
- For all major surgery, ensure that blood has been cross-matched or serum-grouped, and retained.

Correct patient

If you are the surgeon who is going to operate, talk to every patient beforehand. Do not allow the patient to be anaesthetized until you are satisfied. It is your sole responsibility.

Correct operation (p. 8)

Ensure that the correct procedure is done by marking the appropriate side (in cases of hernias, varicose veins, lumps, etc). See the patient before anaesthetic and confirm the site of operation in all cases.

Theatre routines

- Use standard aseptic scrub-up techniques.
- Do orthodox procedures at the start of your career.
- Do not be overconfident. Know your limitations and be honest about them. Seek help if necessary.
- Use large swabs and as few as possible at a time.
- Do not bury packs in the abdomen—rather attach a clip to the tape outside the abdomen.
- Check the wound and abdomen yourself, even though the count is correct.
- Remember that instruments may have been left in the wound. Check these assiduously before closure.

There are no new mistakes.
Try to avoid the time-honoured ones!

Consent

Informed. Parental or guardian if under 16 years. Get consent for orchidectomy in recurrent hernias in elderly males or in orchidopexy in adults.

Instructions to nurses

Removal of dentures, jewellery, etc. The site should be shaved by an orderly, e.g. upper abdomen (cholecystectomy), abdomen and groins (aortic bifurcation), groin (hernia repair). Preop diet—nil orally for patient from 12 midnight on day before surgery. Passage of nasogastric tube in GI surgery. Bowel preparation. Catheter. Premedication. Prophylactic antibiotics (p. 112).

Inform

- Laboratories re frozen section
- Anaesthetist re specific problems
- Physician re diabetes, thryoid disease, etc. (if required by anaesthetist)

Chest X-rays and surgical patients

Preoperative

'Routine' preoperative chest X-rays will detect significant abnormalities like cardiomegaly, tuberculosis, pulmonary infection, and collapse in 10 per cent of patients. The yield increases with age, and 40 per cent of patients over 65 years will show abnormalities.

Effect of X-ray findings on patient management

The chest X-ray findings do not significantly influence the decision to operate. 96 per cent of patients with a normal report and 92 per cent with an abnormal report still proceed to operation, and inhalational anaesthesia is used in greater than 96 per cent. In 25 per cent of patients the operation is carried out without the X-ray report being available.

Baseline chest X-rays

These are carried out to provide comparison with postoperative chest films. They may be of value in older patients but are unhelpful in young adults and children. They should always be done before general anaesthetic in patients over 50 years old.

Costs

Chest X-rays are costly, especially at the bedside.

Guidelines for taking X-rays in surgical patients undergoing elective non-cardiopulmonary surgery

- The 'routine' preoperative chest X-ray is unjustified in terms of detection of abnormalities, effect on patient management and cost.
- Preoperative chest X-rays should be considered if:
  —the patient has acute respiratory symptoms.
  —metastases are suspected.
  —there is established or suspected cardiorespiratory disease in patients who have not had a chest X-ray in the last 12 months.
  —the patient is an immigrant from a country where TB is still endemic and has not had an X-ray in the last 12 months.
- Postoperative chest X-rays should be taken for diagnostic reasons.
  N.B. Always give your clinical reasons on the request form for a chest X-ray; 'Routine—preoperative assessment' is not acceptable.

Reference

Fowkes, F. G. R. (1986). The value of routine preoperative chest X-rays. *British Journal of Hospital Medicine*, **35**, 120–33.

Preoperative biochemical testing

Indications

- To confirm suspected abnormalities before surgery
- As a screening procedure to detect unsuspected abnormalities

Common conditions detected biochemically

- *Renal disease*: The effects of muscle relaxants may return once the action of a reversing agent like neostigmine wears off. Renal failure may also be precipitated.
- *Hepatic disease*: Hepatic failure may follow surgery. If there is jaundice the hepatorenal syndrome may develop postoperatively.
- *Diabetes mellitus*: Unrecognized hypoglycaemia which can be sudden is a danger whilst the patient is under anaesthesia. Diabetic ketoacidosis may develop in undiagnosed diabetes and cause death during prolonged surgical procedures.
- *Abnormalities of serum, potassium and sodium*: Hyperkalaemia predisposes to cardiac arrest, hypokalaemia to cardiac arrhythmias. These effects may be exacerbated by anaesthetic drugs like suxamethonium (muscle relaxant) and halothane.

Incidence of unsuspected abnormalities

In patients under 50 who appear clinically well, screening will detect abnormalities of U&Es in around 0.06 per cent. Abnormalities of creatinine or urea will be found in less than 1 per cent, but abnormal glucose concentrations will be found in around 8 per cent. Therefore, in patients who are clinically well, most surgically significant abnormalities can be detected by testing the urine with potential savings in laboratory costs and staff time.

Guidelines

- Take a careful history. Biochemical investigations are justified in patients with unexplained symptoms.
- Try to avoid excessive 'routine' biochemical testing. Remember the 'no-lose' philosophy.
- Test the urine of all patients. Specifically test for glucose, protein, and bilirubin.
- If an abnormality is found then carry out specific tests, e.g. blood glucose, serum creatinine, etc.
- In patients under 50 years, routine urine testing is an adequate screening method.

Reference

Campbell, I. T. and Gosling, P. (1988). Pre-operative biochemical screening. *British Medical Journal*, **297**, 803.

Nutrition in surgical patients

Malnutrition is linked to postoperative morbidity and mortality. There is reduced resistance to infection, and poor wound healing. These complications are observed in patients undergoing radio- or chemotherapy and in diabetes, older patients, and those who are immunosuppressed or on long-term steroids.

Examine and weigh the patient. Is there evidence of weight loss?—loose skin folds, gaunt appearance, 'I've lost 1½ stones in 3 months'. Ask how the weight loss came about, e.g. recurrent vomiting, pain on eating, diarrhoea, etc. Weigh the patient (loss of 10 per cent is accepted as evidence of malnutrition). What is the weight related to the patient's height?

Benefits of nutritional support

Immune competence increases. Wound and anastomotic healing improve when albumin levels approach normality, and survival from surgical complications is more likely. Postoperative nutritional support reduces the length of stay in hospital.

When should nutritional support be used?

- Preoperatively in malnourished patients
- Patients with exacerbations of inflammatory bowel disease
- Patients with sepsis and major surgical complications
- Patients with external GI fistulas
- Patients with extensive burns or trauma
- Patients undergoing chemotherapy or radiotherapy for certain cancers

Which route?

If the GI tract is available, use it.

Oral Food preparations or supplements can be given if the GI tract is functioning. Give through fine-bore nasogastric tubes, e.g. Silk and Corsafe (E.Merck Ltd.), Clinifeed.

Enteral Fine-bore nasogastric tubes, gastrostomy, and feeding jejunostomy in patients with functioning small bowel. Give a commercial feed containing 1 kcal/ml and 5 g nitrogen/l. Start by continuously infusing at 25 ml/h, gradually increasing to 100 ml/h.

Parenteral
- *Peripheral vein* feeding is useful in the short term (up to 5 days). Patients with malnutrition for malignant disease may benefit from preoperative nutrition of this kind.
- *Total parenteral nutrition (TPN)* via a central vein is indicated when other methods of feeding are either impossible or unsuitable. Most patients require about 2500 kcal daily, but more in severe burns or sepsis. Carbohydrate as glucose and fat are used as energy sources. Nitrogen requirements are about 10 g/24 h, except in hypermetabolic states when 25–30 g/day may be required. Most parenteral feeding systems are now presented in a 2.5–3 l bag format. Each bag contains all

the nutrients required for a 24-hour period. They are easy to use for nursing and medical staff, and are safe for the patient. Control of the rate of administration is essential, and counting devices or infusion pumps should be used. Maintain a separate peripheral line to correct fluid and electrolyte losses or administer drugs.

Monitoring TPN

- Daily blood sugar, U&Es and 6 hourly dip stick test for blood sugar. Correct elevations with insulin infusion at a rate of 1 unit/h until normal levels are reached.
- Twice weekly LFTs and FBC, and blood calcium, magnesium, and phosphate levels
- Blood cultures if the patient is pyrexial
- Dress the insertion site of the catheter every 7 days under aseptic conditions.
- Use the feeding line only for that purpose.
- Remove the central venous catheter if the patient is septi-caemic. Culture the tip.

Complications

Catheter-related
- *Air embolism*: Avoid by keeping the patient supine when manipulating the line.
- *Arrhythmias*: ECG monitoring may be necessary.
- *Haemo/pneumothorax*: Chest X-ray; insert a chest drain if necessary.
- *Haemopericardium*: Both occur due to perforation by the catheter and wire. The flexible J-tip introducer wire reduces the risk. Pericardiocentesis may be indicated.
- *Central venous thrombosis*: Heparin 1000 units/l of infused fluid may be prophylactic.

Metabolic
- Deficiencies of folate, magnesium, zinc, and other trace elements. Blood sugar abnormalities. Electrolyte abnormalities. Hyperosmolar diureses.
- Prevent by careful monitoring (see above).

References

Clarke, P. J., Ball, M. J., Turnbridge, A., and Kettlewell, M. G. W. (1988). The total parenteral nutrition service: an update. *Annals of the Royal College of Surgeons of England*, **70**, 296–300.

Shenkin, A., Fraser, W. D., McLelland, A. J. D., Fell, G. S., and Garden, O. J. (1987). Maintenance of vitamin and trace element status in intravenous nutrition using a complete nutritive mixture. *Journal of Parenteral and Enteral Nutrition*, **11**, 238–42.

Examples of feeding regimens

Intravenous feeding via a central line

Regime 1—standard

	Volume	Total energy	Nitrogen (g)	Na (mmol)	K (mmol)	Ca (mmol)	Mg (mmol)	P (mmol)
Vamin 9 glucose	1000	650	9.4	50.0	20.0	2.5	1.5	
Addamel	10					5.0	1.5	
15% Potassium chloride	10				20.0			
Glucose 10%	1000	400						
Addiphos	15			22.5	22.5			30.0
Intralipid 20%	500	1000						7.5
Vitlipid N	10							
Solvito N	1 vial							
Total	2545	2050	9.4	72.5	62.5	7.5	3.0	37.5

Infusion protocol (*see opposite*)

Time (h)	Line A		Line B	
8	Vamin 9 glucose	500 ml	Glucose 10%	500 ml
	Addamel	10 ml		
	15% KCl	10 ml		
8	Vamin 9 glucose	500 ml	Intralipid 20%	500 ml
			Vitlipid N	10 ml
			Solvito N	1 vial
8	—		Glucose 10%	500 ml
			Addiphos	15 ml

- Addamel (10 ml) and potassium chloride (10 ml) are added to 500 ml of Vamin 9 glucose which is then infused over 8 h in *Line A*. Glucose 10 per cent (500 ml) is infused simultaneously in *Line B*.
- The contents of 1 vial of Solvito are dissolved by the addition of 10 ml of Vitlipid N and the reconstituted mixture aseptically transferred to 500 ml Intralipid 20 per cent. This is then infused over 8 h in *Line B*.
  Vamin 9 glucose (500 ml) is infused simultaneously in *Line A*.
- 15 ml Addiphos is added to 500 ml glucose 10 per cent which is infused over 8 hours in *Line B*.

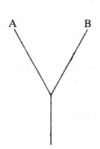

Illustration of feeding system (when 3L bags are not available)

Regime 2—increased requirements

	Volume	Total energy	Nitrogen (g)	Na (mmol)	K (mmol)	Ca (mmol)	Mg (mmol)	P (mmol)
Vamin 9 glucose	1500	975	14.1	75.0	30.0	3.75	2.25	
Addamel	10					5.00	1.50	
15% Potassium chloride	20				40.0			
Glucose 20%	1000	800						
Addiphos	15			22.5	22.5			30.0
Intralipid 20%	500	1000						7.5
Vitlipid N	10							
Solvito N	1 vial							
Total	3055	2775	14.1	97.5	92.5	8.75	3.75	37.5

Infusion protocol (*see opposite*)

Time (h)	Line A		Line B	
8	Vamin 9 glucose	500 ml	Glucose 20%	500 ml
	Addamel	10 ml		
	15% KCl	20 ml		
8	Vamin 9 glucose	500 ml	Intralipid 20%	500 ml
			Vitlipid N	10 ml
			Solvito N	1 vial
8	Vamin 9 glucose	500 ml	Glucose 20%	500 ml
			Addiphos	15 ml

- Addamel (10 ml) and potassium chloride (20 ml) are added to 500 ml of Vamin 9 glucose which is then infused over 8 h in *Line A*. Glucose 20 per cent (500 ml) is infused simultaneously in *Line B*.
- The contents of 1 vial of Solvito N are dissolved by the addition of 10 ml of Vitlipid N and the reconstituted mixture aseptically transferred to 500 ml Intralipid 20 per cent. This is then infused over 8 h in *Line B*.
  Vamin 9 glucose (500 ml) is infused simultaneously in *Line A*.
- 15 ml Addiphos is added to 500 ml glucose 20 per cent which is infused over 8 hours in *Line B*.
  Vamin 9 glucose (500 ml) is infused simultaneously in *Line A*.

Nutrition in surgical patients (*cont.*)

Enteral feeding regime
- *Always use a fine-bore tube*, e.g. Clinifeed, and use the following method for *continuous drip feeding*.
- Check tube is in the stomach by X-ray or aspiration.
- Introduce 50 ml water; aspirate after 1 hour to check stomach emptying.

- Send diet request form to dietitian.
- Order feed from pharmacy (Ensure). Feeding equipment: Clinifeeding System 3 (1 l reservoir and drip line) and Clinifeeding nasogastric tube are also available from pharmacy.
- Clinifeed System 3 is presented as a sterile package, and Ensure is in sterile tins. IT IS IMPORTANT THAT BOTH FEED AND EQUIPMENT REMAIN AS CLEAN AS POSSIBLE. Feeds must be made up immediately prior to use and NOT IN ADVANCE. Wear sterile gloves when filling the reservoir with feed. If the feed needs to be diluted, USE STERILE WATER.
- Administer feed according to regime given by dietitian. The following is given for guidance only:
  - *Day 1*: 3 tins Ensure + 300 ml water given over 12 hours, then 4 tins Ensure given over next 12 hours
    Total for day = 2000 ml fluid; 1750 kcal; 61.6 g protein
  - *Day 2 onwards*: 8 tins Ensure given over 24 hours
    Total for day = 2000 ml fluid; 2000 kcal; 70.4 g protein
  In both cases 2000 ml fluid is given over 24 hours, which gives an *approximate* drip rate of 15 drops per minute. This may be sufficient for the patient's daily needs, or may need to be increased according to the dietitian's instructions. Discard reservoir and run 50 ml water to clean tube every 24 hours.

Hazards of enteral feeding

Diarrhoea may be due to:
- *Bolus feeding*: *always* feed continuously by gravity drip, *never* give as bolus feeds.
- *Not enough fluid*: keep fluid balance chart to check that fluid intake is approximately equal to output.
- *Antibiotics*: commonly provoke diarrhoea. If the antibiotic cannot be withdrawn the diarrhoea may be controlled with loperamide (Imodium).
- *Giving a hyperosmolar feed*. Ensure has a low osmolarity. Any other liquid given via the tube will cause diarrhoea if it has a high osmolarity, e.g. Hycal.
- *Feed contamination*. This should not occur if care is taken in preparing feeds and feeds are *not* made up in advance.

Nausea This is usually preventable if the feed is given slowly by the continuous drip method described over 24 h. *If nausea persists, consult dietitian.*

Monitoring enteral feeding It is important to monitor the patient's progress if enteral feeding is to be successful and adequate for the patient's needs.

- *Feed administration*: Check drip rates at intervals to ensure even flow. Aspirate periodically to check stomach emptying.
- *Fluid balance*: Fluid balance charts must be strictly maintained throughout enteral feeding. Include all oral fluids taken as well as the enteral feed.
- *Laboratory tests*: Serum urea, electrolytes and glucose must be monitored daily until feeding is well established. Serum albumin should be measured initially and then at weekly intervals.
- *Weight*: Where possible, weigh patient daily as an index of nutritional status.

Changing from enteral to oral feeding The change from enteral to oral feeding must be gradually phased to ensure that adequate nutrition is maintained. Records of food taken orally must be kept. Consult dietitian when tube feeding is likely to be discontinued.

If phasing from enteral to oral feeding is impractical, e.g. if patient can no longer tolerate the tube, the dietitian must be consulted to ensure the patient receives adequate nutrition.

N.B. It takes six sachets of Build Up and three pints of milk OR 500 g Complan to provide 2000 kcal.

Body water and electrolytes

There are 42 l of water in a 70 kg man, made up of intracellular fluid (ICF = 40 per cent of body weight = 28 l) and extracellular fluid (ECF = 20 per cent of body weight = 14 l). Extracellular fluid further comprises interstitial fluid (15 per cent body weight = 10.5 l) and plasma (5 per cent body weight = 3.5 l).

Composition

The cell membrane is relatively impermeable to Na^+ and K^+, but despite this the osmolalities of ICF and ECF are similar. This is because water can pass freely from compartment to compartment along any osmotic gradient. In the ICF the principal cation is K^+; in the ECF it is Na^+. Therefore ICF osmolality is K^+-related and ECF osmolality is Na^+ related; total body water is proportional to total body Na^+ and K^+.

Loss or gain of electrolytes from either compartment will lead to a proportional contraction or expansion, but because of the distribution of body water (40 per cent ICF, 20 per cent ECF) loss of a given volume of saline from the ECF will produce a much greater circulatory disturbance than loss of water alone.

The main anions in ECF are chloride and bicarbonate. In ICF the main anions are organic phosphates and proteins. ICF osmolality is about twice the plasma Na^+ concentration, but correction must be made for elevated plasma urea or glucose which will alter ECF osmolality.

Plasma and interstitial fluid

Protein plays a major part in determining plasma volume because of its osmotic effect, and is in higher concentration in plasma than in interstitial fluid. Otherwise their composition is similar.

Water balance

Intake (ml)		*Output* (ml)	
Water as fluid	1200	Urine	1500
Water in food	1000	Lungs	400
Oxidation	300	Skin—sweat —insensible loss	500
		Faeces	100
	2500		2500

Control of water balance

This is effected through osmoreceptors in the hypothalmus which are sensitive to changes in plasma osmolality and stimulate or inhibit thirst or the secretion of antidiuretic hormone (ADH).

Thirst is induced by increased plasma osmolality (from water loss), hypovolaemia, and angiotensin.

Antidiuretic hormone controls water excretion. Its secretion from the posterior pituitary is stimulated by increase in osmolality, hypovolaemia, trauma, and pain.

Reference

Fleck, A. and Ledingham, I., McA. (1988). Fluid and electrolyte balance In *Jamieson and Kay's textbook of surgical physiology*, 4th edn. (eds I. McA. Ledingham and C. McKay), pp. 35–54. Churchill Livingstone, Edinburgh.

Disorders of water and electrolyte balance

Water depletion

	Mild	Moderate	Severe
Deficit	2 l	2–4 l	> 4 l
Signs and symptoms	Thirst, oliguria, dry tongue	As before, + weakness, tachycardia, increase in plasma volume	Decreased plasma volume, hypotension, GFR down, raised urea + raised plasma Na^+

Water excess

More than 5 l leads to headaches, confusion, and convulsions.

Sodium depletion

Deficit	Mild (300 ml)	Moderate (300–600 mmol)	Severe (600–1300 mmol)
Signs and symptoms	Lassitude, fainting, postural hypotension	ECF falls, haematocrit rises, marked postural hypotension, GFR down, urea raised	Shock, severe hypotension, urea raised, Na^+ down, confusion, coma

Sodium excess

This is seen clinically in postoperative and post-traumatic states when equal amounts of Na^+ and water are retained. The body weight may increase by 15 per cent but there is no oedema because of equal retention of Na^+ and water. In congestive cardiac failure and cirrhosis there is excess aldosterone secondary to inadequate renal perfusion. The interstitial compartment expands, leading to clinical oedema.

Potassium depletion

Deficit	Mild (200–400 mmol)	Moderate (600 mmol)	Severe (1000 mmol)
Signs and symptoms	None	Early symptoms are weakness and anorexia; ECG changes appear	Myasthenia, paralysis, ileus, cardiac arrhythmias, metabolic acidosis, dilute acid urine

Potassium excess

More than 7.5 mmol/l leads to muscle paralysis, cardiac arrhythmias with peaked T waves on the ECG.

Clinical examples

Mixed abnormalities are commonly seen in practice, usually related to diarrhoea, fistula losses, or vomiting.

Example 1: Patient with vomiting due to pyloric stenosis

	Na	K	Cl	Urea	pH	PCO$_2$ (mmHg)	Base excess	PCV	Urine pH
Electrolytes (mmol/l)	130	3.0	63	15.2	7.5	55	+16	60%	5.0

Comment Na$^+$ is low, indicating that Na$^+$ loss is greater than water loss. There is a metabolic alkalosis leading to K$^+$ loss in urine. Chloride is low due to loss of H$^+$ and Cl$^-$ in the vomit. The metabolic alkalosis is due to H$^+$ loss in vomit and K$^+$ depletion. The ECF is contracted as a result of Na$^+$ loss in digestive secretions, so the PCV is elevated. The acidic urine in the presence of an alkaline plasma is diagnostic of K$^+$ depletion and is due to a renal mechanism which acts to preserve K$^+$ in the presence of severe fluid losses and hypokalaemia by excreting H$^+$ in the urine.

Principles of treatment
- Restore ECF volume with 0.9 per cent saline.
- Give KCl to restore K$^+$. The metabolic alkalosis will self-correct with restoration of K$^+$ and fluid balance.
- Repeat the U&E at regular intervals during treatment.

Example 2: Intestinal obstruction

High obstruction Lesions below the duodenal papilla lead to severe effects as the patient may vomit all gastric, biliary and pancreatic secretions.

	Na	K	Cl	Urea	pH	PCO$_2$ (mmHg)	Base excess	PCV
Electrolytes (mmol/l)	135	3.2	90	14.0	7.46	40	+5	60%

Low obstruction Sufficient reabsorption of water and electrolytes delays salt and water depletion.

	Na	K	Cl	Urea	pH	PCO$_2$ (mmHg)	Base excess	PCV
Electrolytes (mmol/l)	138	3.8	100	10.0	7.34	35	−5	52%

Principles of treatment Restore the ECF volume with 0.9 per cent saline and correct K$^+$ depletion as above.

Example 3: Patient losing 1–2 l of alkaline fluid per 24 hours through an intestinal fistula

	Na	K	Cl	Urea	pH	PCO$_2$ (mmHg)	Base excess
Electrolytes (mmol/l)	128	5.6	100	16.2	7.30	28	−10

Comment There is depletion of Na$^+$ and K$^+$ with a metabolic acidosis. Plasma K$^+$ is evaluated because acidosis displaces K$^+$ from the cell.

Principles of treatment 0.9 per cent saline + sodium bicarbonate. Plasma K$^+$ will fall as the acidosis is corrected. K$^+$ must then be added to the IV fluids. Monitor the U&E levels regularly.

Acid–base balance

Normal regulation of changes in pH is by:

- *Paired buffer systems*
  - carbonic acid and sodium bicarbonate
  - plasma protein and sodium proteinate
  - sodium dihydrogen phosphate and disodium hydrogen phosphate

- *Pulmonary excretion of carbon dioxide*. Carbon dioxide is transported in solution in the plasma as carbonic acid; in the red cells as carbaminohaemoglobin (25 per cent) where it forms carbonic acid under the action of carbonic anhydrase then dissociates to H^+ ions and HCO_3^- ions. H^+ ions combine with haemoglobin. HCO_3^- ions exchange with Cl^- ions outside the red cells, leading to chloride shift. At pulmonary level the reverse process occurs and carbon dioxide is excreted.
- *Renal excretion of acids and bases*. Most are excreted as neutral bases, formation of which depends on the presence of carbonic anhydrase in the renal tubules. Under its action phosphates are excreted; NH_4^+ is substituted for Na^+ which is reabsorbed from the tubular urine. $NaHCO_3$ is also reabsorbed.

Disorders of acid–base balance

- Respiratory acidosis/alkalosis relates to retention or excessive loss of carbonic acid.
- Metabolic acidosis/alkalosis relates to retention or loss of H^+.

These can be uncompensated, the initial phase, or compensated when regulatory physiological mechanisms come into play to restore the status quo.

Respiratory acidosis (PCO_2 greater than 6.0 KPa)
Causes: Hypoventilation due to drugs, surgery, or faulty ventilatory systems.
Compensation: H^+ is excreted in the renal tubules. HCO_3^- is reabsorbed.

Metabolic acidosis
Causes: Prolonged tissue hypoxia, e.g. cardiac arrest, abdominal aortic surgery (following release of the aortic clamp), small-bowel fistulas, colonic bladder (hyperchloraemic acidosis).
Compensation: Excretion of CO_2 by hyperventilation. Excretion of an acid urine. Retention of HCO_3^-.

Respiratory alkalosis (PCO_2 less than 4.7 kPa)
Causes: Hysteria, high altitude.
Compensation: Renal excretion of HCO_3^- leading to production of alkaline urine.

Metabolic alkalosis
Causes: Chronic alkali ingestion, potassium depletion, vomiting (see pyloric stenosis).

Reference

Fleck, A. and Ledingham, I. McA. (1988). Fluid and electrolyte balance in *Jamieson and Kay's textbook of surgical physiology*, 4th edn (eds I. McA. Ledingham and C. MacKay) pp. 35–54. Churchill Livingstone, Edinburgh.

Stopping smoking before surgery

Cigarette smoking compounds the risks of anaesthesia and surgery. There is a sixfold increase in postoperative respiratory morbidity in patients smoking more than 10 cigarettes per day.

Effects of smoking

Smoking affects several aspects of the immune response with reduction in neutrophil chemotaxis, immunoglobin concentrations, and natural killer cell activity.

Platelet aggregability is also increased, but paradoxically there is a lower risk of isotopically-detected postoperative deep vein thrombosis in smokers compared with non-smokers. Nevertheless, there is no evidence that stopping smoking increases the risk of postoperative thrombotic problems and routine precautions against deep vein thrombosis will lower the risk of thrombo-embolism.

Tissue oxygenation is reduced by carbon monoxide, by the formation of carboxyhaemoglobin. This results in an increased risk of cardiovascular ischaemia in susceptible people.

Effects of stopping

After 24 hours abstinence, nicotine and CO are eliminated reducing the risk in patients with coronary artery disease. One week's abstinence improves pulmonary alveolar monophage performance, and after six weeks immunoregulatory T-cell activity returns to normal in heavy smokers. A minimum of 6 weeks' abstinence is needed before there is any beneficial influence on postoperative respiratory morbidity.

Advice to patients

Ideally, patients should be given firm advice in the hospital environment to stop smoking. If this is repeated at follow-up clinics, some patients may succeed. They should stop smoking as long as possible before surgery because of the undoubted short- and long-term benefits, but at least 6 weeks before an elective operation.

For those who are unsuccessful the hospital environment and prospect of surgery provide prime opportunities to give up. If, however, they cannot be persuaded to stop altogether, considerable benefit will result from stopping 12 to 24 hours before theatre. This is of particular importance in ischaemic heart disease.

Reference

British Medical Journal (1985). **290**, 1763–4.

Anaesthetic agents

Objectives of anaesthesia

- Unconsciousness, analgesia, skeletal muscle relaxation (general anaesthetic)
- Regional or local anaesthesia
- Haemodynamic and myocardial stability pre- and post-operatively
- Rapid induction and recovery
- Minimal side-effects and toxicity

Drugs used in anaesthesia

Inhaled anaesthetics

Drug	Comment
Halothane Enflurane Isoflurane	Halothane is the most pleasant to use, especially in children. Dosage can be reduced by combination with hypnotics or opioids. All three may cause myocardial depression, and their use should be limited in hepatic and renal disease
Nitrous oxide	Rapid uptake and elimination. Its use reduces the dose requirements of volatile agents, opioids, hypnotics. May cause marrow depression. Contra-indicated in pneumothorax or in gaseous distension of the gut
Muscle relaxants	*Comment*
	Their use reduces the dosage requirements of inhaled agents. Their major side-effect is postoperative ventilatory depression.
Depolarizing agents	*Comment*
Succinylcholine	May increase intraocular and intragastric pressure. Postoperative myalgia, malignant hyperpyrexia, hyperkalaemia, and prolonged paralysis due to cholinesterase deficiency are possible side-effects
Non-depolarizing *d*-Tubocurarine Pancuronium Vecuronium	Variable onset and duration. Their action is prolonged in renal and hepatic failure. Drug interactions can affect the onset and duration of paralysis, e.g. aminoglycosides potentiate the action of *d*-tubocurarine

Intravenous anaesthetics

Opioids

Fentanyl (primary anaesthetic) Morphine Pethidine Buprenophine (anaesthetic supplements and analgesics) Naloxone (antagonists) Nalbuphine	Used widely in the care of surgical patients. They are usually given as premedication, and may be used intraoperatively to supplement regional or general anaesthesia. Postoperatively they are used as sedatives in patients with endotracheal tubes who are being mechanically ventilated and as systemic or spinal analgesics. The main side-effect is respiratory depression.

Hypnotics

Thiopentone	Commonly used induction agent. Is protective to the brain in patients at risk from cerebral ischaemia.
Midazolam	Short acting. Similar to diazepam. May cause respiratory depression
Etomidate	Not widely used because it may cause increased muscle tone.
Ketamine	This is analgesic in subanaesthetic doses. Causes disorientative anaesthesia (patient is unconscious but looks awake). It can cause postoperative delirium and myocardial depression.

Balanced anaesthesia

Implies the use of multiple drugs to achieve the objectives of anaesthesia. The patient usually receives a premedication. Anaesthesia is induced by thiopentone or other induction agent and maintained by a combination of inhalational, intravenous drugs and the use of muscle relaxants. Non-anaesthetic drugs are also used to maintain vital functions, e.g. beta-blockers control drug-induced tachycardias.

Reference

Hug (Jr), C. C. (1988). New perspectives on anaesthetic agents. *American Journal of Surgery*, **156**, 406–15.

Prophylactic antibiotics in surgery

Most prophylactic antibiotics are given to prevent wound infection. Some are indicated in instrumental procedures in potentially infected sites to prevent septicaemia. The aim is to achieve a high tissue level at the time of surgery, so most are given on induction of anaesthesia. Usually 1–3 doses of the antibiotics are given within 24 hours.

Established indications for prophylaxis

Type of surgery	Recommended regime	No. of doses
Gastrointestinal surgery	Cefuroxime 1500 mg IV	1 dose
	Cefuroxime 750 mg IV	8 and 16 h post-operatively
	Metronidazole 500 mg IV	1–3
Gastrectomy/ oesophageal surgery	Gentamicin 1500 mg IV	1 immediate pre-operative dose (3 in high-risk patients)
	Cefuroxime 1500 mg IV	1 immediate pre-operative dose (3 in high-risk patients) (see below)
	Metronidazole 500 mg IV	(1 dose, 3 doses in increased risk patients, e.g. carcinoma/ obstruction)
Elective biliary surgery	Cefuroxime 1500 mg IV	1 immediate pre-operative dose
Appendicectomy	Metronidazole 1 g suppository (Give more +/− gentamicin in perforation or gangrene)	Continue post-operatively 8 hourly; in cases of sepsis 3–5 days
Elective colorectal surgery	Cefuroxime 1500 mg IVI	1 immediate pre-operative dose
	Cefuroxime 750 mg IV	8 and 16 h post-operatively
	Metronidazole 500 mg IV	1–3
Genito-urinary instrumentation	Gentamicin 80 mg IVI	1 dose
	Amoxycillin 500 mg IVI	1 dose
ERCP	Gentamicin 80 mg IV	1 dose pre-operatively
	Amoxycillin 500 mg IV	1 dose/examination
Vascular surgery	Flucloxacillin 1000 mg IV	4 doses pre-operatively 8, 16, and 24 h post-operatively
	Gentamicin 80 mg IV	
Insertion of prosthesis e.g. breast, repair of abdominal wall with mesh	Cefuroxime 1500 mg IV	1 immediate pre-operative dose
	Cefuroxime 750 mg IV	8 and 16 h post-operatively
Joint replacement	Flucloxacillin 1–2 g IV	1 immediate pre-operative dose
	Flucloxacillin 500 mg IV	8, 16, and 24 h post-operatively
Vaginal	Metronidazole 500 mg IV	1 immediate pre-operative dose
	Metronidazole 1 g suppository	4 h preoperatively

Prophylaxis of endocarditis*	Amoxycillin 3 g orally	1 h before surgery and 6 h post-operatively
	or	
	Amoxycillin 1 g IM	1 dose before surgery
	Amoxycillin 3 g IM	6 h postoperatively

*Use Vancomycin 1 g IV over 1 hour + gentamicin 80 mg IV immediately preoperatively in penicillin-sensitive patients.

N.B. Patients with chronic bronchitis and compound skull fracture should also receive antibiotics (ampicillin or cotrimoxazole for the former, ampicillin or flucloxacillin for the latter) commenced before surgery and maintained for several days.

Patients at risk from gas gangrene (benzylpenicillin 1 hour preoperatively then 6 hourly for 3–5 days) or tetanus (1 ml toxoid + 1 g penicillin) should also be given prophylaxis, as should immunosuppressed patients undergoing surgery.

Patients with a prosthesis in place should be given longer prophylaxis during contaminated procedures (as in GI surgery).

References

The Association of the British Pharmaceutical Industry (1988–89). *Data Sheet Compendium*. Datapharm Publications Ltd., London.

Pollock, A. V. (1988). Surgical prophylaxis—the emerging picture. *Lancet*, **i**, 225–30.

Antibiotic therapy in surgery

Antibiotics can be costly. Their prescribing is often inappropriate, and it is impossible for most clinicians to keep pace with all new preparations. However, a small number of antibiotics are indicated for most surgical problems. Get to know them well. When in doubt, seek advise from your microbiologist. Maintain wound surveillance through an infection control programme.

Decision to prescribe

In most situations this is clinical, and a 'best guess' policy based on the probable pathogens will get results. Cultures should also be taken, and treatment reviewed on their results. Do not, however, change antibiotics on the basis of resistance on microbiological culture if the clinical response is good. Specific microbial isolation is rarely necessary except in situations such as infection with *Mycobacterium tuberculosis* and *Salmonella typhi*.

Try to avoid the general use of broad-spectrum antibiotics. Rather treat obvious problems, e.g. wound abscess, etc. by surgical drainage.

Route of administration

Bolus parenteral therapy is preferred for sick patients. When the patient improves, drugs should be given orally.

Length of therapy

Assess each individual's response. When a clinical cure has been achieved, stop treatment. Remember that most courses of antibiotics are too long: 5–7 days should suffice.

Choice of antibiotic

In many cases of sepsis the patient needs surgical drainage and not antibiotics. Avoid inappropriate use for the usual infections (amenable to surgery).

Infection	Common organisms	Choice of antibiotic
Skin and soft tissues	*Staph. aureus*	Flucloxacillin (250 mg 6 hourly oral, IM or IV) MU IMI
	Strep. pyogenes	Penicillin (0.6–1.2 g daily IM in divided doses), ampicillin (250 mg 6 hourly orally), amoxil (250 mg 2 hourly) Erythromycin
Postop wound infection ('clean')	*Staph. aureus* *Strep. pyogenes*	Flucloxacillin, erythromycin
Contaminated surgery (e.g. colorectal surgery)	Anaerobes	Metronidazole (500 mg IV 8 hourly, 800 mg, stat + 40 mg 8 hourly) plus Gentamicin (80 mg IMI 8 hourly) or cefuroxime (750 mg IM or IV 8 hourly)

Urinary tract	*Escherichia coli*	Oral trimethoprim (2 tabs twice daily) Oral cotrimoxazole (2 tabs twice daily) Oral ampicillin
Pyelonephritis		Parenteral gentamicin, cefuroxime, or cotrimoxazole
Biliary tract	*E. coli* *Klebsiella* spp.	Ampicillin + gentamicin *OR* cefuroxime alone
Respiratory tract	*Strep. pneumoniae* *Haemophilus influenzae*	Physiotherapy ± amoxycillin or erythromycin Septrin types are very cheap (+ a penicillin)
Septicaemia/	Organisms unknown: check the history for probable pathogen	Gentamicin and metronidazole Cefuroxime and metronidazole
Bone and joint	*Staph. aureus* (common) *H. influenzae* (children) (do gram stains)	Flucloxacillin + fusidic acid, erythromycin amoxycillin or cefuroxime (beta-lactamase producers)

New broad-spectrum antibiotics

Imipenem (Primaxin) is a beta-lactam antibiotic which inhibits bacterial cell wall synthesis. It is indicated for the prevention and therapy of mixed infections caused by faecal flora and it is effective against *Bacteroides fragilis* which is usually resistant to aminoglycosides, penicillins, and cephalosporins.

Ciprofloxacin (Ciproxin) is a synthetic 4-quinolone antibiotic indicated for the treatment of infections caused by organisms which are resistant to other antibiotics. It inhibits bacterial DNA gyrase, is bactericidal and of value in severe systemic infections, infection in immunosuppressed patients, and infected burns.

Ceftazidime (Fortum) is a bactericidal cephalosporin indicated for the treatment of infections caused by single or two or more susceptible organisms. It acts by inhibiting cell wall synthesis and because of its broad spectrum may be used as a first-line agent in severe infections until bacterial sensitivity has been established.

Postoperative complications: Gastrointestinal problems

Paralytic ileus

After abdominal surgery normal bowel sounds disappear for about 48 hours usually returning on the third or fourth day. This postoperative ileus is due to paralysis of the myenteric plexus and is of two types; intestinal ileus (the commonest) and acute gastric dilatation. Ileus will also occur following peritonitis, abdominal trauma, and immobilization. It may be prolonged in hypoproteinaemic and hypokalaemic patients.

Symptoms Abdominal distension and vomiting. Absent or 'tinkling' bowel sounds.

Prognosis Intestinal ileus usually settles with appropriate treatment (see below). Acute gastric dilatation is an emergency. A nasogastric tube must be passed immediately to prevent inhalation of gastric juices, hypovolaemia, or gastric rupture.

Treatment Pass a nasogastric tube to empty the stomach of fluid and gas. Maintain continuous aspiration. Ensure adequate hydration by IV infusion ('drip and suck') and maintain the electrolyte balance. When ileus persists more than 5–7 days, institute TPN.

Nausea and vomiting

These are common symptoms in surgical patients. They may be the result of anaesthetic agents, analgesics, cytotoxic drugs, intestinal ileus, and mechanical obstruction.

Treatment Ensure nil orally for 4–6 hours before surgery. Exclude mechanical causes and ileus. Give central antiemetics, e.g. metoclopramide (Maxolon) 10 mg orally 8 hourly, 10 mg slow IV or prochlorperazine maleate (Stemetil) 12.5 mg IV, 25 mg orally. When persistent pass a nasogastric tube, correct hydration and electrolytes.

Diarrhoea

Diarrhoea may occur following an ileus or in association with continuing sepsis, e.g. pelvic abscess. It may also follow specific procedures such as ileo-anal anastomosis, ileal pouch, truncal vagotomy, and right hemicolectomy. Remember too, the possibility of infection; pseudomembranous colitis is commonly forgotten.

Treatment Identify the cause:
- Rectal and ultrasound examination to exclude pelvic abscess
- Stool samples for bacteria, protozoa
- Microbiology of stools for *Clostridium difficile*
- Erect/supine abdominal films (obstructive ileus)

Treat the patient by:
- Drainage of a rectal abscess
- Specific antibacterial, antiprotozoal chemotherapy
- Vancomycin 125 mg orally every 6 hours for 7–10 days
- Drip and suck for ileus

General measures Replace lost fluids with codeine phosphate 10–60 mg orally 4 hourly or diphenoxylate hydrochloride (Lomotil) 10 mg orally initially then 5 mg 6 hourly until controlled.

Constipation

117

Commonly encountered after elective surgery, particularly in the elderly.

Treatment Ensure adequate hydration. Limit the use of constipating analgesics after the first two days. Ensure adequate dietary fibre. Suppositories or enemata may be indicated, depending on the surgery performed. Manual evacuation is rarely necessary, but nevertheless a rectal examination should not be forgotten.

Postoperative complications: haemorrhage

Bleeding after surgery may be arterial or venous. Arterial blood is bright red and spurts in time with the pulse. Venous blood is darker in colour and flows steadily. There can be massive (1 l in 5 minutes) blood loss if it is from the large veins.

Types

Primary haemorrhage Occurs during surgery and continues.

Reactionary haemorrhage Occurs in the first 24 hours (usually 4–6 hours). It may follow primary haemorrhage or result from the slipping of a ligature or the removal of primary clot as a result of coughing or increasing postoperative blood pressure.

Secondary haemorrhage Due to infection. It occurs about 7–14 days after operation.

Clinical features

There may be visible blood loss into a drainage bottle, on to the bedclothes, etc.

The patient is restless, cold, and clammy with an increasing pulse rate. There is pallor, and in continuing bleeding there is air hunger (sighing, gasping respirations), thirst, and even blindness.

Blood pressure measurements

Pulse and blood pressure should be recorded every 15–30 minutes. In haemorrhagic shock a fall in blood pressure occurs late, and fit young people can maintain near normal levels for some time in spite of blood loss. The trend of pulse and blood pressure is the key to blood loss. A rising pulse rate or a 'thready' low-volume pulse is an indication of possible blood loss.

Treatment

- Raise the legs, establish an IV infusion or CVP line.
- Apply pressure to obvious external bleeding points.
- Replace lost blood volume with Haemaccel or dextran 70 until whole blood is available. Coagulation problems are minimized if the blood used is as fresh as possible.
- Arrange for the patient to return to theatre if necessary:
  —to deal with sepsis
  —to ligate bleeding source
- Carry out a coagulation screen, and treat specific defects in liaison with the haematologist. Clotting factors may be replaced with fresh frozen plasma. Screening is advocated for those patients who have been transfused around 5 units (rule of thumb).
- Measure urine volume (catheterize the patient).

The wound

Wound infection

Is related to the type of surgery carried out, with an incidence ranging from 1–50 per cent without prophylaxis depending on whether the procedure is classified as clean, potentially contaminated, or contaminated. Haematoma formation, diabetes mellitus, poor nutrition, reduced immunity, and drugs (e.g. steroids) increase the risk.

Clinical features The distinction between normal healing and mild infection may be difficult, for in both there may be reddening (erythema). Signs of cellulitis are suggested by more widespread erythema and early induration. Frank pus, purulosanguinous discharge with or without constitutional upset are confirmatory of wound infection.

Treatment

- Antibiotics are indicated when there is cellulitis, or immunodeficiency. Foreign bodies, e.g. suture in the wound should be removed (see soft tissue infection).
- Release pus, if present, by removing sutures or opening the wound to provide adequate drainage. Irrigation with dilute aqueous antiseptics like povidone or chlorhexidine followed by packing impregnated with the same or silastic foam dressings are indicated if the cavity is large.
- Recognize aggressive specific infections, e.g. gas gangrene (Foumier's) or Meleney's synergistic bacterial gangrene and follow on with aggressive therapy.
- Prevent infection by using specific antibiotic prophylaxis (page 112).

Wound sinus

Usually caused by retained foreign material such as a suture. (Suture abscesses are especially common with silk.) Perineal sinuses after excision of the rectum for inflammatory bowel disease are not uncommon. They can become chronic.

Treatment

- Remove the irritating cause.
- Curette the tract to remove debris and insert a wide or light pack (see above).
- Treat underlying pathology (e.g. Crohn's).

Hypertrophic and keloid scars

The two are very different. In hypertrophy the wound becomes broad and raised. It usually settles or even begins to regress at about 6 months. Keloid extends beyond the wound itself and continues to increase beyond 6 months. Extensor surfaces and burned skin are the commonest sites. Negroes and young people are more often affected.

Treatment
- Allow the wound to mature to at least 6 months before instituting further measures. Many scars will begin to fade at this stage.
- Injection of triamcinolone can be successful. (2–3 mg depending on the size of the lesion. No more than 5 mg should be injected.) Repeat at 1–2 week intervals.
- Prevent keloid formation by making incisions along the natural skin creases or Langer's lines. Pressure devices are also said to be effective in preventing keloid formation.
- Avoid excising keloids.

Surgery and the diabetic patient

Admit the patient at least 48 hours before operation. Inform the surgeon, anaesthetist, and diabetologist.

Postoperative risks

- Protein depletion, unrecognized hypoglycaemia, keto-acidosis
- Uncontrolled diabetes predisposes to infection—vulval candidiasis in women and non-clostridial gas-forming organisms.
- Small-vessel disease is a complication of diabetes mellitus. This predisposes to gangrene and ulceration due to overgrowth of anaerobes and microaerophilic organisms.

Assess

The age of onset Is the diabetes 'fragile'—patient psychologically incapable of controlling his disease, or 'brittle'—prone to ketoacidosis?

The disease severity More than 40 units insulin/day indicates moderate disease and increases postoperative risk.

Current disease status Give the patient his normal regime if the diabetologist advises it. Do blood sugars and urine analysis (3–4 may be necessary). If there is poor control postpone the operation.

Action

Major surgery This patient is first on the list. Do ECG, chest X-ray, FBC, U&E. Assess and stabilize diabetes. Halve long-acting insulin 24 hours before operation. Reduce by 25 per cent medium-acting insulin at evening meal before the morning of operation. Omit normal dosage in the morning and start IV dextrose 5 per cent with 15 units/l soluble insulin 2 hours before surgery. Infuse at about 1 1/8 hours. Check blood sugar before surgery. Let the anaesthetist know the result.

Postoperative

- Measure blood sugars 2 hourly until stable (6–10 mmol/l—upper end of normal is ideal), then 6 hourly.
- Measure U&E twice daily.
- Give IV insulin by infusion pump piggybacked into dextrose 5 per cent IV as follows:

Blood glucose	Soluble insulin
< 4	0.5 units/h
4–15	2.0 units/h
15–20	4.0 units/h
> 20	get help from the diabetologist

When the patient is eating, give insulin half an hour before breakfast and supper. Adjust the dose to achieve normoglycaemia.

Minor surgery
- The patient goes first on the list.
- Follow the above preoperative regime. Omit insulin on the day of operation. Do preoperative, immediate postoperative and 4 hourly postoperative blood sugars.
- Give oral food: 30 g carbohydrate every 4 hours.
- Begin insulin: 50 per cent usual dose, give by noon.
- Give normal evening insulin.
- Use soluble insulin regime if blood sugar > 15 mmol/l (above).

Emergency surgery
- Do preoperative U&E and blood sugar.
- Start IV dextrose 5 per cent 1 1/8 h.
- If blood sugar greater than 20 mmol/l correct as for keto-acidosis. Get help.
- Aim for blood sugar < 20 mmol/l bicarbonate > 16 mmol/l preoperatively.
- Take expert advice.

References

Hope, R. A. and Longmore, J. M. (1985). Diabetic patient undergoing surgery In *Oxford handbook of clinical medicine*, p. 130. Oxford University Press.

Smiddy, F. (1976). In *The medical management of the surgical patient*, pp. 10–15. Edward Arnold, London.

Surgery and myocardial infarction

Surgery increases the risk of peri- or postoperative myocardial infarction. In previously asymptomatic men over 50 years the incidence is around 0.5 per cent. If the patient has had a previous infarct the incidence increases almost 40-fold with a mortality of 70 per cent and a recurrence rate of 50 per cent.

Adverse factors

- Preoperative hypertension
- Evidence of preoperative congestive cardiac failure
- Long operations (more than 3 hours)
- Intraoperative hypotension
- Angina pectoris, especially at rest
- Myocardial infarction in the last 6 months

Advice for surgeons

- Elective procedures should be deferred for at least 6 months. Before 3 months more than 35 per cent will reinfarct, after 3 months 15 per cent, after 6 months only 4 per cent.
- Preoperative anticoagulation and antiplatelet drugs may be of value.
- Give appropriate medical treatment to patients with angina or congestive cardiac failure.
- In patients who need emergency surgery or whose cardiac status is unstable or deteriorating the risk of surgery is justifiable.

Reference

Portal, R. W. (1982). Elective surgery after myocardial infarction. *British Medical Journal*, **284**, 843–4.

Surgery and the contraceptive pill

The risk of deep venous thrombosis is doubled after major surgical procedures in women taking oestrogen-containing oral contraceptives and is greatest following major orthopaedic, abdominal, or cancer surgery and hypotensive anaesthesia, although it has been suggested that the risk is over-rated.

Risks in minor surgery

Minor surgery, e.g. dental or laparoscopic with mobilization on the same day, has *virtually* no risk at all. However, it is recommended that the combined oral contraceptive should be discontinued for patients having minor varicose vein surgery or sclerotherapy.

The progestogen-only pill does not require to be discontinued before surgery—minor or major.

When should the pill be discontinued?

The risk of deep venous thrombosis returns to normal within 4 weeks of stopping the pill. Therefore women should stop their oral contraceptives 4 weeks before major surgery and restart 2 weeks after, provided they are mobile again.

Advice for surgeons

- Give the patient adequate advance warning that the pill is to be discontinued. This should be done at SOPD. Ideally, a definite date for surgery should be given so that at least 1 month elapses before admission.
- Give advice on alternative methods of contraception.
- Ensure that the patient is not pregnant when she is admitted.
- Give advice on reducing weight. Take prophylactic measures if there is a family history of deep venous thrombosis.
- Do a clotting screen in those patients with risk factors in their history.
- In emergencies give the patient prophylactic subcutaneous heparin.

Reference

Surgery and the Pill (1985). *British Medical Journal*, **291**, 498–9.

Deep venous thrombosis

Deep venous thrombosis (DVT) is highest in patients over 40 years who undergo major surgery. Postoperative increase in platelets coupled with venous endothelial trauma and stasis all contribute (Virchow's triad). If no prophylaxis is given, 30 per cent of surgical patients will develop deep venous thrombosis and 0.1–0.2 per cent will die from pulmonary thrombo-embolism (PTE).

High-risk groups
- Patients undergoing pelvic surgery
- Patients undergoing hip replacement surgery
- Patients with malignant disease
- Patients on the contraceptive pill (pregnancy)
- Previous history of DVT or PTE
- Older patients. The increase in DVT is almost linear with advancing age.
- Other factors—obesity, diabetes mellitus, polycythaemia, varicose veins, cardiac, and respiratory disease

Diagnosis

Clinical Systemic pyrexia at 7–8 days postoperatively is often the first sign. It may be associated with pain, swelling of the leg and rise in skin temperature.

Investigations Bilateral ascending venography and lung scanning with radioactive technetium—labelled microaggregates of albumin (^{99}Tcm MAA) are most useful. Other techniques include ^{125}I—fibrinogen uptake, Doppler ultrasonic flow detector, strain gauge, impedance plethysmography, and radionuclide venography. All are useful for DVTs occurring at different levels. Venography is conclusive in making the diagnosis but it is also risky.

Treatment

Confirm the site and extent of the DVT by venography and lung scan.
- Calf vein thrombosis may be treated by compression bandage alone.
- All other thrombi should be treated by heparin for 4–7 days (40 000 units/24 hours) (5000–10 000 units by bolus then 1000–1500 units/hour IV alone) or followed by oral anticoagulation for 6–12 weeks (Warfarin).
- Lytic therapy—urokinase, streptokinase—is most effective within 24–36 hours of onset of DVT. This may be effective in life-threatening pulmonary embolus.
- Surgical thrombectomy or embolectomy is used when there is massive PTE or thrombus confined to the femoral vein or below in one leg. The superficial femoral vein may also be ligated. In bilateral iliofemoral DVT embolism may be prevented by inserting an 'umbrella' filter into the vena cava at a level below the renal vessels. An example is the Mobin-Uddin umbrella filter which is introduced in a capsule by

catheter through the internal jugular vein under local anaesthetic. Inferior vena cavography is carried out to determine the levels of the renal veins. The filter is released from the capsule below the lowest renal vein. Postoperatively, anticoagulation is instituted.

Prevention

Identify the high-risk patient (see above).

Methods

- *Mechanical*
  Prevention of venous stasis and increased venous return may be achieved by:
  —Use of elastic stockings, e.g. Bayer's Decopress stockings
  —Intermittent perioperative pneumatic calf compression
  Electrical stimulation of the calf muscles, e.g. Thrombophylacter
- *Low-dose heparin*
  —Subcutaneous calcium heparin—5000 units 2 hours postoperatively then 8 hourly until mobile
  Ultra low dose by infusion pump—1 unit per kg per hour for 3–4 days
- *Dextran 70*
  Started on induction of anaesthetic. Infuse 500 ml perioperatively. Repeat postoperatively and daily thereafter if necessary.
- *Alternative agents*
  —Subcutaneous ancrod (Arvin). Give 280 units after surgery then 70 units daily for 4–8 days.
  —Aspirin (300 mg orally per day), dipyridamole (Persantin 75 mg orally four times daily)

Venous gangrene

Treat by venous thrombectomy and superficial femoral vein ligation. Give ancrod for 2 weeks in late cases.

Advice for surgeons

Low-dose heparin is the method of choice for most general surgical procedures. Patients with fractures, e.g. neck of femur, femur, are best covered with dextran 70 or ancrod (for 4 days).

References

Kakkar, V. V. and Scully, M. F. (1978). Thrombolytic therapy. *British Medical Bulletin*, **34**, 191–9.
Scurr, J. H., Coleridge-Smith, P. D., and Hasty, J. H. (1988). Deep venous thrombosis: a continuing problem. *British Medical Journal*, **297**, 28.

Pain relief after abdominal surgery

Essence

Effective control of postoperative pain is difficult. After surgery abdominal muscle spasm may elevate the diaphragm, rapid breathing may lead to microatelectasis, and the inability to cough may predispose to retained secretions with segmental collapse and chest infection, pyrexia, reduced P_{O_2}, and elevated P_{CO_2}. Effective pain control can reduce these complications.

Methods of pain control

Psychological preparation Can reduce narcotic requirements by 50 per cent. Presurgery anxiety correlates with the severity of postoperative pain.

Intermittent intramuscular opiate analgesia This is the most commonly used, but because the prescription is usually 4 hourly more than 50 per cent of patients undergoing upper adbominal or thoracic surgery still complain of pain.

Continuous opiate infusions Provide effective control but may be associated with postoperative pulmonary dysfunction due to respiratory depression.

Patient-operated analgesic systems Allow administration of small preset measurements of IV analgesics in response to patient demand. Expensive, but early studies indicate efficacy.

Use of selective opioids Permit analgesia with less respiratory depression than, e.g. morphine. Examples include pentazocine, meptazinol, and buprenorphine which, administered sublingually, may reduce the need for parenteral narcotics.

Extradural blockade Patients receiving epidural analgesia (bupivacaine 0.5 per cent) intraoperatively and for 12 hours postoperatively had less pain than those receiving intermittent IM morphine with a lower incidence of chest infection (cholecystectomy patients). Opiates are also effective but cause itching.

Intercostal nerve blocks Does not provide relief of visceral pain and there is a risk of pneumothorax.

Direct perfusion of surgical wounds with long-acting agents, e.g. bupivacaine, may reduce narcotic requirements but offers no additional benefit.

Transcutaneous electrical stimulation (TENS) Reduces narcotic requirements in up to 70 per cent of patients but not respiratory complications after abdominal surgery.

Prospect

Individualization of postoperative pain relief is the ideal. Intermittent IM analgesia does not provide adequate relief although it is conventional. Epidurals are too demanding of time. Continuous IV self-administration of analgesic drugs appears to hold the key to adequate relief. It is, as yet, expensive to administer but as computer technology develops it may become less so.

Reference

Cuschieri, R. J. (1988). Management of post-operative pain after abdominal surgery *Scottish Medical Journal*, **33**, 227–8.

Shock

The clinical features of this complex syndrome are well recognized. There is tachycardia with hypotension and peripheral shut-down associated with pallor, sweating, clammy skin, reduced urinary output, tachypnoea, agitation, restlessness, and eventually coma. These features are secondary to general and inadequate tissue perfusion from whatever cause and are related to hypoxia and anaerobic tissue metabolism. Once established they become irreversible, leading to progressive cellular failure and death.

Types

- Hypovolaemic (haemorrhage, trauma, burns)
- Cardiogenic (myocardial infarct, arrythmias)
- Distributive (due to changes in peripheral resistance, e.g. sepsis, anaphylaxis)
  Often combinations of all three

Recognize conditions which can produce a shock-like picture:
- Pulmonary embolism: fat, gas, clot from a peripheral vein
- Restricted venous return: 'stove-in' chest, pneumothorax, cardiac tamponade
- Loss of sympathetic tone: head injuries, spinal cord injury
- Adrenocortical insufficiency, pain, panic

Management of haemorrhagic shock

Aims:
- Restoration of intravascular volume
- Improved cardiac output
- Improved pulmonary gaseous exchange

Action:
- Establish an airway. Give 100 per cent oxygen by mask. Start positive pressure ventilation if necessary.
- Place the patient in the lateral position. A useful first aid measure is to raise the legs. The head-down position should not be used. There is no evidence for its value and it may lead to cerebral or pulmonary damage.
- Set up an IV infusion. Cross-match blood. Take blood for electrolytes, blood gases, haematocrit, and pH. Set up a CVP line. Give up to 2 l of 0.9 per cent NaCl solution rapidly according to the CVP in combination with plasma expanders—dextran, Haemaccel, or reconstituted plasma. Give blood when available. Oxygen transport to the tissue becomes ideal at a haematocrit of 30 per cent.
- Correct acid–base disturbances. Hyperkalaemia is especially common. Infuse insulin—25 units/l of dextrose 5 per cent.
- Correct calcium when infusing large amounts of blood.
- Start antibiotics if indicated.
- Carry out corrective surgery as soon as the patient's condition permits.

Reference

Ledingham, I. McA. (1988). Shock. In *Jamieson and Kay's textbook of surgical physiology*, 4th edn, (eds. I. McA. Ledingham and C. McKay), pp. 471–88. Churchill Livingstone, Edinburgh.

5 General surgery and gastrointestinal disease

5 General surgery and gastrointestinal disease

135

Thyroglossal cyst, sinus, and fistula

This remnant of the thyroglossal duct is the commonest midline neck cyst. The track runs from the thyroid, passes through, in front or behind the hyoid bone and ends in the foramen caecum of the tongue at the apex of the V-shaped sulcus terminalis which separates the anterior two-thirds of the tongue from the posterior one-third.

Clinical features

The mean age of presentation is 5 years but it may be seen from 6 months to 70 years. 90 per cent are midline. 10 per cent occur laterally, most frequently on the left side. 75 per cent are prehyoid. 25 per cent are at the level of the thyroid or cricoid cartilage. Some are suprahyoid.

95 per cent of patients present with a painless cyst which moves on swallowing, is mobile, transilluminates, and usually fluctuates.

Thyroglossal duct carcinomas may rarely develop. These are papillary cancers which are treated by excision and suppressive thyroxine therapy.

Treatment

By excision of the cyst, sinus, or fistula including the body of the hyoid bone. Inadequate excision of the complete remnant leads to fistula formation.

Operation

Incision Elliptical over external opening or collar incision over cyst.

Procedure Raise subplatysmal skin flaps above and below. Identify the track or cyst and divide the fascia overlying it in the midline between the sternohyoid muscles. Follow it proximally. It is adherent to the hyoid in the midline. Remove about 1 cm of the body of the hyoid. Follow the track to the mylohyoid muscles. Ask your assistant to depress the posterior part of the tongue. Core out the track in that direction and divide the tissue below the mucous membrane of the tongue. The proximal part of the track is only rarely patent but should be cored out if possible as it may be associated with a lingual thyroid.

Closure
- Secure haemostasis
- Close the mylohyoid and sternohyoid muscles
- Insert a suction drain
- Close the skin with interrupted nylon

Postoperative care
- Remove the drain in 3–5 days
- Remove the sutures in 5–7 days
- Discharge the patient between days 3 and 5 postoperatively

Branchial cyst, sinus, and fistula

These represent remains of the branchial clefts, usually the first or second. The complete lesion is a fistula: the tract extends from skin to the posterior pillar of the fauces. A cyst occurs when the central part of the cleft remains patent, a sinus or tract when the lower part remains patent but does not open into the pharynx.

Pathology

The cyst wall is lined by stratified squamous epithelium and contains lymphoid tissue in 80 per cent. They contain straw-coloured fluid and cholesterol crystals. Branchiogenic carcinoma arising in the wall of the cyst is a rare complication.

Clinical features

3 males: 1 female. Peak incidence is in the third decade, but they may present at any age. 66 per cent occur on the left side. 2 per cent are bilateral. 70 per cent are anterior to the sternomastoid in its upper third. The remainder occur in the lower neck, parotid region, posterior triangle, and pharynx. 80 per cent of patients have a neck swelling. Pain and infection are also features in up to 30 per cent.

Treatment

Surgical excision of the cyst, sinus, or fistula tract including its internal opening (if present).

Operation (see p. 702)

Complications

- Division or damage of the hypoglossal nerve
- Division or damage to the cervical branch of the facial nerve
- Damage to the carotid artery (rare)
- Damage to the cervical cutaneous nerve leading to parasthesia or analgesia of the affected area

Branchial cyst, sinus, and fistula

Salivary calculi

Salivary gland calculi occur most commonly within the sub-mandibular gland (80 per cent) and to a lesser extent the parotid. They are composed of calcium phosphate and carbonate and may be related to sialadenitis. They are more common in adults.

Clinical features

Pain and swelling of the affected gland on eating or drinking. If a stone becomes impacted in the duct, the ensuing inflammation and fibrosis leads to persistent swelling of the gland. The patient may also experience colicky pain in the duct when eating or drinking.

Points in the examination of the submandibular gland

- Examine the gland from behind. Feel the swelling by running the fingers backwards under the jaw. If you cannot feel a lump, ask the patient to suck a sour sweet and re-examine him.
- Examine the duct orifice from the front. Ask the patient to open his mouth wide and point his tongue upwards. The ducts lie near the midline at the root of the tongue. Are they red? Is there pus? Can you see an impacted stone?
- Examine the gland bimanually from the front. Wear gloves. Place the fingers of one hand over the gland. The index of the other hand is placed on the mucosal surface of the mandible, and the gland palpated between the two.

Treatment

- Stones in the intra-oral part of the duct can be removed under general anaesthesia. Steady the stone with a Babcock forceps and incise directly over it. Remove the stone. Leave the duct marsupialized.
- Stones in the gland itself are an indication for removal of the submandibular gland (p. 698).
- Parotid gland disease should be treated conservatively with sialogogues and intermittent massage of the gland towards the duct.

Acute parotitis

The commonest infecting organism is *Staphylococcus aureus*. Predisposing factors include advanced age, debility, poor oral hygiene, dehydration, the postoperative period, and patients who have had anticholinergic drugs.

Clinical features

Acute parotitis is usually unilateral, unlike mumps. The patient has severe pain on the affected side and trismus. There is intermittent pyrexia and constitutional upset. The gland is enlarged, tense, and tender. Occasionally pus exudes from the duct. If so, a bacteriology swab should be taken.

Treatment

- Surgical decompression is indicated if symptoms persist without improvement over 48 hours despite the administration of broad spectrum antibiotics (e.g. flucloxacillin 500 mg orally or IM every 6 hours) or if there is abscess formation. Make a transverse incision over the gland parallel to the body of the mandible and decompress the foci of pus with sinus forceps. Pack loosely with gauze soaked in aqueous hibitane. Complications include damage to the facial nerve (VII) (so care should be taken with the forceps), and fistula formation (if there is duct obstruction).
- Prevention is achieved by adequate hydration and attention to oral hygiene, especially in high-risk patients. This involves the use of antiseptic mouthwashes and treating established oral fungal infection with mystatin mixture 1 ml after food four times daily. Advise the patient to retain the mixture in contact with the lesions for as long as possible. Continue for 48 hours after clinical resolution.

Recurrent parotitis

Is often related to duct obstruction. *Streptococcus viridans* or *Pneumococcus* are usually implicated. Recurrent bouts of pain and swelling are characteristic.

Management

- Bacteriology swabs of saliva and pus (if present).
- Adequate advice on oral hygiene.
- Do the dentures fit? If not, advise a dental appointment.
- Clinical examination and sialography. Sialectasis is a characteristic feature on X-ray. The gland ductules are tortuous and dilated.
- In persistent cases consider parotidectomy. Be warned that this can be very difficult after sepsis. Such patients are best referred to a specialist centre.

Upper gastrointestinal endoscopy

Indications

- Investigation and diagnosis of upper GI disorders by direct examination and/or sampling of tissue for histology, cytology, biochemistry, or bacteriology
- Diagnosis of biliary and pancreatic disease by the retrograde injection of dye and X-rays after cannulation of the ampulla of Vater (ERCP)

- Screening for upper GI cancers, especially stomach (cf. common practice in Japan)
- Assessment of therapeutic measures for upper GI disorders, e.g. drugs, post thermocoagulation of ulcers, etc.
- Investigation of the symptomatic patient who has had upper GI surgery
- Therapy as in the removal of foreign bodies, dilatation of oesophageal strictures, removal of polyps, dealing with bleeding varices by injection, dealing with upper GI bleeding by lasers, heat probe or injection of adrenalin, removal of common bile duct stones by papillotomy.

Advantages

- Well tolerated. As there is no need for general anaesthetic, old and ill patients may be examined.
- Permits direct visual examination and target biopsy. The view may be further enhanced by video techniques.
- More accurate than radiology in experienced hands
- Safer than rigid endoscopy

Disadvantages

Cost: Each instrument is very expensive
- *Danger*: Perforation by the instrument is a real risk. Therefore, supervised training is vital.
- *Biopsy size*: Tissue samples are usually small, and larger biopsy channels are being developed.
- *Sterility*: Complete sterility of the instrument and its channels is important especially where there is a risk of AIDS or hepatitis B (p. 68).

Complications

- Perforation, pulmonary aspiration, haemorrhage due to the endoscope
- Cardiovascular complications related to medication

Guidelines for trainees

- Acquire a sound knowledge of gastroenterology.
- Watch many examinations carried out by a competent endoscopist.
- Familiarize yourself with normal and abnormal appearances.
- Learn to control the instrument at first outside the patient then within.
- Never use excessive force during examination.

- If your view becomes obscured draw back the instrument a little before proceeding again.
- Carry out supervised endoscopy regularly (at least 100 examinations) before embarking on the procedure unsupervised.

Special risk groups

- Elderly seriously ill patients
- Patients undergoing emergency endoscopy
- Patients undergoing endoscopy at the hands of an inexperienced endoscopist

Mortality rate

0.01–0.025 per cent.

Achalasia of the cardia

This condition has two characteristic features: absent or poor oesophageal peristalsis, and failure of the lower oesophageal sphincter to relax. It affects men and women equally between 30 and 60 years and occurs in 1 in 100 000 of the population per annum.

The cause is unknown, but degeneration of Auerbach's plexus occurs. Chagas' disease (caused by *Trypanosoma cruzi*) produces identical changes.

Clinical features

- Dysphagia is the main symptom. Often intermittent to begin with, it becomes progressive and may be worse with liquids.
- Regurgitation of foodstuffs occurs, especially during sleep
- Retrosternal pain

Complications

- *Nutritional*: progressive weight loss
- *Respiratory*: repeated bouts of aspiration of oesophageal contents may lead to pulmonary fibrosis.
- *Oesophageal erosions*
- *Carcinoma*: may develop even after treatment.

Diagnosis

- Barium swallow may show a dilated gullet above a smooth distal narrowing (the rat-tail appearance).
- Oesophagoscopy should be done to exclude either benign stricture or carcinoma. There is frequently decomposing food debris which should be cleared to allow an adequate view.

Treatment

- Oesophageal bouginage will give relief of symptoms in about 70–80 per cent of patients, but complications occur in 5 per cent and this is unacceptably high
- Heller's operation. This is now the recommended procedure for fit patients. The procedure involves an extramucosal oesophagocardiomyotomy and is analogous to Ramstedt's procedure for pyloric stenosis. It can be performed through the chest or abdomen and a longitudinal incision is made in the lower 10–12 cm of the oesophagus extended about 1 cm on to the stomach (see pp. 706, 708).

Pharyngeal pouch

Pharyngeal pouch (*syn.* pharyngo-oesophageal diverticulum, pharyngeal diverticulum) is a false diverculum lined only with stratified squamous epithelium but without a muscle coat. It protrudes through the Killian–Jamieson dehiscence—a weak point posteriorly between the cricopharyngeus muscle and the inferior constrictor of the pharynx above. Its development is associated with neuromuscular unco-ordination of the pharynx and the upper oesophageal sphincter. Initially small it may lie to the right or left but as it enlarges it becomes central to fall between the oesophagus and vertebral column causing compression of the oesophagus.

Clinical features
- Most patients are elderly, suggesting an acquired condition.
- There is dysphagia due to the pouch filling, gurgling on swallowing, and regurgitation of oesophageal contents especially at night with risk of pulmonary aspiration.

Complications
- Nutritional: weight loss due to dysphagia
- Respiratory: due to inhalation of pouch contents
- Squamous carcinoma: rare, but may occur due to persistent exposure to food carcinogens

Diagnosis
- There may be a palpable soft swelling in the neck
- Barium swallow will demonstrate the pouch
- Oesophagoscopy: this is dangerous if a pouch is unsuspected and may lead to mediastinitis if the pouch is ruptured. It may be carried out once the pouch has been delineated by barium swallow to exclude carcinoma.

Treatment
Excision combined with cricopharyngeal myotomy (see p. 704).

Hiatus hernia

In this condition, which is common especially in later life, part of the stomach herniates through the oesophageal hiatus of the diaphragm.

Types

- *Sliding or axial hiatal hernia* (phreno-oesophageal membrane intact). The majority of hiatus hernias are of this type.
- *Rolling or para-oesophageal hiatal hernia* (defect in phreno-oesophageal membrane. True peritoneal sac in thoracic cavity).

Clinical features

Sliding (many do not produce symptoms)
- When reflux oesophagitis occurs there is heartburn especially after meals, made worse by stooping or lying down.
- Waterbrash due to excess salivation
- Regurgitation of food may occur at night when asleep and lead to aspiration pneumonia.
- Stenosis may occur as a result of reflux oesophagitis and lead to shortening of the oesophagus and dysphagia.
- Gallstones and diverticular disease often co-exist with hiatus hernia (Saint's triad).

Rolling
- Elderly patients are affected. There is also usually a degree of sliding hernia present.
- Intermittent dysphagia
- Pain after eating due to distension of the intrathoracic portion of the stomach by food
- Cardiac symptoms due to pressure effects on the heart
- Hiccough due to phrenic nerve irritation

Complications

Sliding
- Oesophagitis, stricture formation
- Stricture formation and dysphagia
- Chronic blood loss, anaemia
- Inhalation pneumonitis

Rolling
- Incarceration
- Gangrene
- Gastric volvulus
- Gastric ulceration

Diagnosis

Sliding
- Barium swallow and meal
- Upper GI endoscopy

Rolling
- Chest X-ray may reveal a fluid and gas level in the thorax.
- Barium meal

Conservative Treatment

Sliding
- No symptoms, no treatment
- Reflux oesophagitis: reduce weight, stop smoking, elevate the head of the bed, avoid stooping, etc.
- Antacids and/or H_2 antagonists (Gaviscon 1–2 tablets chewed after meals and at bedtime/Tagamet 800 mg at bedtime)
- Frequent small meals

Rolling
- Surgical treatment is indicated because of the risks.

Surgical treatment

This is indicated if medical treatment fails or if complications of anaemia or severe dysphagia develop. The procedure most favoured is the Nissen fundoplication which prevents reflux by suturing the mobilized fundus around the gullet. Some advocate the Belsey mark IV procedure through the bed of the eighth rib. The acute angle of the oesophagogastric junction is re-established with mattress sutures and the right crus sutured behind the oesophagus.

Rolling hernia is repaired by reduction of the hernia and closing the diaphragmatic defect with sutures. If strangulation has occurred emergency gastric resection may be indicated.

Peptic ulceration

Peptic ulceration develops when there is a breakdown in the mucosal defence systems of the stomach or duodenum and may be associated with increased or inappropriate acid or pepsin secretion. Cytoprotective systems like production of gastric mucosal prostaglandin E_2 and mucosal bicarbonate secretion are both reduced in patients with duodenal ulcer. These factors plus an acid environment provide optimal circumstances for the proteolytic enzyme pepsin to cause mucosal ulceration—no acid, no ulcer. In gastric ulceration hypersecretion of acid is not commonly a feature. In fact some patients are achlorhydric.

Possible causes and associations

- Smoking
- Blood group O
- Hyperparathyroidism
- Zollinger–Ellison
- Non-steroidal anti-inflammatory drug
- Reserpine
- Steroids
- Stress

Sites of ulceration

- *Common*: Stomach and duodenum
- *Uncommon*: Oesophagus, jejunum, Meckel's diverticulum containing ectopic gastric or pancreatic tissue.

Gastric ulcers

- *Type I* arise on the lesser curve associated with gastritis
- *Type II* arise secondary to duodenal ulcers due to gastric stasis
- *Type III* arise in pyloric and prepyloric region in association with normal or increased gastric secretion (cf. duodenal ulceration)

Duodenal ulcers

Usually occur in the first part of the duodenum. In the Zollinger–Ellison syndrome, ulcers can occur in any part of the duodenum.

Symptoms

Epigastric pain. Often described as 'deep', 'throbbing', or 'colicky'.

Features

Periodicity, pain related to hunger and food, vomiting.

Periodicity

- In DU the symptoms last up to 10–15 days and recur at 3–6 monthly intervals. They tend to be worse in spring and autumn.
- In GU the symptoms can last weeks and periodicity is less well marked.

Pain

- In DU, pain is related to hunger and often relieved by food. As a result patients often look well nourished. Nocturnal pain waking the patient at 2–3 a.m. is classic.

- In GU, hunger-related pain is not so marked. Food may precipitate pain and the patient may even be afraid to eat. Weight loss may be a feature.

Vomiting

- Not a feature of DU unless there is pyloric obstruction, when relentless large-volume vomiting may occur
- In GU vomiting is a marked feature and may even relieve the pain. Therefore some patients induce the vomiting themselves.

Additional symptoms of heartburn, water brash, postprandial bloating, and abdominal discomfort are common. Clinical examination may reveal deep tenderness in the RUQ or epigastrium.

Diagnosis

By contrast radiography or endoscopy.

Treatment

Most patients respond to medical treatment in the form of antacids and antagonist drugs. Antacids (e.g. magnesium trisilicate mixture 10 ml 8 hourly orally) relieve pain. H_2 antagonists (cimetidine 400 mg twice daily or 800 mg at night orally, ranitidine 150 mg orally twice daily or 300 mg orally at night) for 6–8 weeks will cure 90 per cent of peptic ulcers. One or other of these drugs is the treatment of choice.

Surgical treatment is indicated when there are:

- Persistent symptoms despite medical treatment
- Complications such as haematemesis, perforation, or stenosis
- Stress ulceration as a result of operation or injury
- Long-term complications as a result of previous ulcer surgery

Recurrence

Surgery has a lower recurrence than medical treatment (less than 10 per cent long term *vs*. 37 per cent at 1 year).

References

Robert, A. (1979). Cytoprotection by prostaglandins. *Gastroenterology*, **77**, 761–7.

Gledhill, J., Buck, M., and Paul, A. (1983). Cimetidine or vagotomy: Comparison of the effects of proximal gastric vagotomy, cimetidine and placebo on nocturnal intragastric acidity in patients with cimetidine resistant duodenal ulcer. *British Journal of Surgery*, **70**, 704–6.

Surgical procedures for peptic ulcer

Gastric ulcer

The procedure of choice is partial gastrectomy of the Billroth I or II variety depending on the site of the ulcer. These operations have a low recurrence rate of less than 0.5 per cent and a mortality rate of less than 1 per cent.

When the ulcer is in the pyloric or prepyloric regions, truncal vagotomy and antrectomy may be performed.

Perforated gastric ulcer

In most situations partial gastrectomy should be undertaken as there is a 10 per cent risk of malignancy. However, vagotomy and pyloroplasty with wedge excision of the ulcer has been shown to be safe in elderly patients. If the ulcer proves to be malignant a second operation may be required.

Complications of gastrectomy

Early complications such as haemorrhage, pancreatitis, acute gastric dilatation, anastomotic leak, or duodenal stump leakage should be anticipated, recognized, and treated (see post-operative complications).

Late complications may include:

- *Dumping syndrome* (early = unpleasant symptoms while eating, late = symptoms up to 2 hours after a meal). This is caused by hyperosmolar solutions passing rapidly into the small bowel which exert osmotic pressure attracting fluid and electrolytes, especially potassium, into the gut lumen. Symptoms include tiredness, epigastric discomfort, sweating, diarrhoea. Early symptoms resolve by reducing fluids during a meal, eating more frequent smaller meals and avoidance of heavy carbohydrate meals. Late symptoms may be resolved by sucking sweets or revisional surgery.
- *Recurrent ulceration* or stomal ulceration. These may be implied by the return of original symptoms. Confirm by endoscopy. Suspect Zollinger–Ellison.
- *Diarrhoea* may be related to dumping and often responds to dietary measures.
- *Biliary gastritis* often responds to Metoclopramide (10 mg three times daily by mouth.) Severe cases may require revisional surgery (Roux-en Y).
- *B_{12} deficiency*. Check the patient's blood annually. Give B_{12} replacement therapy (cyanocobalamin 250–1000 μg IM monthly as maintenance).

Duodenal ulcer

The most commonly performed operations are:

- *Truncal vagotomy and drainage* either by gastroenterostomy or pyloroplasty
- *Highly selective vagotomy* (HSV) (synonym: proximal gastric vagotomy). Recurrence is 1–10 per cent but side-effects after HSV are few compared with up to 10 per cent in patients with truncal vagotomy and drainage (diarrhoea, dumping, gastric bile reflux).

Mortality rates are low, less than 2 per cent for the former and 0.3 per cent for the latter.

HSV should only be performed by surgeons well experienced in the technique.

Perforated duodenal ulcer (p. 726)

Reference

Koo, J., Lam, S. K., and Chan, P. (1983). Proximal gastric vagotomy, truncal vagotomy with drainage and truncal vagotomy with antrectomy for chronic duodenal ulcer: a prospective randomised controlled trial. *Annals of Surgery*, **197**, 265–71.

Roux-en-Y procedures

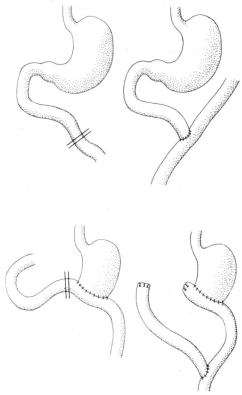

Roux-en-Y procedures

Gastrointestinal bleeding

Haematemesis (vomiting blood) and melaena (passing altered blood, usually black or tarry, per rectum) signify bleeding high in the gastrointestinal tract. The passage of fresh blood pr may signify a high bleed with rapid gut transit or lower GI bleeding (jejunum to rectum).

Causes

- *Upper GI tract*: Peptic ulceration, oesophageal varices, reflux oesophagitis, Mallory–Weiss tear (oesophageal tear due to vomiting), gastritis, bleeding cancers
- *Lower GI tract*: Colonic diverticular disease, AV anomalies (angiodysplasia) ulceration in a Meckel's diverticulum, bleeding cancers

Associations

Non-steroidal anti-inflammatory drugs, multiple injuries (stress ulceration), steroids, warfarin, alcohol ingestion.

Diagnosis

- *Upper GI tract*: Endoscopy or barium meal. Endoscopy is more accurate (greater than 90 per cent).
- *Lower GI tract*
  —Angiography if bleeding is continuing and more than 4 units of blood have been transfused in the first 24 hours
  —Colonoscopy if bleeding is slowing down and the patient has required less than 4 units in 48 hours. Accuracy is 90 per cent but repeated examination may be necessary.
  —Barium enema, although frequently carried out, is seldom diagnostic.

Management

- Put the patient to bed. Monitor pulse rate and blood pressure 4 hourly or less. Establish a CVP line and infusion.
- If shocked, efficient speedy blood replacement is necessary.
- If not shocked, take a history and examine the patient carefully.
- Take blood for haemoglobin, haematocrit, coagulation studies, grouping and cross-matching. Give 2–3 units. Store 2–3 in case the patient has to go to theatre.
- Catheterize the patient. Pass a nasogastric tube if it is your consultant's policy.
- Establish the diagnosis (see above).
- Control oesophageal varices with balloon tamponade and pitressin infusion (p. 168).
- Give IV Cimetidine for upper GI bleeding (200 mg/6 hours). There is little evidence that it stops bleeding, but it is cytoprotective in acute gastritis.

Indications for surgery

Exsanguination. Rebleeding with endoscopic stigmata of recent haemorrhage. If more than 4 units of blood are required in 24 hours, there is an increased probability that surgery or therapeutic endoscopy will be needed.

Procedure

157

Non-operative Therapeutic embolization, laser coagulation of bleeding vessels, local cautery with a heat probe, and application of adrenalin to the bleeding site.

Operative

- *Peptic ulceration*: Stop bleeding by oversewing the ulcer. If this is unsuccessful in patients with deep ulcers, isolate and ligate the gastroduodenal artery. Prevent recurrence by truncal vagotomy and drainage.
- Some gastric ulcers may be locally excised. Others will require formal gastrectomy.
- *Multiple erosions*: Vagotomy and drainage or gastric resection
- *Oesophageal varices*: Treat by injection sclerotherapy or oesophageal transection. Emergency selective shunt procedures, e.g. distal splenorenal (Warren) are worth while although an emergency portocaval shunt may be as effective. The prognosis depends on hepatic function and is excellent in well-nourished patients with bilirubin less than 25 mmol/l, albumin more than 30 g/l, no ascites or encephalopathy. Below these levels the prognosis deteriorates (Child's criteria).
- *Lower GI bleeding*: Treat by local resection of affected bowel, colonoscopic electrocoagulation (angiodysplasia) or subtotal colectomy (angiodysplasia, diverticular disease).

Risk of rebleeding

The incidence varies between 5 and 30 per cent depending on the pathology and the procedure performed.

Jaundice

Bilirubin is formed by the breakdown of haemoglobin. It is then made water-soluble by conjugation in the liver cells and excreted via the bile into the gut where it is broken down by bacteria to stercobilinogen. This is mostly reabsorbed and re-excreted by the liver (enterohepatic circulation). Some is excreted as urobilinogen in the urine, some in the faeces giving them their dark colour. Thus if bile is present on testing the urine the jaundice must be cholestatic (see below).

Jaundice is the yellow pigmentation of tissue, especially skin and sclera, due to deposition of bile pigments associated with increased circulating bilirubin (more than 35 μmol/l).

Causes

- *Haemolysis (acholuric jaundice, bilirubin unconjugated)*: Hereditary spherecytosis is surgically the most important as splenectomy is a method of treatment. Rhesus incompatibility, trauma, and incompatible blood transfusion are also causes.
- *Intrahepatic cholestasis (hepatocellular disease)*: Hepatitis A and B, drugs, toxins, defects of bilirubin transfer
- *Extrahepatic cholestasis (obstructive)*: Gallstones, bile duct strictures, carcinoma of pancreas or bile ducts, extrinsic compression of bile ducts by tumour

Diagnosis

Take an accurate history. Look for spider naevi, liver palms. Ask about drugs, alcohol, recent foreign travel, hepatitis, sexual behaviour, and recent anaesthesia.

Biochemistry

Plasma bilirubin is high in extrahepatic cholestasis. Fluctuation in levels suggests calculous disease. High alkaline phosphatase levels also imply extrahepatic obstruction. Plasma transaminase levels (alamine aminotransferase, aspartate aminotransferase) and lactate dehydrogenase reflect hepatocellular damage (intrahepatic cholestasis).

Other tests

Establish hepatitis A surface antigen (HBsAg) status, 'Monospot' (if available), FBC, U&E, chest X-ray. Dark urine and pale stools indicate cholestasis.

Management

Intrahepatic cholestasis is treated medically, extrahepatic surgically.

- Demonstrate gallstones, liver metastases, dilatation of intra/extrahepatic ducts by US scan
- Inject contrast medium via a needle inserted percutaneously into the intrahepatic ducts, if dilated, to demonstrate level of obstruction (percutaneous transhepatic cholangiography, PTC). Cannulation of a duct at PTC can also be used to decompress obstructive jaundice as a form of palliation.

- Use endoscopic retrograde cholangiopancreatography (ERCP) to define the lower edge of a lesion when obstruction is complete. Contrast is injected via a cannula endoscopically introduced into the ampulla of Vater. The technique can also be used to extract stones or insert a stent into the ampulla, e.g. in malignant disease to permit biliary drainage. Prophylactic antibiotics (p. 112) are administered at PTC and ERCP.

159

- Use CT scan/isotope liver scan to identify intrahepatic lesions.
- Liver biopsy is useful in diagnosis especially if targeted under US screening or laparoscopy on to specific lesions. Always group and save blood, get consent. Do full blood count, platelets, and partial thromboplastin time (PTT).

Preoperative preparation of jaundiced patients

- Ensure blood is available. Do PTT, full blood count, platelets, U&E.
- Give preoperative vitamin K 15 mg IM daily for 3 days.
- Give mannitol 25–50 g IV or IV fluids the day before surgery to prevent hepatorenal syndrome.
- Give prophylactic antibiotics (cefuroxime or gentamicin) at induction and 8 and 16 hours postoperatively.

Operative implications of obstructive jaundice

Patients with obstructive jaundice have a high postoperative morbidity and mortality. Three preoperative factors, if present, are associated with a 60 per cent postoperative mortality. If absent the figure is less than 5 per cent.

Poor preoperative factors

- Haematocrit less than 30 per cent (prognosis remains poor even after blood transfusion)
- Plasma bilirubin greater than 200 μmol/l
- Malignant disease

In addition, jaundiced patients may develop endotoxaemia resulting from increased gut absorption of endotoxins and reduced hepatic clearance.

Complications associated with endotoxaemia

Renal failure occurs in 9 per cent and has a mortality greater than 50 per cent. Gut-derived endotoxins lead to peritubular fibrin deposition and acute tubular necrosis. The most important precipitating factors are cholangitis and operation.

Coagulation disorders Disseminated intravascular coagulation is a risk. Biliary tract infection may also be a precipitating factor.

Gastrointestinal haemorrhage Occurs in up to 14 per cent. Stress ulceration is a feature with microvascular thrombosis leading to gastric erosions, a possible result of endotoxin-related disseminated intravascular coagulation.

Delayed wound healing Wound dehiscence occurs in 4 per cent, and incisional hernia in 12 per cent. The incidence can be reduced by closing the wound with deep tension sutures. Malnutrition, malignancy, and infection are also associated.

Aim of treatment

Neutralize the potential effects of endotoxaemia. The following have been demonstrated to be effective:
- Oral bile salts: sodium deoxycholate and chenodeoxycholate maintain renal function
- Oral lactulose reduces endotoxamia and presents a fall in renal function postoperatively
- Prophylactic antibiotics (p. 112) reduce the risk of post-operative infection and may also reduce endotoxaemia. The second-generation cephalosporins such as cephazolin and cefuroxime are most commonly used. 3–5 days of therapy are indicated in cases of cholangitis.
- Mannitol and adequate preoperative hydration are very important in the maintenance of renal function. They prevent tubular obstruction and endothelial cell swelling. Give a test dose of 200 mg/kg by slow intravenous injection. Infuse 20–50 g IV over 24 hours beginning the day before surgery.

Reference

Pain, J. A., Cahill, C. J., and Bailey, M. E. (1985). Peri-operative complications in obstructive jaundice: therapeutic considerations. *British Journal of Surgery*, **72**, 942–5.

Management of gallstones

Gallstones are common. They are present in 10 per cent of people over the age of 50 years in developed countries and are a frequent post-mortem finding. They are probably diet-related. Most are cholesterol stones. Pigment stones (calcium bilirubinate) account for 30 per cent and are associated with haemolytic disease, alcoholic cirrhosis, and bile infected with *E. coli*.

The incidence of gallstones increases with age and biliary stasis. Thus they are more frequent in multiparous women, the obese, during drug therapy (the Pill, clofibrate), haemolytic disease, TPN, post-truncal vagotomy, and disorders or resection of the terminal ileum (e.g. Crohn's disease).

Pathological effects

Less than 20 per cent cause complications or lead to surgery. 80 per cent remain silent.

Stones in the gallbladder:
- Biliary colic is common. Symptoms are due to impaction in Hartmann's pouch.
- Acute and chronic cholecystitis } may lead to empyema,
- Mucocoele of gallbladder (rare) } gangrene, or perforation

Stones in the ducts Obstructive jaundice, ascending cholangitis, acute pancreatitis.

Cholecysto-enteric fistula Stones in the gallbladder may pass into the small bowel via a fistula, impact and cause gallstone ileus.

Other associations Carcinoma of the gall bladder is more common in long-standing stone disease.

Clinical features

Around 20 per cent have symptoms of flatulent dyspepsia or fat intolerance. Biliary colic leads to transient but recurrent right upper quadrant and epigastric pain associated with vomiting. Acute-cholecystitis is more severe. The pain localizes in the right upper quadrant with tenderness over the gall bladder especially on inspiration (Murphy's sign) and a swelling may be felt. Attacks last up to 5 days and recur. Complications include empyema and jaundice.

Diagnosis

Straight X-ray (10 per cent are radio-opaque), oral cholecystography, US scan, PTC, isotope scanning (useful in acute cholecystitis)

Indications for surgery

- Patients with biliary colic, acute cholecystitis, jaundice, pancreatitis, carcinoma of gall bladder
- Failure of conservative management: Gallstones not responding to dissolution, perforation of the gall bladder, persistent jaundice. 50 per cent of patients with gallstones will develop symptoms within 10 years.

Early or delayed cholecystectomy?

Early cholecystectomy is performed within the first 7 days of illness. It is not technically more difficult and is economically viable in that only one hospital admission is necessary. There is no increase in mortality or morbidity compared to delayed cholecystectomy, in which the acute inflammatory process is allowed to resolve before undertaking surgery. Early cholecystectomy also reduces the risk of perforation, empyema, and failure of conservative treatment but the diagnosis must be accurately made early in the disease. This is not possible in many areas because of long waiting lists for radiological procedures and ultrasound. After 7 days, however, it is more difficult with increased morbidity and mortality.

Reference

Fowkes, F. G. R. and Gunn, A. A. (1980). The management of acute cholecystitis and its hospital cost. *British Journal of Surgery*, **67**, 613–18.

Non-surgical treatment of gallstones

Many patients with gallstones wish to avoid surgery or are unfit for it. Management options for some of these patients have widened (Bouchier 1988).

Dissolution therapy

Requirements Functioning gall bladder. Small or medium-sized radiolucent stones, 15 mm in diameter. Patience. Less than 30 per cent of patients fulfil the criteria.

Drugs used Chenodeoxycholic acid, ursodeoxycholic acid, combinations of the two, chenodeoxycholic acid + Rowachol.

Doses
- Chenodeoxycholic acid 13–15 mg/kg/day (18–20 mg/kg/day in the obese)
- Ursodeoxycholic + ursodeoxycholic acid in doses of 7.5 mg/kg and 6.5 mg/kg respectively by mouth
- Rowachol one capsule/day + chenodeoxycholic acid 7–10.5 mg/kg/day by mouth

Side-effects Diarrhoea, pruritus, transient rise in serum transaminases.

Results
- By carefully selecting patients, dissolution rates of 70–75 per cent are obtained.
- Stone recurrence rates are 25 per cent at three years of stopping therapy, 50 per cent at seven years, 64 per cent at 12 years.
- Low-dose therapy does not prevent recurrence.

Extracorporeal shock-wave lithotripsy

Requirements Functioning gall bladder. Medium-sized stones. Less than 30 per cent of patients are suitable.

Contra-indications Very large and very small stones. Calcified stones.

Technique Intravenous analgesia. Shock waves are generated by electrostatic spark discharge, electromagnet, or pulsed piezoelectric shock. The patient is immersed in a tank, or a compressible water bag is held against the body.

Additional therapy Oral bile salts assist fragmentation.

Results Still being evaluated. Stone recurrence may be a problem.

Therapeutic endoscopy

Common bile duct stones, retained, and recurrent stones may be managed by endoscopic sphincterotomy and stone extraction.

Results 90 per cent success rate.

Percutaneous transhepatic catheterization and MTBE infusion

This is the domain of the radiologist. The gall bladder is catheterized and the cholesterol solvent MTBE (methyl tertiary butyl ether) is instilled. Gallstone dissolution occurs within a few hours, but it is not suitable for pigment stones.

Side-effects Nausea, pain, vomiting, haemolysis, duodenal erosions. These can be reduced by using a controlled infusion system.

Reference

Bouchier, I. A. D. (1988). Non-surgical treatment of gallstones: many contenders but who will win the crown? *Gut*, **29**, 137–42.

Common bile duct stones

Stones are found in the common bile duct (CBD) of 10 per cent of patients undergoing cholecystectomy.

Preoperative diagnosis

Suspect if there is a history of jaundice. Confirm their presence by grey-scale US, PTC, or ERCP. If the patient is not jaundiced, stones in the CBD may be suggested by these diagnostic investigations.

Operative diagnosis

Operative cholangiography should be carried out as a routine procedure in all cholecystectomies before making the decision to explore the CBD (Gunn 1982). A normal cholangiogram shows a normal calibre duct (10–12 mm diameter) without filling defects. There is free flow of contrast into the duodenum and easy filling of the intrahepatic ducts. The terminal narrow segment of the CBD is seen clearly (Hand 1976).

Technique

Identify the cystic duct, CBD, and common hepatic ducts. Tie the cystic artery. Ligate the proximal end of the cystic duct. Pass a ligature distally but do not tie it. Open the cystic duct close to the CBD. Pass a cholangiogram catheter filled with saline attached to a syringe to exclude bubbles. Aspirate the syringe to obtain bile. Now make the distal ligature secure to keep the cannula in place. Remove all swabs and instruments. Cover the wound. Position the X-ray films and centre the X-ray machine. Replace the 20 ml syringe with 10 ml of Hypaque. Inject 2–3 ml initially and X-ray. Repeat with a further 6–8 ml. Interpret the films as above. If there are stones, exploration is indicated. If there are no stones, remove the cannula and tie the cystic duct flush with an absorbable ligature with the CBD and remove the gall bladder. This procedure may also be carried out under direct screening with an image intensifier.

Exploration of CBD (see operative surgery)

Residual stones

If the common bile duct has been explored it is closed over a T-tube and T-tube cholangiography is carried out 7–10 days postoperatively. This may reveal stones.

Management

- If the stones are small, IV ceruletide may be effective. This relaxes the sphincter of Oddi and coupled with saline flushing of the CBD via T-tube may allow the stones to pass into the duodenum.
- The T-tube is left for 3–4 weeks until a tract is established. This may subsequently be dilated and the stones removed by various methods—Dormia basket, Fogarty catheterization—under screening.

- Endoscopic sphincterotomy with removal of the gallstones is possible using therapeutic ERCP.
- Shock-wave lithotripsy. The stones may be fragmented by ultrasound and subsequently removed.
- Dissolution techniques. Many solutions have been passed through the T-tube to dissolve stones. Monooctanoinin is reasonably effective.
- Re-exploration is indicated if these techniques fail.

References

Gunn, A. A. (1982). The management of gallstones. In *Recent advances in surgery*, **11** (ed. R. C. G. Russell), pp. 183–96, Churchill Livingstone, Edinburgh.

Hand, B. H. (1976). Presentation and management of stones in the common bile duct. In *Current Surgical Practice*, Vol. 1 (eds J. Hadfield and M. Hobsley), pp. 114–31. Edward Arnold, London.

Portal hypertension

Increased portal venous pressure has serious clinical implications when it exceeds 20 mmHg (normal level 5–10 mmHg).

Causes
Prehepatic (20 per cent)
Congenital portal vein atresia
Portal vein thrombosis due to:

- neonatal umbilical sepsis
- pyelophlebitis after acute appendicitis
- exchange transfusion
- trauma
- tumour
- thrombosed porta caval shunt

Hepatic (80 per cent)
Cirrhosis (alcoholic in West)
Other causes elsewhere—p. 170
Schistosomiasis

Posthepatic (rare)
Tricuspid valve incompetence
Budd–Chiari syndrome
Constrictive pericarditis

Effects
- *Dilatation of portasystemic collaterals* leading to oesophageal varices, haemorrhoids and rectal varices, epigastric veins (caput medusae), retroperitoneal or diaphragmatic collaterals. Oesophageal varices can produce massive bleeding.
- Ascites, hypersplenism and splenomegaly, encephalopathy

Management of bleeding oesophageal varices (acute bleeding)

Conservative treatment 60 per cent of patients respond. *Resuscitate* with fresh blood. Monitor CVP, urine output, pulse, blood pressure. Prevent coagulopathies by advice from haematologist on the use of fresh frozen plasma and platelets. Give aperients, cimetidine, and oral neomycin to reduce the risk of portal pyaemia.
- *Upper GI endoscopy* establishes the diagnosis.
- *Balloon tamponade*: Use the Sengstaken–Blakemore triple-lumen tube or Minnesota four-lumen tube (superior). The tube is passed into the stomach. The lumina permit inflation of the gastric balloon, tamponade of the varices with air, aspiration of gastric and oesophageal contents so preventing aspiration pneumonia. Use with continuous IV infusion of vasopressin to reduce portal pressure, 0.4 units/minute. Add glyceryl trinitrate to reduce cardiovascula side-effects. Keep the tube in position for 24 hours then deflate to prevent oesophageal/gastric ulceration. 90 per cent of patients respond but almost all will rebleed unless definitive treatment is carried out.

Surgery
- Injection sclerotherapy is effective in 90 per cent. Repeat injections to prevent rebleeding.
- Oesophageal transection and re-anastomosis with the stapling gun is effective in 60 per cent. Operative mortality is 10 per cent. Recurrence 5 per cent.
- Emergency shunts are effective provided correct patient selection is used in accordance with Child's criteria of hepatic function (p. 157).

Definitive treatment after variceal bleeding

Investigations Spleno-portography, CT, and US scan to assess the portal vein.

Shunt operations have variable results, increasing the risk of both encephalopathy and liver failure and reducing the incidence of rebleeding. Long-term survival is poor. Portacaval, mesocaval, and splenorenal are described. Those with the lowest incidence of encephalopathy are the distal splenorenal and the left gastric-vena caval shunt, although there is controversy that they may be no better than non-selective shunts.

Chronic injection sclerotherapy The varices are injected every 2–3 weeks until obliterated with follow-up endoscopy at 3-monthly intervals. 60 per cent occur within 1 year.

References

McDougal, B. R. D., Westaby, D., Theodossi, A., Dawson, J. L., and Williams, R. (1982). Increased long term survival in variceal haemorrhage using injection sclerotherapy. *Lancet*, **i**, 124–7.

Warren, W. D., Millikan, W. J. Jr, Henderson, J. M. *et al.* (1982). Ten years portal hypertensive surgery at Emory: Results and new perspectives. *Annals of Surgery*, **195**, 530–42.

169

Cirrhosis of the liver

The term refers to a group of chronic hepatic diseases characterized by necrosis, fibrosis, and nodular regeneration. It involves the whole organ. The main effects are portal hypertension and progressive hepatic failure.

Causes

- *Children*: Bile duct atresia, congenital, viral hepatitis, kwashiorkor, Galactosaemia
- *Adults*: Alcohol, viral hepatitis B, biliary tract obstruction, primary biliary cirrhosis, haemochromatosis, schistosomiasis in endemic areas, chemical toxins, idiopathic (around 50 per cent)

Morphological types

- *Micronodular*: Every lobule is destroyed and < 4 mm
- *Macronodular*: Irregular lobules; some survive; size varies
- *Mixed*: Both of the above

Complications

Portal hypertension Results from obstruction of the blood through the liver. Its main effects are oesophageal varices, distended anterior abdominal wall veins (caput Medusae), splenomegaly and hypersplenism, haemorrhoids.

Liver failure
Features include:
- Jaundice
- Spider naevi: Overgrowth of end arteries of the skin of the face, neck, and upper arms. They blanch when their centre is depressed. Normal people can have up to five.
- Liver palms: Erythema of the thenar and hypothenar eminences. The hands are warm.
- Bleeding varices occur in around 40 per cent of patients.
- Failure to synthesize albumin leading to hypoproteinaemia and oedema
- Ascites is related to hypoproteinaemia, fluid retention (due to secondary hyperaldosteronism), and portal hypertension.
- Gynaecomastia and testicular atrophy, due to failure to inactivate oestrogens
- Clotting disorders
- Mental deterioration and coma are final events which are due to the liver's inability to detoxify ammoniacal breakdown products (porta systemic encephalopathy).

Primary liver carcinoma occurs in around 15–20 per cent of patients dying of cirrhosis.

Treatment

Cirrhosis is progressive. Patients should, however, give up alcohol and increase their carbohydrate intake. Give blood transfusions for GI haemorrhage or varices. If there are signs of hepatic failure, maintain fluid balance and correct electrolyte abnormalities especially hypokalaemia. Reduce the effects of nitrogen products in the bowel by restricting dietary intake, giving enemas and oral lactulose to remove blood from the bowel and non-absorbed antibiotics like neomycin (1 g 6-hourly orally).

Prognosis

30 per cent of patients die within a year of diagnosis.

Intestinal ischaemia

Causes of small intestinal vascular insufficiency:
- Fall in cardiac output: cardiac failure, myocardial infarct, shock
- Vascular occlusion: emboli (atrial fibrillation, post-myocardial infarct), thrombosis (atheroma, polycythaemia, oral contraceptives)
- Microvascular damage: hypersensitivity reactions, vaso-constriction (pressor drugs), endotoxins (*Cl. perfringens*)
- Reduced circulation: Dehydration, herniation, portal hyper-tension, idiopathic necrotizing enterocolitis

Acute occlusion

Superior mesenteric artery occlusion from embolus or throm-bosis (commonest) is the commonest cause.

Clinical features

The onset is sudden, with severe colicky abdominal pain associated with a soft, silent non-tender abdomen and the development of shock. Vomiting and diarrhoea, with occult blood, may occur. The physical signs are out of all proportion to the severity of the patient's illness and tend to be intermittent. This is because necrosis progresses from the mucosa to the serosa. Peritonitis does not develop until all layers are involved.

Treatment

Treat causal factors first. Correct shock with Hartmann's solution, proteins and plasma. Give IV antibiotics (metroni-dazole 500 mg/8 hourly and gentamicin 80 mg/8 hourly or cefuroxime 750 mg/6 hourly). In intravascular coagulation give continuous heparin infusion 10 000 units/6 hourly. Reverse with protamine sulphate before surgery.

Laparotomy

Viable bowel (intestinal vessels and arcades pulsating). Consider superior mesenteric artery embolectomy or bypass with side-to-side ileocolic to right common iliac artery. Have a 'second look' 24 hours later.

Dead bowel Reconstruction or embolectomy is useless. Resect dead bowel and restore continuity with end-to-end anastomosis if possible.

Prognosis

75 per cent mortality.

Chronic ischaemia

Associated with colicky postprandial pain lasting for about 1 hour. The patient may lose weight and be afraid to eat (intes-tinal angina). An epigastric bruit conducted to the right iliac fossa is an unusual finding. Treatment is with jump grafting or ileocolic to right common iliac artery anastomosis.

Ischaemic colitis

Associated with inferior mesenteric artery occlusion. It is classified as gangrenous, stricturing, and transient. The patient has
pain, rectal bleeding, vomiting, and fever with peritonitis and
shock in the gangrenous form when colonic resection with
exteriorization of the bowel and delayed anastomosis should be
performed. In other forms barium enema may indicate 'thumb
printing' or stricture, most often of the splenic flexure.
Carcinoma and inflammatory bowel disease must be excluded.

Acute intestinal obstruction

Carry out erect and supine abdominal films if you suspect this common general surgical problem.

Classification

- *Adynamic*, e.g. paralytic ileus

```
                          gallstone ileus (rare)
                         /                          high small gut
           simple ————————adhesions
          /              \                          low small gut
         /                carcinoma ———— small bowel, colon
- Mechanical
         \
          strangulation—intussusception, infarction,
                         volvulus, internal/external
                         hernia, bands
```

                              (after Le Quesne 1976)

Adhesions and hernia are the commonest causes. 'The more developed a country, the more likely the cause is adhesions' (Ellis' Law).

Pathophysiology

- Distension of bowel above the level of obstruction with gas
- Accummulation of intestinal secretion above the obstruction
- Progressive depletion of ECF
- Multiplication of bacteria especially coliforms, *Strep. faecalis*, *Clostridia perfringens*, *Bacteroides* spp

Clinical presentation

Abdominal colic and distension (often slight in small bowel but marked in colonic obstruction), anorexia, nausea. Vomiting is an early feature of high small-bowel obstruction but may be absent in large-bowel obstruction. Absolute constipation. Bowel sounds may be loud enough to hear but may also be tinkling. Toxaemia and peritonitis occur late and in paralytic ileus pain is absent, but before this diagnosis is made closed-loop obstruction as a result of intestinal strangulation or carcinomatous stricture should be excluded.

Investigations

Erect and supine abdominal films are a key to the diagnosis. Look for gas-filled distended loops of bowel with multiple horizontal fluid levels. The appearances of small- and large-bowel obstruction are very different (see below) and may be diagnostic of the level.

Key decisions

Is the patient obstructed? Is there vomiting, colic or distension? Yes. Carry out an abdominal X-ray. Is the bowel distended? Are there fluid levels? Yes = obstruction.

Is the obstruction simple or due to strangulation? Look for external hernia? (umbilical, inguinal, femoral). Are there abdominal operation scars? (internal, hernia related to adhesions). Is the patient fibrillating? (?infarction). Is there a sentinel loop on the abdominal X-ray (?internal hernia). Is the pain continuous? Is there local peritoneal irritation? These latter findings imply obstruction due to strangulation. If you are in doubt it is better to look and see than wait and see.

What is the level of obstruction? Is there vomiting with minimal obstruction? (high). Are colic and distension features with vomiting developing later? (low). Is distension the single feature? (low colonic obstruction).

Look at the abdominal X-ray (erect and supine). There are radiological differences between distended jejunum, ileum, and large bowel. In jejunum there are transverse markings running right across the bowel; in ileum the markings are absent; in the large bowel the haustral markings do not go right across the bowel.

Management

Conservative Simple obstructions like paralytic ileus or adhesive small-bowel obstruction. Pass nasogastric tube and aspirate continuously. Give IV fluids—drip and suck. Many simple obstructions will settle on this regime. But consider surgery if the patient does not show signs of settling.

Surgery Patients with strangulation or evidence of closed-loop obstruction need urgent laparotomy. Give large volumes of IV fluids with CVP line and urine output monitoring to correct deficits, and operate as quickly as possible after admission.

Guidelines for success

- Suspect if there is vomiting, colic and distension.
- Examine hernial orifices. Note abdominal scars.
- Note the bowel patterns on abdominal X-ray.
- Laparotomy sooner rather than later if there is doubt about the cause.

Reference

Le Quesne, L. P. (1976). Acute intestinal obstruction. In *Current surgical practice*, Vol. 1, (eds J. Hadfield and M. Hobsley), pp. 168–84. Edward Arnold, London.

Pseudo-obstruction

Patients in medical, general surgical, or orthopaedic wards may develop symptoms of mechanical bowel obstruction often related to the large bowel but without an obvious mechanical cause (Ogilvie's syndrome).

Associations

Lumbar spinal fracture, fractured hip or pelvis, retroperitoneal irritation (haematoma, etc.), hypoxia, uraemia, disorders of water and electrolyte balance.

Clinical features

The patient is usually elderly with renal or cardiac disease, a recent history of illness or injury, and is confined to bed. Abdominal distension gradually develops but bowel sounds are heard which may become obstructive in character. The abdomen can become massively distended and is tympanitic to percussion.

Diagnosis

- Think of pseudo-obstruction as a possibility in the elderly.
- Water-soluble contrast enema to establish the absence of mechanical obstruction.
- Colonoscopy

Treatment

Laparotomy carries a high mortality rate.
- Correct disorders of fluid and electrolyte balance.
- Correct hypoxia and uraemia.
- Intravenous nutrition may be necessary.
- Colonoscopy and decompression of the distended colon is very successful.
- Correct the underlying cause.

Complications

Massive colonic distension can be confused with closed-loop obstruction with the caecum becoming worryingly dilated. Perforation is a risk and this is an indication for surgery. Fortunately it is rare. Usually the caecum is involved, the patient presenting with caecal tenderness, localized pain or guarding, and evidence of pneumoperitoneum.

Treatment

By caecal exteriorization. Caecostomy is dangerous and carries a high mortality.

Reference

Dudley, H. A. F. and Patterson-Brown, S. (1986). Pseudo-obstruction. *British Medical Journal*, **292**, 1157–8.

Meckel's diverticulum

Meckel's diverticulum occurs in 2 per cent of people, affects 2 males:1 female, is 2 feet from the ileocaecal valve on the anti-mesenteric border of the small intestine and is 2 inches long (mnemonic—all the 2s). It is a true diverticulum with a mucus membrane and muscular coat and may be connected to the umbilicus either by a fibrous band or as a complete fistula—remnants of the vitello-intestinal duct.

Clinical features

2 males:1 female. Many cause no symptoms but 20 per cent have heterotopic gastric or pancreatic mucosa. The following complications may be associated with Meckel's diverticulum.

Diverticulitis This may mimic appendicitis. If the appendix is normal in a patient with suspected appendicitis examine the terminal 2 feet (60 cm; from ic valve, 5 cm long) and more of the ileum in search of a Meckel's.

Intussusception Again intermittent right-sided abdominal pain may mimic appendicitis but an intermittent mass may be present. The initiating factor is the inflamed mucosa at the mouth of the diverticulum, not inversion of the diverticulum itself.

Rectal bleeding Due to peptic ulceration in the diverticulum. The blood is usually bright red or slightly altered in colour. Think of this possibility in cases of fresh rectal bleeding.

Peptic ulceration The pain is felt around the umbilicus and is usually related to meals.

Intestinal obstruction The presence of congenital bands between the tip of the diverticulum and anterior abdominal wall can lead to volvulus or internal herniation.

Perforation Leads to localized or generalized peritonitis.

Diagnosis

Most are silent. They are also difficult to diagnose radiographi-cally when symptomatic due to oedema of the mouth. If bleeding occurs, technetium scanning may localize the site.

Littre's hernia

This is an inguinal or femoral hernia which contains a Meckel's.

Diverticulectomy

If a Meckel's is found coincidentally it should be left alone if uninflamed with a wide neck.

Indications for diverticulectomy Inflammation, ulceration, narrow neck, peritoneal bands, perforation. Diverticulectomy may be achieved by resection and end-to-end anastomosis of the segment which contains the diverticulum.

If it is broad-based the tip should be grasped by a Babcock's forceps, the base clamped transversely, and excised. The clamp should then be oversewn with continuous vicryl and removed before tightening the suture. A further layer may be added to invaginate the first. Closing the intestine transversely prevents stricture and is analogous to the Heineke–Mikulicz pyloroplasty.

Appendicitis

This is the commonest cause of an acute abdomen in the UK. The presentation is variable and the diagnosis essentially clinical. It most often affects teenagers and young adults. Consider the diagnosis in ill neonates, infants with diarrhoea, anorexic schoolchildren, the pregnant, and the aged. If in doubt re-examine the patient. Remember that as many as 500 people may die each year of acute appendicitis in the UK.

Symptoms

Colicky periumbilical pain which shifts to the right iliac fossa and localizes there. Loss of appetite and nausea are common. Alteration of bowel habit may lead to the erroneous prescription of aperients for constipation. Diarrhoea is a feature of pelvic appendicitis. Vomiting is uncommon.

Signs
General
Fetor, fever, flushed appearance
Furred tongue
Tachycardia
Coughing hurts
White blood cells $> 12\,000$ mm^3 in 25 per cent

Rectal examination
Tender anteriorly to the right in 30 per cent

Right iliac fossa
Tenderness, guarding maximal over McBurney's point
Rebound tenderness
Hyperaesthesia

Other
Pressure in left iliac fossa produces pain in right iliac fossa (Rovsing's sign)

Differential diagnosis (includes amongst others)

Mesenteric adenitis
Meckel's diverticulitis
Terminal ileitis
Ureteric colic
Acute renal disease
Acute salpingitis
Mittelschmerz
Ruptured ectopic pregnancy
Perforated viscus
Acute pancreatitis
Non-specific abdominal pain

Treatment

Appendicectomy. Give 500 mg/1 g metronidazole as suppository with premedication at diagnosis to reduce the risk of wound infection. In uncomplicated appendicitis one is adequate.

If perforatiion is present resuscitate the patient then carry out appendicectomy. Pass a nasogastric tube, start IV fluids, give metronidazole and gentamicin IV. At operation take bacteriology swab for culture and sensitivity. If peritonitis is widespread carry out peritoneal lavage before closing the wound. Continue antibiotic therapy for 3–5 days. Drain local abscesses.

Appendix mass

This is omentum and small bowel adherent to the inflamed appendix. Treat expectantly with IV fluids, analgesia, and antibiotics. Postpone appendicectomy until after resolution has occurred and carry out interval appendicectomy 6–8 weeks later. Wound infection is the commonest complication. Failure of resolution implies abscess formation.

Complications

Paralytic ileus, abscess formation (appendix, subphrenic, and pelvic all need drained), overwhelming sepsis (immunosuppressed patients), portal pyaemia (rare).

Acute pancreatitis

After an attack of acute pancreatitis the gland usually returns to functional and anatomical normality provided the predisposing cause is removed. Some patients develop recurrent attacks, enjoying normal health between attacks. In chronic pancreatitis there are anatomical or functional abnormalities of the gland.

Incidence

Incidence varies geographically, but is estimated at 54 per million of the population in the UK and 100–115 per million in the USA.

Mortality

Acute pancreatitis still causes some deaths in those under 50 years but the highest mortality is in older patients when it may approach 20 per cent.

Aetiology

Certain factors predispose:
- Gallstones (54 per cent) and alcohol (more than 20 per cent) are common.
- Mumps, Cocksackie B infection, steroids, trauma, neoplasia, scorpion bites are less common.
- Idiopathic

Theories of causation

- Obstructive hypersecretion (e.g. gallstones impacted at ampulla)
- Duodenal reflux into pancreatic duct (e.g. in afferent loop obstruction)
- Bile reflux into pancreatic duct
- Acinar cell derangement

All may play a part in conjunction with predisposing conditions.

Pathology

The inflammation ranges from mild pancreatic oedema to necrosis and haemorrhage. These result from the release of proteolytic enzymes leading to autodigestion. Vasoactive kinins—kallikrein and bradykinin—add to the circulatory changes leading to fluid and electrolyte loss, oedema, and exudate. Fat necrosis is a feature which may be seen at laparotomy.

Clinical features

Pain is the most clamant feature. It is epigastric, often radiating to the back. Nausea and vomiting are common. Shock is a feature of severe acute pancreatitis. The onset of symptoms may develop some 24 hours after alcohol ingestion. There may also be a past history of dyspepsia, biliary colic, or transient jaundice.

Complications

- Renal failure associated with shock
- Latent hypoxia is common in severe disease. Acute respiratory distress syndrome (ARDS) may develop due to denaturization of surfactant.
- ECG changes due to 'myocardial depressant factor'
- Consumptive coagulopathy
- Multiple organ systems failure
- Psychosis
- Local complications—pseudocyst, abscess, pleural effusions, fistula formation

Diagnosis

Serum amylase greater than 1200 IU. Mesenteric ischaemia, perforated duodenal ulcer, and acute cholecystitis may also lead to transiently elevated levels. Amylase level is not an indicator of severity.

Treatment

- *Mild*: Nasogastric suction, intravenous fluid replacement, analgesia
- *Severe*: Patients risk cardiovascular, renal/respiratory complications. They need monitoring of blood pressure, CVP, blood gas profile for PaO_2 and urine output. Peritoneal dialysis or plasmaphoresis may be indicated.

Criteria of severity

- White blood cells $> 1500 \times 10^6/l$, blood sugar more than 10 mmol/l
- LDH > 500 IU/l, SGOT more than 200 IU/l, urea > 16 mmol/l
- $Ca^{2+} < 2$ mmol/l, albumin less than 32 g/l
- $PaO_2 < 60$ mmHg

If three of the above criteria obtain the patient has SEVERE ACUTE PANCREATITIS. However, the criteria are not 100 per cent accurate.

Laparotomy is indicated when there is uncertainty of the diagnosis or when the patient fails to improve by conservative measures. Depending on the findings, pancreatectomy or pancreatic necrosectomy may be necessary with drainage of the pancreatic bed. Peritoneal lavage in patients with severe pancreatitis has not been shown to improve the prognosis.

References

Corfield, A. P., Williamson, R. C. N., McMahon, M. J., *et al.* (1985). Prediction of severity in acute pancreatitis: Prospective comparison of three prognostic indices. *Lancet*, ii, 403–7.

Mayer, A. D., McMahon, M. J., and Corfield, A. P. (1985). A randomised trial of peritoneal lavage for the treatment of severe acute pancreatitis. *New England Journal of Medicine*, **312**, 399–404.

Chronic pancreatitis

This may follow incomplete resolution of acute pancreatitis. It is commoner in men than women and is especially associated with chronic alcohol intake. Other causes include obstruction of the ampulla of Vater or pancreatic duct by tumour or calculus, post-traumatic, cystic fibrosis, familial, hyperparathyroidism, malnutrition in infancy, and primary sclerosing cholangitis.

Pathology

The pancreas becomes diffusely hardened. Histologically there is glandular atrophy, duct ectasia, calcification, and stone formation. Duct occlusion leads to cystic changes. The whole or only part of the gland may be affected.

Clinical features

Upper abdominal pain (more than 90 per cent of patients), weight loss, malabsorption, and diabetes mellitus (30–40 per cent).

Investigations

- Plain abdominal X-ray may show calcification.
- US scan demonstrates cystic change and duct dilatation.
- ERCP establishes the extent of the disease by assessing duct dilatation. It is a useful investigation preoperatively.
- Examination of the stool: steatorrhoea occurs in more than 30 per cent.
- Serum amylase is elevated during painful exacerbations (acute on chronic pancreatitis).

Treatment

Medical

- Stop alcohol.
- Give analgesics: Coeliac plexus block may be necessary for pain.
- Increase dietary intake of carbohydrate.
- Give pancreatic supplements, e.g. pancreatin according to patient needs:
  - *Adults*: Pancrex forte 6–10 tabs 6 hourly 30 mins before meals
  - *Children*: Pancrex V capsules 2–6 6 hourly swallowed whole
  - *Infants*: Pancrex V contents of 1–2 capsules mixed with feeds

    Neonates: Pancrex V 125 contents of 1–2 capsules mixed with feeds

    (Other preparations include Cotazym, Pancrease, Creon, Nutrizym.)
- Treat diabetes mellitus.

Surgical

- To relieve pain—distal pancreatico-jejunostomy—the tail is transected and a Roux loop of jejunum sutured to it to drain the duct retrogradely.
- Longitudinal pancreatico jejunostomy. The main pancreatic duct is laid open and the cut duct is enclosed by a Roux segment of jejunum sutured over it.
- Varying degrees of pancreatic resection may be indicated if the ducts are not dilated, but the procedures have high morbidity.
- Endoscopic sphincterotomy may relieve symptoms if there is main duct sphincter stenosis.

Prognosis

Patients usually live more than 20 years. Death is often due to another disease.

Peritonitis

Peritonitis is an inflammation of the lining of the abdominal cavity, the peritoneum. Although it may be limited to one area in the form of an abscess, it usually extends to both the visceral and parietal peritoneum to become generalized. Untreated, it is usually fatal.

Causes

Chemical, e.g.

- Perforated duodenal ulcer, perforated gall bladder, pancreatitis
- Ruptured ectopic pregnancy, Mittelschmerz
- Barium sulphate in X-ray diagnosis

Bacterial, e.g.

- Penetrating foreign body. Direct spread from the female urinary tract—primary peritonitis. Perforation of uterus by IUD.
- Primary pneumococcal peritonitis post-splenectomy. Complication of septicaemia.
- Most common cause is from the viscera, e.g. appendicitis, perforation of bowel in Crohn's disease, diverticulitis or carcinoma of colon.

Organisms

Extremely variable. When introduced from without *Staphylococcus aureus*, *Streptococcus* spp., gonococcus are implicated. However the most common infecting organisms are from the bowel itself—Bacteriodes, *Escherichia coli*, *Clostridium perfringens*, *Pseudomonas* spp., and *Klebsiella*. Remember that TB can be a cause.

Signs and symptoms

Local features

- Sudden onset with severe pain at first localized, then generalized
- Reflex contraction of abdominal muscles (guarding) with rigidity when generalized
- Tenderness to palpation
- Bowel sounds are initially present but may become absent due to ileus

Systemic features

- These may be shock with increase in pulse rate and temperature and falling blood pressure.
- The patient lies on his back, knees forward, looks pale and anxious with shallow breathing.

Early diagnosis is essential

Diagnosis Clinical findings are important. Erect chest and abdominal X-rays may demonstrate free gas. Routine haematology and biochemistry especially lipase or amylase to exclude pancreatitis. If the diagnosis is difficult and the patient's condition worsening, exploration is indicated.

Treatment

Resuscitation Give IV fluids, plasma, or plasma expander quickly until the blood pressure starts to rise. Check and correct electrolytes, blood gases. Pass a catheter to monitor urine output. Consider a CVP line. Pass a nasogastric tube. Give adequate analgesia—narcotic agents.

Removal or repair of the causative lesion When the cause is unknown explore the abdomen through a right paramedian incision centred on the umbilicus. Identify and repair or remove the cause. Drain the abdomen.

Treat infection and toxaemia This can be achieved by:
- Peritoneal washout at laparotomy with saline containing antibiotics (e.g. tetracycline 1 g/l), or antiseptics (e.g. noxythiolin which also prevents adhesions)
- IV antibiotics—aminoglycoside or cephalosporin with metronidazole 8 hourly

References

Anonymous (1979). Antibiotic lavage for peritonitis. *British Medical Journal*, **2**, 691–2.

Stewart, D. J. and Matheson, N. A. (1978). Peritoneal lavage in appendicular peritonitis. *British Journal of Surgery*, **65**, 54–6.

Perforated peptic ulcer

This is usually an anterior duodenal ulcer which perforates leading to acute diffuse peritonitis. Less often subacute perforation may lead to subphrenic or perigastric abscess formation.

Most gastric perforations are on the anterior surface near the lesser curvature. They are less common on the posterior surface when perforation occurs into the lesser sac. Ulcers at this site usually present with bleeding as they are extraperitoneal and involve the gastro-duodenal artery.

Associations

- Recent dyspepsia. Known ulcer. About 20 per cent of patients have had no symptoms.
- Recent ingestion of non-steroidal anti-inflammatory drugs
- Trauma. Major surgery. Burns. Other forms of stress.
- Carcinomas of the stomach may present as perforation.

Clinical features

Pain is sudden, severe, and sometimes accompanied by vomiting. It becomes generalized as the gastric juices spread throughout the peritoneal cavity. This is a chemical peritonitis which becomes secondarily infected by bacteria. Before the onset of bacterial peritonitis there may be an apparent degree of recovery or latent period. The patient has a rigid abdomen, almost board-like, and shallow breathing. He lies still, preferring not to move. He may report shoulder-tip pain. Bowel sounds are absent and the liver dullness may be obscured by intraperitoneal gas. Shock and collapse may develop.

Investigations

- Erect chest X-ray. There is free subdiaphragmatic air in 70 per cent.
- White blood cell level is increased, amylase is elevated but usually less than 1200 IU. There is haemoconcentration.

Treatment

- Resuscitate the patient.
- Laparotomy and oversewing of the ulcer perforation is indicated for patients with a short or no history of dyspepsia (less than 1 month). Postoperatively such patients should receive H_2 antagonists for a variable period up to 4 months. 75 per cent of patients will develop further problems without this regimen.
- Patients with chronic dyspepsia or longstanding duodenal ulcer should have a definitive procedure, e.g. truncal vagotomy and drainage, HSV + suture of the perforation.
- Gastric perforations may be treated by local excision with truncal vagotomy and pyloroplasty or antrectomy. Always check for gastric cancer.

References

McKay, A. J. and McArdle, C. S. (1982). Cimetidine and perforated peptic ulcer. *British Journal of Surgery*, **69**, 319–22.

Tanphinphat, C., Tanprayoon, T., and Na Thalang, A. (1985). Surgical treatment of perforated duodenal ulcer: A prospective trial between simple closure and definitive surgery. *British Journal of Surgery*, **72**, 370–2.

189

Abdominal abscesses

An intra-abdominal abscess is a localized collection of pus within the peritoneal cavity.

Aetiology

Sepsis can be
- Primary
- Secondary to intra-abdominal disease
- Due to contamination at surgical procedures or trauma
- Secondary to blood-borne organisms
- Secondary to splenectomy

Sites of abscess formation

They can occur anywhere, but are commonly located in the sub-phrenic and subhepatic spaces in the upper abdomen. Below this level the paracolic gutters, especially the right, are prime sites. They also occur commonly in the right and left iliac fossae in relation to a perforated appendix or diverticular disease respectively. Pelvic abscess may be a complication of diffuse peritonitis, as are collections between loops of bowel. Organ abscesses, e.g. pancreatic, perinephric present a specific clinical picture.

Clinical features

Localizing signs are absent in as many as 30 per cent of patients and these may be further confused by antibiotics or in immuno-compromised patients. The presentation is usually of an inter-mittent pyrexia and leucocytosis. This gives a classical notched appearance to the temperature chart. A mass is palpable in 10 per cent, and rectal or vaginal examination should always be carried out when pelvic sepsis is suspected.

Imaging and localization of abdominal abscesses

- US examination is the first choice.
- CT scanning is of more value in detecting retroperitoneal, pancreatic, mesenteric, and psoas abscesses but its avail-ability is not universal.
- Radionuclide imaging with ^{67}Ga citrate or ^{111}In labelled leucocytes is valuable for small collections especially around the bowel.

Treatment

- Percutaneous drainage under US or CT guidance. An initial diagnostic tap is followed by the insertion of a pigtail catheter or trocar catheter. These drain under gravity into a drainage bag until resolution is complete. The catheter may be used to irrigate the cavity with antibiotics.
- Operative drainage is infrequently necessary and carries a 40 per cent re-operation rate. Extraserosal approaches to avoid the peritoneal or pleural cavities should be utilized, e.g. the posterior approach through the bed of the 12th rib for subphrenic abscess.

5 General surgery and gastrointestinal disease

Abdominal abscesses

- Antibiotics have limited value, and continued change of antibiotics has a risk of pseudomembranous colitis. They may also increase the population of resistant organisms and mask clinical signs. They are indicated only in cases of septicaemia or gas gangrene.

Reference

Joseph, A. E. A. (1985). Imaging of abdominal abscesses. *British Medical Journal*, **291**, 1446–7.

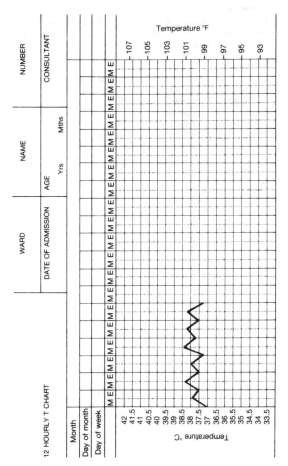

Acquired megacolon

This is caused by faecal impaction usually due to chronic constipation in childhood. It does not occur at birth and may be seen in the elderly.

Examination

- The abdomen is dilated and faeces are palpable in the colon.
- Rectal examination reveals a dilated faecally impacted rectum. There is often a painful fissure.
- Sigmoidoscopy may show 'melanosis coli', a sign of chronic aperient ingestion.

Diagnosis

Is both clinical and by barium enema which confirms dilatation to the level of the anus. Ensure that there is no anorectal pathology, e.g. carcinoma.

Treatment

Aim is to maintain an empty rectum and colon and permit the return of muscle tone. This is achieved by:
- Treatment of the anal fissure or fistula
- Manual disimpaction of faeces
- Regular colonic washouts
- Investigation of the family environment
- Re-establishment of toilet training

Volvulus

Volvulus results from rotation or torsion of the mesenteric axis of a portion of the alimentary tract.

Effects

Partial or complete intestinal obstruction. Ischaemia leading to gangrene and perforation.

Types

Volvulus neonatorum. Gastric volvulus. Small-intestinal volvulus. Caecal and sigmoid volvulus.

Gastric volvulus

Occurs as a complication of a rolling hiatus hernia.

Small-bowel volvulus

Usually occurs in the ileum and is related to the presence of adhesions on the antimesenteric border around which the bowel rotates.

Treatment is laparotomy, untwisting of the volvulus, and division of the adhesion. Resection of the bowel is indicated if it is non-viable.

Caecal volvulus

Occurs when the caecum and ascending colon are excessively mobile. The twist is usually clockwise and obstructs the ascending colon first. If a second twist occurs the ileum may obstruct. Delayed diagnosis has a 30 per cent mortality.

Presentation Vomiting, abdominal pain, constipation, occasional palpable tympanitic mass in right iliac fossa or lower abdomen.

Diagnosis Plain abdominal X-ray shows gas-filled ileum and occasionally a gas-filled caecum. Barium enema.

Treatment Decompression and resection or fixation by caecostomy prevent recurrence.

Sigmoid volvulus

This is the most common site, usually affecting middle-aged or elderly males. A redundant pelvic or sigmoid colon favours its occurrence.

Presentation is usually sudden with severe abdominal pain often when the patient is straining at stool. The abdomen subsequently becomes massively distended and there is absolute constipation. There may also be a history of recurrent attacks of left-sided pain as a series of partial volvulus occurs with subsequent untwisting followed by the explosive passage of large amounts of faeces and flatus.

Diagnosis Plain abdominal X-rays shows massive distension of the colon. Barium enema may delay surgery in acute cases.

Treatment
- Preoperative decompression with a sigmoidoscope may be successful. A rectal tube should be left in place for 48 hours.
- Surgical resection and end-to-end anastomosis can be carried out electively after decompression in patients who have had recurrent episodes.
- If decompression is unsuccessful a laparotomy should be performed with exteriorization of the affected segment (Hartmann's procedure). On-table colonic lavage may be followed by resection and anastomosis obviating the need for the second stage of the procedure.

195

Reference

Ellis, H. (1982). *Intestinal Obstruction*. Appleton Century Crofts, New York.

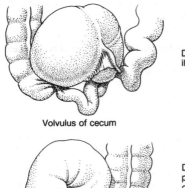

Distended ileum

Volvulus of cecum

Distended proximal colon

Volvulus of sigmoid colon

Volvulus

Crohn's disease

Crohn's disease is a chronic inflammatory granulomatous disorder which can affect any part of the alimentary tract from the mouth to the anus. The cause is unknown. It is premalignant in small and large bowel.

Pathology

The rectum is often spared. 30 per cent involve distal small bowel, 30 per cent colon, 40 per cent both. There are 'skip' lesions between unaffected intervening bowel. Pseudopolyps are unusual (cf. ulcerative colitis). The inflammation is transmural, the bowel wall and mesentery greatly thickened and crypt abscesses unusual (cf. ulcerative colitis). Ulceration is usually submucosal and the disease is characterized by fistulas, fissures, and non-caseating granulomas.

Histology

Confirms transmural inflammation, widespread granulomas in the bowel, and lymphadenopathy adjacent to the bowel. In the late stages fibrosis and cicatrization are common.

Clinical features

Fever, abdominal pain, diarrhoea, moderate anaemia, mass often in right iliac fossa, perianal abscess, and fistual formation (suspect in patients with recurrent peritoneal abscesses). Malabsorption. Acute or chronic intestinal obstruction. May mimic acute appendicitis. Internal fistulas.

Diagnosis

Sigmoidoscopy, biopsy. Barium and small-bowel enemas show thickened, stenosed areas with deep ulcers (raspberry thorn, rose thorn) and cobblestone appearance. Note skip lesions, internal fistulae and the characteristic string sign of Kantor when the terminal ileum is constricted.

Treatment
Medical
- As for ulcerative colitis with oral Salazopyrine 1 g three or four times daily by mouth. Many patients attend as outpatients.
- More severe attacks require hospitalization with bed rest.
- Steroids and/or azothiaprine. Anaemia should be corrected and TPN may be required but it does not lead to a cure as was previously thought.

Surgical treatment is indicated in the presence of complications:
- Drainage of abscesses
- Resection of stricture, fistulae, and perforated bowel with end-to-end anastomosis
- Intestinal obstruction
- Associated conditions like uveitis and polyarthritis if Crohn's disease is active
- The development of and prevention of carcinoma

Ulcerative colitis

Definition

A diffuse inflammation of the mucus membrane of the colon and rectum, which normally starts in the rectum (proctitis) and extends to involve the distal or whole colon.

Proctoscopic pathology

Mild disease inflammatory oedema, touch bleeding, watery faeces often bloodstained or purulent. The rectal wall is more rigid than usual. Presence of pseudopolyps and shallow ulcers.

Severe disease early age of onset extensive colonic involvement, large irregular ulcers, ulcer base necrosis. When extensive the bowel may be denuded. Areas between take the form of polypoidal tags. Basal fibrosis.

Histology

Suppurative acute inflammation with polymorphs, crypt abscesses, ulceration, granulation, loss of goblet cells, decreased crypts of Lieberkuhn. In chronic cases there are lymphocytes, plasma cells, but little evidence of repair.

Aetiology

Unknown. 10 per cent of patients have allergies to milk proteins. The theory that colonic bacteria may be causative remains not proven.

Clinical features

Characterized by remissions and exacerbations but may be mild distal disease (proctitis) to more severe disease with incessant diarrhoea, fever, malabsorption, anaemia, and dehydration. May also occur as a fulminating, debilitating attack. Patients may develop toxic dilatation or perforation (both life-threatening).

Diagnosis

Proctosigmoidoscopy plus biopsy. Characteristic barium enema with loss of haustration and ulceration.

Management

Medical or surgical.

Medical
Acute attack
5-day regime: replacement therapy with blood, iron electrolytes, TPN
Drugs: Prednisolone 60 mg in 24 h, IV
Hydrocortisone 400 mg in 24 hours, IV
Steroid retention enemas when diarrhoea is settling start with 50–100 ml graduating to 200–300 ml twice daily
Moderately severe
Replacement therapy
Systemic prednisolone 10 mg orally four times daily
Salazopyrin (may be enteric coated) 1 g three or four times daily by mouth

Distal colitis/proctitis
Replacement therapy
Oral Salazopyrin or Predsol suppositories thrice daily (5 mg prednisolone)
Retention enemas at night
   N.B. Check with your pharmacist first regarding prescriptions for patients with ulcerative colitis.

Surgical treatment is indicated in:
Perforation
Acute toxic dilatation—immediate
Failure to respond to medical treatment—semi-elective
Chronic disease
Unremitting/recurring symptoms
Severe polyarthritis
Stricture or risk of cancer (risk rises with recurrent symptoms over 10 years and with early-onset unremitting widespread disease).

Operations

Panproctocolectomy with ileostomy. Colectomy with J or W pouch and pouch-to-rectum anastomosis. This procedure retains continence and is combined with rectal mucosectomy. The procedure should be carried out by specialized surgeons.

Diverticular disease of the colon

Colonic diverticula are acquired outpouchings of colonic mucosa through the bowel wall associated with increased intraluminal pressure. They occur in rows between the taenia coli at the points of penetration of blood vessels in the colonic wall and are associated with hypertrophy of the circular and longitudinal muscle layers which predates their formation. Almost all occur in the sigmoid colon, but the rectum with its complete layers of muscle is unaffected. The condition is rare before 35 years and uncommon in countries where a high-fibre diet is eaten.

Clinical features

Diverticulosis, the state of having diverticula, may be asymptomatic.

Patients may complain of left-sided colicky abdominal pain, an alteration of bowel habit, or rectal bleeding.

Diverticulitis, inflammation in diverticula, may produce persistent left iliac fossa pain, localized (more common) or widespread peritonitis in a patient with fever or malaise. There may be a history of diarrhoea, constipation, and abdominal distension relieved by passing flatus. There is left-sided tenderness, and the sigmoid colon may be palpable. Urinary symptoms, like the passage of flatus through the urethra (pneumaturia), implies vesicocolic fistula formation.

Diagnosis

Clinical. Barium enema (not in diverticulitis: may cause perforation). Colonoscopy.

Exclude

Carcinoma of colon, urinary tract infection, appendicitis, or Crohn's colitis.

Complications

Stricture formation, severe rectal bleeding, intestinal obstruction, adhesions, perforation and faecal peritonitis, abscess formation, fistula formation especially into bladder or vagina.

Treatment

Symptomatic diverticulosis High-fibre diet—wholemeal bread, flour and rice, fruit and vegetables. Add 2 teaspoonfuls of bran to food at each meal. Use bulking agents like Colofac or Fybogel until the stools are soft and defecation is painless. Warn the patient that flatulence is to be expected and usually passes off in 2 weeks. Advise bed and antispasmodics if there is pain.

Acute diverticulitis is treated by bed rest, nil orally, and antibiotics given IV or IM—cefuroxime 750 mg four times daily and metronidazole (suppositories) three times daily for 10 days. Relieve pain with pethidine 75–150 mg/4 h IM.

Perforation Treat by hemicolectomy or defunctioning transverse colostomy. Give antibiotics as above. If there is widespread peritonitis carry out peritoneal lavage. The mortality rate is 40 per cent.

Bleeding The source can be difficult to find. Treat by blood transfusion. If uncontrollable carry out a restorative procto-colectomy (total colectomy + construction of an ileal pouch and pouch-rectum anastomosis).

Obstruction, stricture, fistula formation are treated by resection of the affected area of bowel with end-to-end anastomosis.

Sigmoid myotomy (division of the circular muscle of affected segments) is rarely performed.

Pseudomembranous colitis

This is watery or bloody diarrhoea associated with abdominal cramps, tenderness, fever, and leukocytosis. Endoscopically the appearances are of proctitis or colitis of the left colon. The mucosa is reddened and oedematous with raised white or yellow areas—the pseudomembranes. Biopsy confirms mucosal inflammation. The condition is associated with antibiotic ingestion (antibiotic-associated pseudomembranous colitis) and high numbers of *Clostridium difficile*.

Diagnosis

- Suspect if there is persistent diarrhoea which occurs 5–10 days after antibiotic ingestion, particularly of ampicillin and broad-spectrum combinations. Diarrhoea is not uncommon with antibiotics and usually stops spontaneously.
- Assay the patient's stool for *Clostridium difficile* or its toxin. Ask the laboratory how the specimen should be taken.
- Carry out endoscopy with biopsy.
- Do routine stool cultures (three specimens on successive days).

Treatment

- Discontinue the antibiotic.
- Eradicate *Clostridium difficile* with oral vancomycin (150–200 mg daily in divided doses) for 7–10 days
- Give oral cholestyramine 12 000–24 000 mg daily in divided doses. This binds the toxin of *Clostridium difficile* and is of value in mild cases.

Relapse

30 per cent of patients may have further problems. Repeat the course of vancomycin for a further two weeks. If there is recurrence taper the dose of vancomycin as follows:
125 mg orally four times daily for the first week, 125 mg twice daily for the second week, 125 mg orally on alternate days for the third week, 125 mg every third day for the fourth week.

Polyps

Any lesion which projects into the lumen of the bowel is a polyp. It is composed of epithelial and connective tissue.

Colonic polyps: Types

- *Neoplastic*: Adenoma, e.g. familial adenomatous polyposis
- *Inflammatory*
- *Hamartomatous*
- *Miscellaneous*: Metaplastic in mucosal prolapse syndrome

Malignant potential

Related to the presence of dysplastic epithelium. Other significant factors are the type of adenoma, size of the lesion, and invasion of the muscularis layer of the mucosa.

Polyposis syndromes

Associated with multiple GI polyps:
- Familial adenomatous polyposis
- Peutz–Jegher's syndrome
- Juvenile polyposis
- Inflammatory polyposis (chronic inflammatory bowel disease)

Familial adenomatous polyposis

Inherited as an autosomal dominant, the polyposis gene being situated on the long arm of chromosone 5. Such patients are strongly predisposed to develop colonic cancer. This strong association is possibly the best known and most frequently taught precancerous condition in medicine. The polyps are thought to develop around puberty.

Symptoms may develop at around 20 years. There is a change of bowel habit, usually diarrhoea associated with the passage of blood and mucus.

Carcinoma develops around 5 years later, the average age being 35–40 years. The average age of death is 40–2 years in untreated patients.

Management

1 *Screening*: Multiple areas of congenital hypertrophy of the retinal epithelium is present from birth and can be picked up on ophthalmoscopy.
2 *Treatment*: Total colectomy + mucosal proctectomy + ileo-anal pouch anastomosis (specialized surgeons). Panprocto-colectomy + ileostomy.
3 *Follow-up* is long term. The use of long-acting prostaglandin antagonists like sulindac may be of value.

Reference

Chapman, P. D., Church, W., Burn, J., and Gunn, A., (1989). Congenital hypertrophy of retinal pigment epithelium: a sign of familial adenomatous polyposis. *British Medical Journal*, **298**, 353–4.

Examination of the anus and rectum

Tell the patient what you intend to do. Ask him to lie on his left side with his knees drawn up towards his abdomen and his buttocks at the edge of the bed. Raising the buttocks on a sandbag facilitates sigmoidoscopy. Stand with your left side next to the patient's back.

Action

Separate the buttocks and inspect the anus. Note whether there are changes of excoriation, erythema, or moistness of the perianal skin. Are there anal skin tags? Is there a prolapsed haemorrhoid? Are there perianal warts?

Digital examination

Put a rubber glove on your right hand. Apply some K-Y jelly (Johnson and Johnson) to the tip of the index finger. Place the pad of your index on the anal orifice posteriorly and apply enough pressure to permit entry. At this stage the patient may complain of pain and discomfort and you may note spasm of the sphincter. This suggests an anal fissure. Do not persist with digital rectal examination but gently separate the buttocks to expose the anal verge and the lower end of the fissure and a sentinel pile usually situated posteriorly. Lubrication of the finger with lignocaine gel may permit digital examination. However, if there is spasm and pain it is better to examine the patient under anaesthetic.

Once the finger is in the anus, change its direction so that the tip 'points' into the rectum. With the pad of the finger feel the posterior and lateral walls. Are there any abnormalities—mass, ulcer, polyp?

Now bend your knees and rotate the finger so that you can feel the anterior rectal wall. Is there a mucosal abnormality? Can you feel the prostate anteriorly in a male patient? Note its consistency. In female patients can you feel the cervix through the anterior wall. It is a knob-like projection. The os is usually palpable as a slit. Note whether movement of the cervix leads to pain or discomfort.

Rigid sigmoidoscopy

Should be carried out at this session, usually prior to proctoscopy. The rigid sigmoidoscope is a calibrated tube consisting of a viewing barrel, obturator for introduction, and an eye piece, air, and light source. The length is variable but usually 20–35 cm. The patient lies in the left lateral position. The lubricated instrument is introduced gently, the obturator removed and the sigmoidoscope advanced under direct vision whilst gently insufflating air. Keep the lumen in site at all times and take care at the level of the rectosigmoid junction where undue force may cause perforation. Note the mucosa—reddened, bleeding? Are there polyps? Is there ulceration, pus, blood, or suspected tumour? Take biopsies.

Proctoscopy

The proctoscope is a short instrument consisting of an obturator enclosed in a viewing barrel with a handle and light source.

Action

Apply K-Y jelly to the tip of the obturator and press it on to the anal orifice until it has penetrated. Once the instrument has been completely inserted remove the obturator, attach the light source and gently but slowly withdraw the instrument. Note the state of the rectal mucosa. Is it pink (normal)? Does it bleed easily? Does it completely fill the barrel (prolapse)? As you withdraw the instrument note whether there are haemorrhoids. They appear as purple cushions at positions 3, 7, and 11. Band or ligate them if indicated (p. 216). Note also whether there are polyps or an anal fissure anteriorly or posteriorly (although proctoscopy is usually painful in the presence of a fissure but may be made possible with local anaesthetic lubricating jelly).

Painful anorectal conditions

Prolapsed thrombosed haemorrhoids (strangulated piles)

This is fairly common. The patient usually has a history of piles for 3 years or more, with episodes of prolapse. All three haemorrhoidal sites may be involved, giving the appearance of gangrene.

Treatment
- *Conservative*. Bed rest, elevation of the foot of the bed, local cooling, and analgesia. The haemorrhoids subside after a few days and further treatment is often not necessary.
- *Manual dilatation* under general anaesthetic (Lord's procedure). Formal treatment may be indicated later.
- *Emergency haemorrhoidectomy*, especially if there have been previous attacks or there is gangrene. Give prophylatic antibiotics to prevent portal pyaemia. This procedure is not commonly performed nowadays.

Thrombosed external haemorrhoid (synonym: external plexus haematoma)

Produces acute anal pain which may last 2–3 days. The pile appears as a tense dark blue swelling at the anal margin.

Treatment
- Reassurance and analgesia
- If large, excise under local anaesthetic (2 per cent lignocaine + 1/200 000 adrenaline). Do not puncture and express the clot: this can lead to anal tag formation.

Fissure in ano

Affects men and women equally, with a peak incidence in 20–30 year olds. It is due to a breach in the squamous epithelium of the lower one third of the anal canal. Most common in the posterior midline in men and the anterior midline in women. Pain is exquisite, burning or shooting, related to defecation and often associated with fresh bleeding. Even sitting down may hurt. It may be transient or last for hours. Often the diagnosis can be made from the history alone. The patient is afraid to pass stools and is extremely apprehensive about rectal examination, so be gentle. Stop and examine under general anaesthetic if pain is extreme. The fissure is exposed by carefully separating the buttocks. Often a sentinel tag or pile overlies it.

Treatment
Acute fissure should be regarded as an urgent SOPD appointment.
- *Conservative therapy* is of value in acute fissures. Use a St Marks No. 2 anal dilator with 2 per cent lignocaine gel. Pass this gently and advise the patient to repeat the procedure twice daily until relief occurs then continue for a further 3 weeks with lubrication. This method is not popular with patients.

- *Surgery* is indicated in very painful fissures, failed conservative treatment, associated piles, or fistula and chronic fissure. The options are manual dilatation (Lord style) of the anus to six or eight digits and lateral sphincterotomy (which is marginally superior to manual dilation). Do not perform a sphincter stretch. It may lead to incontinence.

Perianal abscess

Affects men more often than women. They are perianal, ischiorectal (commonest), intermuscular, or pelvirectal and colonized by gut or skin organisms. They may present insidiously or acutely. Antibiotics should only be prescribed if there are systemic symptoms. Treatment is by surgical drainage under general anaesthesia, although minimal surgery under local anaesthesia has much to recommend it. Incision, curettage with antibiotic cover, and primary closure is also an effective technique but the incidence of fistula formation remains around 30 per cent. The infecting organisms originate from the skin in the case of perianal abscess with fistula and from the gut in ischiorectal abscess with fistula.

References

Jensen, S. L., Lund, F., Neilson, O. V., and Tange, G. (1984). Lateral subcutaneous sphincterotomy versus anal dilatation in the treatment of fissure in ano in outpatients: a prospective randomised study. *British Medical Journal*, **289**, 528–30.

Lock, M. R. and Thomson, J. P. (1977). Fissure in ano, the initial management and prognosis. *British Journal of Surgery*, **64**, 355–8.

Vasilevsky, C. and Gordon, P. H. (1985). Results of treatment of fistula-in-ano. *Diseases of Colon and Rectum*, **28**, 225–31.

Fistula in ano

This is a track of granulation tissue between the perianal skin and the anal or rectal lumen which leads to its constant re-infection and persistence. It occurs commonly after spontaneous or inadequate surgical drainage of a perianal or ischiorectal abscess.

Associations

Inflammatory bowel disease, rectal carcinomas, trauma, tuberculosis.

Types

Anatomical classification
Subcutaneous
Submucous
Low anal (below anorectal ring (puborectalis))
High anal (above or at the level of anorectal ring)
Pelvi rectal (penetrates levator ani)

Park's classification
Intersphincteric
Transphincteric (high or low)
Supralevator

Goodsall's rule

If the skin opening is on the anterior half of the anus the fistula is direct. Fistulas opening on the posterior half of the anus open in the posterior midline internally and may extend laterally in horseshoe fashion on both sides.

Clinical features

Fistulas often begin as abscesses. After drainage a persistent purulent discharge develops, associated with perianal irritation and discomfort. Pain develops only if the orifice becomes occluded.

Examination

There is usually a single external opening near the anus. Sometimes several openings are seen.

Rectal examination may detect an indurated track, and pus can occasionally be expressed from the opening on pressing the track.

Proctoscopy/sigmoidoscopy should be carried out to visualize the internal opening if possible and exclude associated disease. If the internal opening can be seen, pass a malleable probe along the fistula to define its course.

Treatment

When the fistula is secondary to other disease processes these must be managed initially.

When the fistula itself is the only feature, treatment is to lay the track open and allow it to heal by granulation. High fistulas should be treated in the same manner but there is a risk of incontinence if the anorectal ring is divided. In these a seton (ligature of silk or linen) is passed along the line of the fistula which acts as a wick and allows fibrosis to occur so that the track can be excised with gradual removal. In some patients a long-term defunctioning colostomy must be established.

References

Parks, A. G., Gordon, P. H., and Hardcastle, J. D. (1976). A classification of fistula in ano. *British Journal of Surgery*, **63**, 1–12.

Ramanujam, P. S., Prasad, M. L., and Abcarion, H. (1983). The role of seton in fistulotomy of the anus. *Surgery, Gynaecology and Obstetrics*, **157**, 419–22.

Pilonidal sinus

This is a midline tract in the natal cleft of young adults, usually male between 20–40 years. It is lined by granulation tissue, contains loose hairs, and may have associated secondary tracts and external openings. The condition has also been described in the interdigital clefts of barbers.

Aetiology

Although congenital sinuses do occur associated with dermoid cysts or vestigial glands, most are acquired. Hairs are driven into the skin by sitting or by the shearing action of the buttocks, and infection is initiated from faecal contamination. As the sinus develops further, shed hairs are sucked in by the intermittent negative pressure created by the action of the buttocks.

Clinical features

The patient is usually a thickly-haired young man (or woman) who presents with recurrent abscess formation.

Examination

Reveals a tender swelling (pilonidal abscess) or single or multiple external openings which may exude pus or blood-stained sebum. Hairs are often visible at the orifice.

Treatment

Pilonidal abscess incision and drainage.

Pilonidal sinus
- Lay open the track and allow it to heal by granulation.
- Excise the track and its ramifications. Close by primary suture or rotation flaps.
- Wide excision. Allow healing to occur by granulation from below.
- Phenol injection
- Can be brushed out with excision of the skin openings

Recurrence

There is a 10–15 per cent recurrence rate with all forms of treatment. Reduce the risk by advising the patient and spouse (for shaving) to keep the area clean and hairfree.

Rectal prolapse

Protrusion of the rectum or anal canal through the anal orifice may be complete or incomplete (anterior rectal wall only). It is most common in women over the age of 40 years, although men may be affected (peak incidence 20–40 years). Children are rarely affected and there may be an association with cystic fibrosis.

Causes

Denervation of puborectalis muscle, loss of the acute anorectal angle, poor fixation of rectum to the sacrum. Haemorrhoids, anal polyps, warts, and papillae may also prolapse through the anus.

Symptoms

Prolapse may occur with defecation and reduce spontaneously or it may be permanent. Some patients are anxious. Many are unaware that they have a problem. Some degree of faecal incontinence and irregularity of bowel function are both common features.

Clinical findings

Lax anal sphincter. Voluntary contraction is weak. The perineum descends and the prolapse may appear on straining.

Investigations

- *Sigmoidoscopy*: Traumatic proctitis
- *Proctoscopy*: The prolapse fills the lumen.
- *Anorectal manometry*: Intraluminal pressures in the anus and rectum can be measured at different levels with the sphincters lax and maximally contracted. The results are compared to normal controls.
- *Evacuation proctography*: May demonstrate an increased anorectal angle, occlusion of the rectal canal or recto-rectal intussusception.

Complications

Irreducibility, ulceration, bleeding, strangulation, perforation of the bowel.

Treatment

Surgery is necessary to control the prolapse and improve continence.

Procedures

- Circumferential wiring of the anus, usually with prolene or polyamide (Thiersch). Complications (ulceration, infection and faecal impaction) are common. Reserve for unfit patients.
- Anterior prolapse in women usually responds to submucous injection of phenol.
- Perianal rectosigmoidectomy. This is amputation of the prolapse. Recurrence is 50 per cent. Consider in frail patients.

- Delorme operation. The mucosa is excised from the prolapse and the muscle is longitudinally plicated. The results are good.
- Perirectal implantation procedures—Wells' operation involves the suture of a rectangular sponge to the sacrum. The edges of the sponge are then sutured to the anterior of the rectum on each side. In Ripstein's procedure a Teflon prosthesis is sutured anteriorly to the rectum. The flaps formed are then sutured on each side to the sacrum to form a sling. The pelvic floor may be repaired at the same time.
- Rectosigmoidectomy and anterior resection have high recurrence and complication rates.

Prolapse in children

More common in boys, especially if thin, when there is reduced ischiorectal and pararectal fat. The rectum intussuscepts and a plum-coloured doughnut protrudes from the anus.

Treatment

Conservative. The condition usually corrects itself as the child grows. Reassure the parents. Advise them to keep the act of defecation as short as possible, otherwise the child may use the prolapse as a form of attention-seeking. Simple repeated reduction is usually all that is required. Persistent prolapse may be treated by phenol injection or linear cauterization.

References

Bartolo, D. C. C., Rose, A. M., Virjee, J., and McC Mortensen, N. J. (1985). Evacuation proctography in obstructed defaecation and rectal intussusception. *British Journal of Surgery*, **72**, Supplement, 5111–16.

Henry, M. M. (1986). Rectal prolapse and the descending perineum syndrome. *Surgery*, **1**, 682–6.

Haemorrhoids

Haemorrhoids are the commonest of all anal conditions, affecting as much as 40 per cent of the population at some time.

Aetiology

Three theories:
- Varicosities of the internal haemorrhoidal plexus
- Distension of arteriovenous anastomosis near the anal cushions
- Prolapsed anal cushions

Symptoms

Most are asymptomatic. Bleeding and prolapse at defecation are the commonest symptoms.

Pain is the result of thrombosis or associated anal fissure.

Soiling, itching, and perianal irritation are all less frequent.

Classification

Internal or external. They may also be classified as first degree (entirely within anal canal and above dentate line), second degree (prolapse occurs on defecation with spontaneous reduction afterwards), and third degree (piles remain prolapsed after defecation and require manual replacement, descend during exercise, or are outside the anal canal all the time).

Differential diagnosis

Rectal prolapse, inflammatory bowel disease, anal polyp, anal fissures, malignant disease. Biopsy suspicious lesions.

Investigation and examination

Ask the patient to describe the symptoms. Examine the abdomen, groins, and genitals. Inspect the anus carefully before carrying out rectal examination with proctoscopy and rigid sigmoidoscopy. Note anal sphincter tone. Differentiate rectal prolapse which completely fills the proctoscope from the quadrants of haemorrhoids.

Treatment

Reassurance and exclusion of serious disease. Regulation of bowel habit with a high-fibre diet and bulk laxatives is effective in many.

Injection sclerotherapy of the submucosa around the haemorrhoid with up to 20 ml of 5 per cent phenol in arachis oil leads to a chemical thrombosis followed by fibrosis after 2–3 weeks. This is an outpatient procedure which is effective for bleeding haemorrhoids. Review after 6 weeks. Rubber-band ligation is of value especially in patients with bleeding and prolapse. The band is placed 1–2 cm above the haemorrhoid. This is an outpatient procedure. Two bands are placed at a time, with regular review and bulking agents.

Lord's procedure. Under general anaesthetic the anus is gently dilated: four fingers of both hands are inserted. After the procedure the patient is instructed to pass a dilator into the anus.

Formal haemorrhoidectomy is indicated for patients who do not respond to conservative measures or who have large prolapsing haemorrhoids associated with skin tags.

References

Alexander, R. M. (1985). A technique for avoiding mucosal stenosis and secondary haemorrhage after haemorrhodectomy. *Diseases of the Colon and Rectum*, **28**, 271–3.

Corman, M. L. and Veindenheimer, M. C. (1973). The New haemorrhoidectomy. *Surgical Clinics of North America*, **53**, 417–22.

Poon, G. P., Chu, K. W., Lau, W. Y. *et al.* (1986). Conventional Vs triple rubber band ligation of haemorrhoids—a prospective randomised trial. *Diseases of the Colon and Rectum*, **29**, 836–8.

Restorative proctocolectomy

This has been developed to avoid the need for a permanent ileostomy in patients having operations for ulcerative colitis and familial polyposis coli. Total colectomy and mucosal proctectomy are performed (to eradicate the risk of rectal cancer) and the terminal ileum is mobilized by division of the ileocolic, ileal, and if necessary the superior mesenteric artery distal to its first branches. It is then fashioned into a pouch which can be of J, H, or W shape. The advantage of the W pouch is its increased capacity as compared to the others, but if a 20 × 20 cm loop is used for J pouch construction there appears to be little difference at long-term follow-up. The disadvantage of the H pouch is that it is isoperistaltic and catheterization is occasionally necessary.

Indications

- Ulcerative colitis where a conventional proctocolectomy would be necessary
- Familial adenomatous polyposis
- Colorectal cancer provided radical dissection is possible

Contra-indications

- Crohn's disease
- Severe acute colitis ⎱ may be treated by colectomy + rectal
- Patients taking high ⎰ preservation as a primary procedure
  doses of steroids   with reconstruction later

Advantages

- The addition of an ileal reservoir (pouch) after proctocolectomy results in less frequent defecation (around 3–4 times/day) than with direct ileoanal anastomosis (more than 8 times/day).
- The pouch is superior to ileostomy for most patients.
- The risk of rectal cancer is eradicated by mucosal proctectomy. (mortality from rectal cancer is 1 per cent with ileorectal anastomosis).

Disadvantages

- A temporary defunctioning loop ileostomy is usually necessary.
- The serious complication rate is high—pouchitis, anastomotic dehiscence, fistula formation, stenosis—but diminishes with the experience of the surgeon. Therefore most procedures are performed in specialist centres.
- Restorative proctocolectomy gives poor results in Crohn's disease.
- It is contra-indicated in patients over 55 years.

References

Hawley, P. R. (1985). Ileorectal anastomosis. In *Symposium XII—Ulcerative Colitis: Sphincter-Saving Operations. British Journal of Surgery Supplement*, **72**, S75–S82.

6 Vascular surgery

Evaluation of patients with peripheral vascular disease

The diagnosis can usually be made from the history and clinical examination.

History

Most patients' problems relate to occlusive arterial disease. Determine the main complaint, its mode and speed of onset. Has it persisted? If so is it continuous or intermittent? What precipitates attacks? How frequently do they occur, and how is the patient between attacks?

Inquire specially about symptoms relating to the cardiac (myocardial infarction, angina), cerebral (transient ischaemic attacks, stroke), visceral (intestinal angina, ischaemic colitis) and renal hypertension), vascular systems.

Is there local pain, swelling, or sensory loss? What is the state of the venous system?

Determine the family history. Is there hypercholesterolaemia? Are there associated risk factors like diabetes, hypertension, smoking?

Clinical examination

Examine the patient's face (arcus senilis, xanthelasma = subcutaneous cholesterol deposits, anaemia, or polycythaemia), hands and feet (tar on fingers from smoking, loss of pulp at tip of digits due to chronic ischaemia, shiny skin, tapering or frank gangrenous change in Raynaud's disease).

Inspect the local problem, comparing it to the other side. What is the colour? Are there vasomotor or trophic changes? Is there evidence of sepsis, ulceration, or gangrene? Is there an obvious pulsatile mass? Raise the patient's legs to 60° above the examination couch. Watch for pallor of the feet/foot. This happens in seconds if there is extensive disease. Then ask the patient to sit up and let the legs hang down. Record the time for venous filling, which should normally occur in about 10 seconds. In significant disease the veins remain collapsed for some considerable time. Note also the time it takes for reactive hyperaemia to develop in the dependent limbs. There are often marked differences between the limbs when obstructive disease is unilateral only (Buerger's test).

Palpate the skin for temperature changes. Feel the neck vessels. Are there cervical ribs? Feel the arterial pulses on each side, starting from the neck and following the order: common carotid, subclavian, brachial, radial, aortic, common femoral, popliteal, posterior tibial, and dorsalis pedis.

Percuss varicose veins for impulses (see varicose veins).

Auscultate the praecordium, carotid, subclavian, aortic, renal, iliac, and femoral arteries. Are there bruits? Systolic bruits may be heard over aneurysms or stenosis. Machinery murmurs may be heard in arteriovenous fistulas.

Investigations

Do an ECG and chest X-ray.

Non-invasive

- *Cerebrovascular disease*: ocular plethysmogaphy, spectral analysis of Doppler ultrasound flow signals. These tests are not universally available
- *Lower limb* Doppler ultrasound indicates flow patterns in veins and arteries. Ankle Doppler pressure can be compared to brachial pressure to give the ankle/brachial ratio or ischaemic index

Invasive Arteriography. IV digital subtraction angiography.

Preinvestigative, preoperative, and postoperative care for patients with vascular disease

Outpatient clinic procedures

Chest X-ray, ECG, ESR, FBC and platelet count, U&E, MSSU.

Admission to hospital

- A short preoperative stay prevents colonization with multiple resistant staphylococci.
- Clotting screen
- Discontinue antiplatelet or anticoagulant therapy if indicated (e.g. patient undergoing surgery).
- Group and cross-match blood: abdominal aneurysm, aortic bifurcation graft 4–6 units; femoropopliteal bypass procedures, femoral, popliteal aneurysms 2–4 units.
- Treat infections with appropriate antibiotics *before* surgery and delay operation if necessary.
- Special investigations, e.g. CT scan, IVU, abdominal films, arteriography

Patient undergoing arteriography

- *Preinvestigation*: Obtain consent. Nil orally from midnight before investigation. Prepare the groin or axillae by shaving. The patient should urinate before the study. Contrast solutions produce diuresis.
- *Postinvestigation*: Bed rest for 12–24 hours. Check peripheral pulses. Report immediately any change in pattern. Start oral fluids after 12 hours if well. If operation is considered set up an IV of N-saline/dextrose 5 per cent before the investigation.

Complications of arteriography

- Haematoma
- Intimal damage leading to arterial occlusion + embolus or thrombosis
- Allergic reactions to contrast agent

Patient undergoing surgery

- *Preoperative*: Shave the operation site if necessary. Arrange preoperative physiotherapy. Give clear fluids 24 hours before surgery and nil orally after midnight. Start IV fluids in the evening before theatre with N-saline/dextrose 5 per cent 500 ml/6 hourly. 1 hour before induction of anaesthesia give IV Flucloxacillin cloxacillin (500 mg) IV +/− an aminoglycoside. Repeat ×2 at 6 hourly intervals. Catheterize the patient for aortic procedures.
- *Postoperative*: Monitor appropriate pulses and blood pressure ($\frac{1}{4}$, $\frac{1}{2}$ hourly if required) for 24 hours. Monitor fluid input and urine output against CVP or wedge pressures. Ensure adequate blood volume replacement. Check neurological function frequently after carotid surgery which may have delayed complications. Postoperative ventilation may also be required.

6 Vascular surgery

Pre-investigative, preoperative, and postoperative care for
patients with vascular disease

Smoking and peripheral vascular disease

The 1977 report of the UK Royal College of Physicians, *Smoking and Health*, stated that more than 95 per cent of patients with peripheral vascular disease of the legs were smokers. Those who continue to smoke were more likely to develop critical ischaemia or gangrene and suffer subsequent amputation than those who stopped. Of 5540 amputees who attended artificial limb centres in 1978, 3514 had amputations because of vascular disease (63.4 per cent).

Smoking as a risk factor in peripheral vascular disease

Five factors have been shown to be precursors of atherosclerosis in the brain, heart, or legs. These are hypertension, serum cholesterol greater than 7.7 mmol/l, glucose intolerance, left ventricular hypertrophy on ECG, and smoking. Smoking is a greater risk in the production of intermittent claudication than stroke where hypertension is more important. The risk of developing intermittent claudication for a smoker of 15 cigarettes per day in people over 45 years is 9 times that of non-smokers. If other associated risk factors are included such as increased serum triglyceride, raised diastolic blood pressure, or increased serum urate, the risk of developing claudication is multiplied by a further 20 times.

Does smoking cause atherosclerosis?

Smoking does not appear to cause arterial disease, but probably accelerates its progression. An important factor in cigarette smoke appears to be carbon monoxide. It can shift the oxygen dissociation curve to the left, combine with haemoglobin to form carboxyhaemoglobin (CoHb) and also with myoglobin and the cytochrome systems. An association between CoHb and atherosclerosis has been established in humans, and the presence of CoHb in the blood correlates well with the prevalence of intermittent claudication.

Smoking and reconstructive vascular surgery

Continued cigarette smoking after aortic and femoral bypass procedures increases the risk of graft failure (90 per cent graft patency at 2 years) (less than 5 cigarettes/day), cf. 60 per cent graft patency at 2 years (more than 5 cigarettes/day). Even if a reconstruction is not carried out patients who stop smoking and begin to take exercise can double their claudication distance in 2–3 months. Those who continue to smoke are more likely to develop ischaemic rest pain or symptoms in the unaffected leg, or undergo amputation.

Advice to patients

- Patients with intermittent claudication should be advised that smoking is a correctable risk-factor and they should give up.
- Patients should not be offered reconstruction unless they can demonstrate that they have stopped smoking or are prepared to cut down. (Blood samples will reveal CoHb levels, which

are a good indicant of recent smoking.) Options for surgery are reduced if more severe symptoms appear after surgical reconstruction.

- Stopping smoking can be difficult. Refer the patient to the antismoking clinic. Try smoking withdrawal aids such as nicotine chewing gum. Other methods like terror tactics: 'Your legs will drop off', or hypnosis 'Cigarettes make me sick' have reasonable success in the short term but relapse is common.

227

References

Greenhalgh, R. M., Laing, S. P., Cole, P. V., and Taylor, G. W. (1981). Smoking and arterial reconstruction. *British Journal of Surgery*, **68**, 605-7.

Thomas, M. (1981). Smoking and vascular surgery. *British Journal of Surgery*, **68**, 601-4.

Occlusive disease of the aorto-iliac and femoro-popliteal segments

Atherosclerosis is a generalized disease. It may involve the abdominal aorta and its branches either singly or in combination, but usually affects the bifurcation and the iliac vessels. In the lower limb the femoro-popliteal segment is most frequently affected.

Clinical features

Intermittent claudication The patient complains of pain, usually in the calf, after walking 100–400 yards (90–350 metres). The onset is insidious. The pain is relieved by rest only to recur on exercise. Symptoms may remain unchanged or improve if collaterals develop.

Prevalence Claudication is the most frequent manifestation of arterial disease for which patients attend a vascular clinic. 10 per cent of patients over 65 years will have symptoms.

Site of occlusion The commonest site is the superficial femoral artery beyond the origin of the produnda femoris artery.

Associated disease Many patients have a history of myocardial infarction and angina. Cerebral ischaemia is less common. With or without operation these patients have a 20 per cent risk of death within 5 years from stroke or myocardial infarction.

Symptoms of aorto-iliac disease Buttock and thigh claudication +/− impotence (Leriche syndrome), calf claudication, rest pain.

Symptoms of femoro-popliteal disease Intermittent claudication of the calves. At first the patient can 'walk through' the pain.
   With progressive disease the claudication distance reduces.
   Rest pain of the toes and feet, worse in bed at night, relieved by hanging the foot over the side of the bed or standing up indicates serious ischaemia.

Signs

- Ischaemic ulcers, gangrene, infection of the toes, feet, and pressure areas
- Diminished or absent peripheral pulses distal to the arterial lesion
- Atrophic and ischaemic skin changes—loss of limb hair, brittle nails, digital pulp loss
- Bruits and thrills. A bruit is an audible systolic murmur distal to an occlusive lesion which is higher pitched in severe stenosis. A 'thrill' is a palpable bruit.
- Pallor on elevation of the limb. Rubor when returned to the horizontal due to reactive hyperaemia (Buerger's test).

Investigations

(see p. 223).

Treatment

Many patients can modify their lifestyles and will not require reconstructive surgery. They should:

- *Stop smoking*. Doppler ultrasound and treadmill exercise tolerance have been observed to improve in patients who give up. Smoking is the single biggest factor influencing amputation rates.
- *Take exercise*. Average walking distances are doubled in the first three months.
- *Treat risk factors*. Diabetes, cigarette smoking, and hypertension need attention.

Drugs Vasodilator drugs offer no benefits. More complex compounds like naftidrofuryloxalate (Praxilene) give symptomatic relief in patients over 60. Oxpentifylline (Trental) has been shown to increase treadmill walking distances. Treatment should be given a trial of 2–3 months in selected patients with severe symptoms.

Surgical

- *Reconstructive*, e.g. aorto-femoral bypass, femoro popliteal bypass. These procedures rely on healthy artery proximal and distal to the lesion. Success depends on proximal inflow to the graft and distal 'run-off'.
- *Transluminal angioplasty* is of value for the dilatation of localized or short segment occlusions. A balloon catheter is passed through the stenosed area and the balloon inflated under X-ray screening. Excellent results may be obtained but lesions may recur. There is a risk of embolization.
- *Endarterectomy* is the removal of the diseased intima to restore the lumen. It is rarely used in aorto-ilio-femoral disease and is controversial elsewhere (e.g. carotid).

Prognosis

Short-term results are good in over 90 per cent of patients. Smoking relapsers develop graft occlusions more quickly. Patients are at risk of death from the other dangers of systemic arteriopathy, e.g. myocardial infarction, stroke.

Advice to patients

Patients who claudicate need reassurance. Many will not require reconstructive surgery if aetiological factors are corrected—diabetes, hypertension, but especially smoking. Many will not be heading for amputation.

References

Intermittent claudication. *British Medical Journal* (1986), **292**, 970–1.

Jelnes, R., Gaardsting, O., Hougaard Jenson, K., Baekgaard, N., Tonneson, K. H., Schroeder, T. (1986). Fate in intermittent claudication: outcome and risk factors. *British Medical Journal*, **293**, 1137–40.

Acute arterial occlusion of the limbs

Sudden occlusion of an artery is usually due to embolism or trauma. This is an emergency. Treatment is required immediately.

Clinical features

When a major vessel is occluded there is: pain, pallor, pulselessness, paralysis, paraesthesia, and 'perishing' of the limb.

Diagnosis

Usually clinically obvious. Ultrasound may identify abdominal aneurysms.

Embolic sources

Embolic sources are mural thrombus following myocardial infarction (30 per cent), atrial fibrillation, mitral stenosis, and aneurysms. Less commonly atherosclerotic plaques, prosthetic heart valves, bacterial endocarditis, and left atrial myxoma are the sources. In trauma intimal damage or flaps may lead to occlusion.

Treatment

- *General*: Treat cardiac arrhythmias. Start administration of heparin by continuous IV infusion (5000–10000 units/12 hours) to prevent extension of thrombus as a result of distal disease. Dextran 40 500 ml + Praxilene 400 mg may be infused together IV in the first 3–4 hours of treatment. Relieve pain with IV or IM analgesics
- *Specific*: Embolectomy is indicated when occlusion is due to embolism. Patients sustaining trauma require vascular repair or reconstruction with or without embolectomy

Prognosis

Patients with no clinical history of peripheral vascular disease do well at embolectomy, whereas those with a history often have distal disease requiring more than embolectomy, e.g. reconstruction.

Aneurysms

An aneurysm is a dilatation of an artery formed by widening and expansion of its lumen and containing blood or clot. True aneurysms contain all three layers of artery wall in the sac. False aneurysms, due to trauma or infection (mycotic), are periarterial haematomas encapsulated by surrounding tissue. Most aneurysms in developed countries are due to atherosclerosis. All aneurysms may rupture, irrespective of size.

Types

They may be fusiform (asymmetrical dilatation of artery), saccular (symmetrical), or dissecting (when blood splits the arterial media). The underlying cause may be congenital, traumatic, atherosclerotic, mycotic, symphilitic, or related to Marfan's syndrome.

Sites

Thoraco-abdominal, abdominal, femoral, popliteal, visceral, carotid, subclavian, circle of Willis (Berry aneurysms).

Symptoms are due to expansion, thrombosis, embolism, dissection, or rupture.

Thoraco-abdominal aneurysms Involve the descending thoracic and upper abdominal aorta.
- *Features*: Often asymptomatic. May produce back, chest, or abdominal pain on rapid expansion or rupture. There is a pulsatile abdominal mass found above the umbilicus if the infrarenal aorta is involved. X-rays show widening of the mediastinum or calcification in the aortic wall. CT scan may define extent.
- *Complications*: Rupture, embolization
- *Treatment*: Resection and graft replacement
- *Prognosis*: Good in elective cases if the whole aneurysm is excised. Poor in cases of rupture.

Abdominal aortic aneurysms Usually the infrarenal aorta is involved, with or without common iliac aneurysms.
- *Features*: Often asymptomatic. Abdominal pain, back pain, nerve pain may indicate expansion or rupture. There is a palpable abdominal mass lateral and superior to the umbilicus.
- *Diagnosis*: Ultrasound scan. Plain X-rays may show calcification.
- *Treatment*: Excise and replace with prosthetic graft (Dacron).
- *Prognosis*: 95 per cent survive elective procedure. Less than 50 per cent survive emergency resection for rupture.

Surgery is indicated for all other forms of aneurysmal disease.

Ruptured abdominal aortic aneurysm

Abdominal aortic aneurysms present as an emergency when expansion of the aneurysm causes severe abdominal pain or backache. When the aneurysm ruptures there is sudden severe pain and shock. The patient is usually over 60 years old.

Mortality

Without treatment all patients die. With treatment, even in specialist centres, 50 per cent of patients reaching hospital will die.

Diagnosis

History, clinical examination, and plain abdominal X-rays (if time permits in a non-shocked patient).

There is usually an expansile abdominal mass. Chest X-ray will exclude thoraco-abdominal aneurysm. Blood sugar and serum amylase should be done. In cases of real doubt and a stable patient, an abdominal US scan should be done.

Remember that ruptured abdominal aortic aneurysm is one of the causes of an acute abdomen.

Unusual presentation

Peripheral arterial embolism, aorto-caval fistula with severe abdominal pain and congestive cardiac failure, progressive renal failure and lower limb ischaemia, complete aortic occlusion. It may also mimic other acute abdominal pain syndromes.

Preoperative assessment

In shocked patients immediate surgery is indicated. Almost all patients need surgery. Infra-renal aneurysms are rarely inoperable, and the conservative treatment of aneurysms is contraindicated. Age is not a contra-indication to surgery. Relatives should be aware of this, particularly in the elderly.

Unfavourable prognostic factors

Age > 80 years, sustained hypotension with delayed transport, patients with previous myocardial infarctions.

Operation

Control blood loss as quickly as possible. Withhold muscle relaxants until the surgeon makes the incision. Control bleeding either by cross-clamping or Foley catheter impaction proximal to the neck of the aneurysm allowing time for dissection. Occasionally a transdiaphragmatic approach to the aorta is necessary.

Graft

Woven Dacron without heparinization. After insertion, declamp temporarily and 'punch' the groins' (firmly compress the common femoral artery and its branches with the closed fist) to express loose clot before completion of the lower anastomosis. The residual sac should be used to cover the graft.

Postoperative care

Monitor fluid balance *accurately*, using CVP and/or Swan Ganz control as well as clinical evaluation. Beware of *clotting* abnormalities. Ventilation should be used only if necessary, for weaning can be extremely difficult.

Complications

- *Haemorrhage*: Can occur from the aorta peroperatively. Avoid this by achieving proximal and distal control before opening the sac. Oversew lumbar vessels. Avoid major venous bleeding by ensuring minimal or no dissection near the venacava and common iliac vessels.
- *'Trash foot'*: Avoid by minimal dissection. Carry out Fogarty catheter embolectomy immediately the condition is recognized. (See p. 256).
- *Renal failure*: May be related to prolonged hypovolaemia of the site of the aneurysm. Monitor urine output by inserting a catheter before operating.

The management of stroke syndrome

Atherosclerosis of the extracranial vessels is the common pathology with artery-to-artery embolism in 90 per cent of patients. Polycythaemia, thrombocytosis, anaemia, cardiac lesions like atrial myxoma, aortic valve prolapse, and Takayasu's arteritis may also be implicated.

Terms used

- *Transient ischaemia attack*: Transient loss of sensory and/or motor function. Lasts minutes usually, but always < 24 hours.
- *Reversible ischaemia neurological defect*: As for transient ischaemic attack, but > 24 hours
- *Stroke in evolution*: Progressive neurological deficit over hours or weeks
- *Completed stroke*: The stable end result of an acute or progressive stroke

Diagnosis

The patient is usually a hypertensive male > 50 years who smokes and suffers from intermittent claudication. The clinical features are of transient ischaemic attacks including amaurosis fugax (temporary blindness), vertebro-basilar attacks (ataxia, diplopia, drop attacks), or even frank stroke.

Signs

- Motor deficit: carotid bruit on contralateral side
- Amaurosis fugax: carotid bruit on ipsilateral side
- Other bruits at the right base of the neck or ipsilateral supra-clavicular fossa suggest innominate or subclavian or vertebral artery lesions.

Investigations

- Doppler ultrasonographic imaging of the carotids
- Digital subtraction angiography. Contrast is injected either intra-arterially or intravenously. Not all centres have DSA.

Prognosis of patients with TIAs

17 per cent will have a stroke within the first year, 33 per cent within 5 years. The mortality rate is also high—40 per cent from myocardial infarction or stroke within 5 years.

Treatment: Medical or surgical?

- *General*: Identify and treat risk factors—diabetes mellitus, lipid disorders, hypertension. Advise the patient to stop smoking.
- *Medical*:
  —Aspirin and sulphinpyrazone are better than placebo, reducing the incidence of stroke especially in males.
  —Anticoagulant therapy may be of value within the first few months of transient ischaemia attacks, but this remains unproven.
- *Surgical*: Carotid endarterectomy

- *Indications*: Transient ischaemic attacks, stroke in evolution, vertebrobasilar disease, asymptomatic carotid lesions
- *Outcome*: 12–25 per cent of patients develop a stroke which may be delayed in appearing postoperatively

The operation of carotid endarterectomy is widely practised, but it has still to be established that the risk of stroke without surgery is greater than after carotid endarterectomy.

Reference

Murie, J. A. (1987). Carotid endarterectomy: an effective therapy? *Scottish Medical Journal*, **32**, 163–5.

Thoracic outlet compression syndrome

Thoracic outlet compression syndrome is the term used to cover a spectrum of neurological, arterial, and venous disorders resulting from compression of the neurovascular bundle as it leaves the chest to enter the upper limb. The area of compression is enclosed by the first rib, clavicle, and scalenus anterior.

Clinical features

Venous Compression of the axillary vein leads to cyanosis of the skin of the hand and arm. Impaired venous and lymphatic return produce oedema.

Neurological Pain, paraesthesias, weakness, and numbness of the hand and arm are noted especially over the distribution of the ulnar nerve. These features are exacerbated during sleep and when carrying heavy objects. Abduction of the arm with the head turned to the opposite side also makes the symptoms worse.

Arterial Compression of the axillary artery produces pallor or intermittent cyanosis of the hand and fingers. Embolic episodes give rise to pain, pallor, cyanosis, and even gangrene of the fingers.

Differential diagnosis
- Prolonged cervical intervertebral disc
- Carpal tunnel compression syndrome

Diagnosis
- Cervical spine/thoracic outlet X-rays. These may confirm the presence of a cervical rib or protuberant transverse process of C_7.
- Arteriography may confirm subclavian artery stenosis or post stenotic dilatation.

Treatment

Mild cases: Tell patients to avoid heavy lifting. Prescribe shoulder girdle exercises.

Patients with neurological or vascular symptoms need surgical decompression of the thoracic outlet by excision of a cervical rib and first rib combined with cervical sympathectomy through a transaxillary approach. If there is no cervical rib, division of scalenus anterior (scalenotomy) is helpful in over 70 per cent of patients.

Thrombo-angitis obliterans, vasospastic disorders

Thrombo-angitis obliterans (Buerger's disease)

Affects the small vessels of the hands and feet, with associated thrombophlebitis and Raynaud's phenomenon. Most patients are young men. All are smokers.

Effects Foot claudication, severe digital ischaemia progressing to gangrene, loss of ankle and wrist pulses.

Diagnosis Arteriography confirms patches of occlusion with uninvolved intervening segments. Exclude other forms of arteritis.

Treatment Stop smoking, sympathectomy, reconstructive procedures, amputations.

Prognosis The disease will stabilize if the patients stops smoking.

Raynaud's disease

This occurs in otherwise normal individuals who have digital arterial sensitivity to cold. Young women are most often affected. The peripheral pulse pattern and anatomy of the digital arteries are normal.

Effects Intermittent attacks of digital pallor followed by cyanosis and reactive hyperaemia. In some patients the vasospasm may cause arterial obliteration followed by dry gangrene.

Differential diagnosis Chilblains. Thoracic outlet compression syndrome (see p. 238).

Treatment Avoid cold. No smoking. Use vasodilator drugs (naftidrofuryl oxalate, Praxilene; inositol nicotinate, Hexopal). Cervical sympathectomy if symptoms are severe.

Raynaud's phenomenon

Peripheral vasospasm secondary to organic disease. It is especially associated with occlusive arterial disease, neuro-vascular entrapment syndromes, connective tissue disease especially systemic sclerosis, and mixed connective tissue disease. The phenomenon may be provoked by various stimuli, e.g. drugs like ergot, beta-blockers, or heavy metals; cold; trauma; hormones and emotional upset. Raynaud thought that the condition was caused by sympathetic overactivity but this has never been proved. More recent evidence suggests that raised blood viscosity is a factor coupled with the interaction of thromboxane A_2 (vasoconstrictor produced in platelets) and prostacyclin (PGI_2) (vasodilator, produced in vascular endothelium).

Treatment

Medical:
- Advise patient to stop smoking, avoid cold. Review drug prescriptions. Identify and treat underlying disease.

- Give vasoactive drugs, e.g. naftidrofuryl oxalate (Praxilene), inositol nicotinate (Hexopal). These are frequently prescribed by the patient's GP.
- IV dextran 40 500 ml/12 hours
- Guanethidine sympathetic blockade
- Corticosteroids. Try nifedipine.
- Prostacyclin

Surgical:
- Cervical sympathectomy may be necessary in Raynauds' phenomenon, which is not due to collagen disease, but which remains refractory to drug treatment.

241

Acrocyanosis

May be confused with Raynaud's. It affects young women who develop persistent (not paroxysmal) cyanosis of the hands and feet which may be patchy and affect especially the forearms and calves. It is painless.

Treatment Identify and treat the causal lesion.

Post-traumatic vasomotor dystrophy

Can follow trauma usually to an arm. There is pain, cold sensitivity, chronic cyanosis and coldness. Sudek's atrophy may develop (spotty osteoporosis of the bones of the hand or feet).

Causalgia

Causalgia is severe, persistent burning pain following upper extremity injuries precipitated by minor stimuli. Cyanosis and coldness of the extremity is characteristic.

Treatment Sympathectomy.

Rare arterial diseases

Polyarteritis nodosa

Is a primary necrotizing vasculitis which may be due to circulating immune complexes.

Clinical features Young to middle-aged men are affected. It may present as a PUO with malaise, weight loss, abdominal pain, skin rashes, gangrene of fingers, and GI haemorrhage.

Pathology There is obliteration and necrosis of arteries, arterioles, and capillaries with the development of saccular aneurysms. The kidneys, heart, liver, and muscles are most often affected. Neurological involvement may lead to fits, hemiparesis, or psychosis.

Diagnosis Biopsy of affected tissue usually muscle. Visceral angiography. ESR is raised. FBC shows anaemia, eosinophilia, and leucocytosis.

Treatment Prednisolone and cyclophosphamide.

Temporal, occipital and ophthalmic arteritis

Characterized by localized arterial infiltration leading to occlusion.

Clinical features include headache, tender arteries, blindness when ophthalmic artery is involved. There is often low-grade fever and arthralgia. Elderly women are most often affected.

Diagnosis Biopsy is diagnostic, and also often relieves the pain.

Treatment Prednisolone.

Takayasu's arteriopathy ('Pulseless disease')

This is an obliterative arteritis usually affecting females and pursuing a relentless course to death from cerebral ischaemia.

Clinical features One or more vessels arising from the aortic arch become thrombosed leading to cerebral symptoms or upper limb ischaemia and claudication on exercise (upper limb pulses are absent).

Treatment Endarterectomy. Bypass grafting. Sympathectomy. These provide only temporary relief.

Cystic arterial degeneration

In this condition ganglion-like lesions containing clear jelly develop in the externa of an artery, usually the popliteal.

Clinical features The lesion compresses the vessel when the limb is flexed and leads to claudication especially on stasis.

Diagnosis Arteriography

Treatment Decompress. Repair arterial defect with vein or synthetic patch.

Arteriovenous fistula

Communication between arteries and veins may be congenital, when it is usually multiple, or acquired as a result of trauma. It may also be surgically created to permit vascular access for patients who require renal dialysis.

Congenital lesions

Most often occur on the lower limb, female pelvis, and abdominal viscera. They may be entirely asymptomatic, or cause pain and swelling over the site of the fistula. An affected limb may be thicker and longer than its neighbour.

Clinical examination There is a pulsatile mass with a palpable thrill and a constant machinery murmur. Proximal pressure on the artery reduces the swelling and the pulse rate (Nicoladini–Branham sign).

Complications

Pain. Progressive swelling. Massive haemorrhage. High-output cardiac failure.

Treatment Excise small lesions completely. Arteriography should be used to establish the vascular pattern of large lesions, which can be major in type. The vascularity may then be subsequently reduced by therapeutic embolization and the lesion excised within 24 hours. Re-examination may be necessary at intervals of 1 or 2 years.

Traumatic arteriovenous fistulae

May be secondary to penetrating trauma. They can occur iatrogenically during vertebral disc surgery. As they enlarge they cause pressure effects and high-output failure.

Treatment involves early identification with separation and repair of the affected vessels. Multiple proximal and distal ligation with reconstruction may be necessary.

Vascular access for haemodialysis

Regular vascular access is achieved by anastomosis between the radial artery and the cephalic vein. This takes about 3 weeks to mature and become usable. It has a long patency and low infection rate, making it superior to the Scribner shunt which may be used in emergency dialysis. This is an external prosthesis which is inserted into an artery and adjacent vein. Other methods of vascular access include ulnar artery to basilic vein, subcutaneous arteriovenous grafts (using reversed saphenous vein, human umbilical vein, Dacron, or Goretex).

Complications of arteriovenous anastomosis Bleeding, aneurysmal formation, thrombosis, infection, engorgement of the hand. Cardiac failure from high output is very rare.

The anterior tibial compartment syndrome

This is due to compression of structures in the anterior tibial compartment of the leg, which is especially vulnerable because of its rigid boundaries. It is characterized by pain in the antero-lateral compartment and exists in both acute and chronic forms. The incidence is highest in males, 9:1 females. Young men are most often affected.

Aetiology

There are four causes:
- Severe or unaccustomed exercise (30 per cent)
- Trauma in association with soft tissue injuries or fractures of the tibia or fibula (20 per cent)
- Vascular in association with iliac or femoral artery disease, arterial bypass procedures, Buerger's disease, or polyarteritis nodosa (40 per cent)
- Various: Inguinal hernia repair, open heart surgery, IV infusions, nephrosis, eclampsia, leg braces, and lumbar sympathectomy (10 per cent)

Pathogenesis

It is believed that increased pressure in the anterior tibial compartment results from fluid retention (exercise) or oedema (local tissue damage, trauma, vascular disease). Rise in pressure in the anterior tibial compartment leads to further ischaemia and eventually tissue necrosis. In this aspect the anterior tibial artery behaves as an end artery.

Clinical features

There is pain over the anterior tibial compartment, ± weakness in dorsiflexion of the foot with anaesthesia over the first web space due to compression of the deep peroneal nerve.

Early or mild cases

Pain, tenderness, and swelling of the anterolateral aspect of the leg and foot are exacerbated by active or passive dorsiflexion of the toes or ankle. There may be weakness of dorsiflexion. Sensory loss in the first web space is an important sign.

Late or severe cases

The onset is rapid, often with complete paralysis of dorsiflexion and foot drop. There is anaesthesia of the first web space with eventual necrosis of skin and muscle of the anterior tibial compartment.

Differential diagnosis

Infection, phlebitis, tenovaginitis, osteomyelitis, Osgood–Schlatter's tibial stress fracture.

Investigations

In most patients the diagnosis is clinical. In the chronic group, arteriography and pressure measurements are of value.

Treatment

Recognize the problem. Most complications can be prevented by prophylactic fasciotomy, which will save limbs in traumatized patients or in those with revascularization procedures. Chronic symptoms may respond to rest. Some patients will need amputation.

Varicose veins

These are elongated, dilated, tortuous veins with incompetent valves.

Venous systems of the leg

Three groups:
- Superficial: long and short saphenous systems and tributaries
- Deep: between the muscular compartments of the leg
- Perforators: connect the superficial and deep systems

Blood passes from superficial to deep systems at the sapheno-femoral junction, short saphenous-popliteal junction, and via the perforators which are valved.

Classification

- *Primary*
  —familial can affect the superficial and perforating systems
  —exacerbated by fetus, fibroids, ovarian/pelvic tumour, prolonged standing
- *Secondary*
  —to deep venous thrombosis (postphlebitic syndrome. See p. 252)

Complications

Bleeding, thrombophlebitis, thrombosis, venous eczema, itching, unsightly appearance. Pain is rare.

Clinical examination

- Note the patient's general condition.
- Examine the cardiovascular system for cardiac failure, respiratory system for emphysema or bronchitis, abdomen for masses.
- Ask the patient, standing on a platform, to point out the worst areas.
- Note oedema, pigmentation, ulceration (gravitational ulcer usually above the medial malleolus), papery skin appearance.
- Palpate the veins from ankle to groin. Are there fascial defects at the site of 'blow-outs'?
- Cough impulse test for 'thrill' with coughing over a saphena varix or the long saphenous vein below the sapheno-femoral junction.
- Percussion test. Tap a vein and see how far down its length you can feel the repercussions which are normally interrupted by competent valves.
- Brodie–Trendelenberg test determines the competence of perforating veins and the sapheno-femoral junction
  —With the patient supine, elevate the leg to empty the veins.
  —Place two fingers over the sapheno-femoral junction 5 cm below and medial to the femoral pulse.
  —Ask the patient to stand. Maintain pressure with the fingers.
  —Do the veins begin to fill from below? If so the perforators are incompetent.

—Release the fingers. Does the long saphenous fill rapidly? If so, the sapheno-femoral junction is incompetent. If there is no evidence of incompetence elsewhere, disconnection would be indicated.

- Perthe's test. Empty the veins as above. Place a tourniquet below the knee. Ask the patient to exercise rapidly up and down on tiptoe. Do the veins fill? If so, there is perforator incompetence and they may need to be ligated.

Treatment

- Support stockings, exercise, and periodic elevation
- Compression sclerotherapy. Use the minimum dose, otherwise deep thrombophlebitis may be precipitated.
- Ligation, stripping, stab avulsions. Ligation and stripping of the long or short systems may be combined with perforator ligation and avulsion of unsightly veins.

N.B. Do not ligate or strip varicose veins secondary to disease of the deep venous system without first carrying out venography.

249

Compression sclerotherapy

Indications
- Moderate varicose veins of the lower leg
- Residual varicose veins after surgical treatment

Contra-indications
- Obese limbs. Compression is difficult to maintain.
- Veins above the knee
- Acute thrombophlebitis
- Deep venous thrombosis
- Pregnancy and the 'Pill'

Equipment and technique
- Sclerosant—STD or 5 per cent ethanolamine oleate
- Mark the sites of the veins. Apply a Tubigrip stocking to the foot.
- Have available on a trolley several 2 ml syringes with 0.5 ml in each, fine needles (23 gauge), mediswabs or surgical spirit as skin disinfectant, latex pads.
- Ask the patient to lie supine.
- Prepare the skin.
- Start with the varicosity furthest from the heart. Insert the needle and withdraw blood to confirm successful entry.
- Raise the leg to 45° until the vein empties. Slowly inject sclerosant and apply finger pressure. Replace this with a latex pad and draw the tubigrip over it to maintain pressure.
- Continue this procedure for the remaining veins.

Postoperative management
- The patient should walk daily 2–3 miles (3–5 km).
- The reinjection of residual veins and rebandaging is carried out at 2-week intervals. Usually no further treatment is required after 6–8 weeks.

Complications
- Ulceration due to extravasation of sclerosant
- Gangrene from intra-arterial injection
- Skin irritation from the bandage
- Thrombophlebitis and deep vein thrombosis
- Brownish discolouration of the skin along the path of the injected vein

The postphlebitic limb

This syndrome is the result of deep venous thrombosis or deep thrombophlebitis.

Clinical features

At least two of the following are present: oedema of the leg and ankle; venous ulceration; hyperpigmentation due to haemosiderin from extravasated red blood cells; venous eczema, telangiectasia and/or lipodermatosclerosis; pain and aching relieved by elevation of the limb (venous claudication).

Types

Proximal follows iliofemoral deep venous thrombosis. The iliac vein fails to recanalize adequately but the calf muscle pump still functions well. As a result there is venous obstruction to limb outflow above the inguinal ligament. Blood is then forced through extensive collaterals around the upper thigh to the pelvis and lower abdomen. There is swelling and aching of the leg on exercise.

Distal There is destruction of the valves of the popliteal and lower femoral veins due to deep venous thrombosis, so that local venous hypertension occurs especially when erect or walking. Typically the skin of the lower leg is poorly nourished with venous ulceration, fibrosis, and ultimate contracture of soft tissues which may affect the Achilles tendon and the intrinsic muscles of the foot.

Combinations of proximal and distal types occur.

Investigations

Ultrasound and venography may detect the sites of deep venous occlusion and incompetent perforators. Venograms may induce phlebitis but may be necessary for surgery.

Treatment

- *Control venous hypertension*: Use heavy elastic stockings, intermittent limb elevation. Avoid prolonged standing.
- *Treat stasis ulcers*: Elevate the limb to control oedema. Enforce bed rest in serious cases. Clean ulcers with saline. Apply compressive dressings (blue-line bandage). Change weekly. Skin graft large ulcers. Give systemic antibiotics appropriate to bacterial sensitivity.
- *Treat incompetent perforators*: Subfascial ligation of perforators is occasionally indicated.
- *Bypass proximal obstruction*: Venous reconstruction with bypass grafting or angioplasty has mixed results.

Prognosis

The condition is chronic. All operative procedures fail in the long term, so compression and elastic support may be required indefinitely.

Amputations

Indications

To relieve pain; to restore mobility (with a prosthesis); to save life.

To save life

- Moist gangrene, putrefaction, spreading infection to viable tissue leading to systemic infection
- Extensive tissue necrosis after acute thrombosis, embolism or trauma, which may preclude arterial reconstruction
- Tumours, paralysis

To relieve pain Rest pain, ischaemic ulceration (venous and arterial). Gangrene.

To restore mobility The endpoint of amputation is to achieve a healthy stump so that the patient can be fitted with a suitable prosthesis. Unfortunately, not all patients achieve this endpoint.

Investigations

Identify treatable lesions by arteriography. Treat by reconstruction profundoplasty or sympathectomy.

Preoperative preparation

General Restore haemoglobin status, fluid, and electrolyte balance. Treat cardiac failure or arrhythmia. Chest X-ray. Give antibiotics. Cephalosporins + metronidazole in mixed injections. Penicillin for moist gangrene.

Level Primary healing at the chosen level is the ideal. Failure is costly to both patient and surgeon. Below-knee amputations permit easier rehabilitation, but are more likely to break down than above-knee. Try to predict below-knee healing by Doppler ultrasound systolic pressures. Clinically the skin should feel warm and bleed when incised.

Types

- *Digital*: for Raynaud's, Buerger's etc. Amputate through bone proximal to healthy tissue.
- *Ray*: involves part of the metatarsal and associated tendons (also performed as a carpal ray). Indicated in diabetes when infection tracks up the tendon sheath.
- *Transmetatarsal (Lisfranc)* when several toes are involved. Create a long plantar flap for successful healing.
- *Syme's* when there is distal plantar infection and gangrene. The foot is amputated leaving the calcaneum for weight bearing. The heel skin must be healthy.
- *Gritti–Stokes* creates a long stump which makes rehabilitation difficult. The femur is transected at the adductor tubercle. This and the through-knee are unsatisfactory.
- *Below-knee*. Can be done in equal anterior/posterior flaps or long posterior flap. Anterior incision is at 6–8 cm distal to the tibial tuberosity.
- *Above-knee*. The stump should be fashioned 15 cm above the knee joint by either antero-posterior or myoplastic flaps.

Postoperative care

Control pain with opiates. Prevent muscle atrophy and contracture with physiotherapy. Get the patient up on an inflatable limb until the prosthesis is ready. Practice walking between bars.

Complications

Stump pain. Phantom limb. Causalgia. Treat with analgesia, excision of neuromata, sympathetic nerve blocks, and reassurance.

Complications of vascular procedures

These may occur preoperatively, in the early postoperative period, or late.

Preoperative complications

Haemorrhage This can occur during any vascular procedure and if arterial implies that proximal or distal control is inadequate. Other causes include bleeding from lumbar vessels, the inferior mesenteric or tributaries. On occasions haemorrhage can be the result of an atheromatous vessel ruptured by the clamp. In this situation control may be achieved by passing a Foley catheter into the lumen and inflating the balloon proximally.

Major venous haemorrhage can occur from the vena cava or the iliac veins during dissection. This can be virtually eliminated by avoiding routine dissection of these structures.

'Clamp-shock' When an aortic cross-clamp is released, some patients develop profound hypotension. It is therefore important to ensure adequate fluid replacement and to monitor the acid–base status carefully. Slow release of the clamp and replacement of blood volume may prevent this problem.

Ureteric injury This may occur during the repair of an aneurysm, particularly of an inflammatory type when the left ureter may be adherent to the wall.

'Trash foot' This leads to lower limb ischaemia due to embolization of atheromatous material distally upon release of the clamps. Prevention is by minimal peri-aortic dissection and by adequate endarterectomy around anastomotic sites. If it occurs carry out Fogarty embolectomy, heparinize the patient and give IV dextran 40 to minimize the effects of distal digital emboli.

Postoperative complications

Haemorrhage This may result from dehiscence at an anastomosis. The site should be re-explored and repaired.

Hypothermia This can occur after aortic graft procedures and may contribute to postoperative acidosis and cardia arrhythmias. For these reasons patients undergoing major aortic surgery must have their body heat maintained by the use of on-table heating devices, conventional bedclothes, and postoperative space blankets.

Embolism When there is distal ischaemia return to theatre, as embolectomy is mandatory.

Paralytic ileus (p. 116).

Late complications

Sepsis This is a risk whenever foreign material is used. If it occurs the septic graft must be removed and replaced if possible. The use of prophylactic antibiotics has reduced the risk.

Graft occlusion Depends on the site and type. It occurs in 5–50 per cent of patients over 55 years due to late thrombosis, intimal hyperplasia, or atherosclerosis.

Anastomotic aneurysm/false aneurysm formation This may be sepsis-related, and further surgery is indicated.

Death from cardiovascular disease Remember that atherosclerosis is a systemic disease. Patients who undergo vascular surgery have an increased risk of suffering a stroke or myocardial infarction compared to the general population. Remember that the onset of stroke may be delayed after carotid surgery.

Fistula formation This can occur between the aorta and gut (aorta-enteric) and presents as a massive haematemesis or melaena. It may be sepsis-related.

7 Urology

259

Examination of the urine

Test the urine of all patients. Look at it and smell it. If indicated, read its volume in relation to time by means of a fluid chart. For most situations paper strip tests are sufficiently accurate. They are impregnated with an indicator dye specific for the particular test required. When the strip changes colour it is compared to control colours on the bottle label which gives an indication of pH, protein, etc.

Examples

- *pH*: Acid urine suggests uric acid stones. Alkaline urine may indicate infection with *Proteus*.
- *Protein*: The indicator dye is bromophenol blue, which turns yellow in the presence of protein. If protein is discovered the urine should be collected over 24 hours and its protein content established in the laboratory. If the level is more than 150 mg/24 hours, further investigations are required.
- *Glucose*: The dipstick tests (Clinistix) are impregnated with glucose oxidase which lead to a change of colour in the presence of glucose down to a concentration of 5 mmol/l. Always do a blood sugar test if glucose is found on routine testing. This may have to be repeated with the patient fasting.
- *Blood*: The dipstick test is based on the oxidation of orthotoluidine by the peroxidase-like activity of haemoglobin. The tests will detect haemoglobin, free red cells, etc., as few as 10/high power field. False positives can occur with infected urine or with myoglobinuria, so the urine should be centrifuged and a red blood cell count performed.
- *Pus cells*: Pus can be seen, if present in large quantities. If it is suspected a drop of urine should be examined on a microscope slide when pus cells may be seen.
- *Casts*: These derive from the contents of the renal tubules. Centrifuge the urine. Place a drop on a microscope slide and examine it. There are two types of casts—granular consisting of red or white blood cells or both, and hyaline which are of protein.
- *Crystals*: These have characteristic shapes depending on their consistency, e.g. calcium oxalate are diamond-shaped, cystine are hexagonal.
- *Bacteria*: Bacteria can usually be seen, stained, and identified in the centrifuged deposit. Send three consecutive early morning urines to the laboratory for TB detection. The Ziehl–Neelsen stain is used.
- *Ova*: These can also be detected microscopically.
- *Malignant cells*: Can be detected in the urine, but the method must be emphasized. When collected, the urine must be immediately fixed with formalin (so have some ready). Then it is centrifuged and the deposit fixed and stained. Ask your laboratory for details if you wish this examination.
- *Culture*: Try to send the specimen immediately, otherwise growth of contaminants will occur. Storage in a refrigerator, but not at room temperature, is satisfactory. The dip slide

technique is a fairly accurate method. In this technique a plastic slide coated with culture medium is dipped into the urine, the excess drained off, and the slide placed in a sterile bottle in a warm place for 24 hours. The colonies on the slide can be compared to a control slide to estimate the numbers. More than 1×10^5 colonies/ml implies infection.

Urine specimens

Obtain urine when the patient has a full bladder, usually first thing in the morning (except for urine cytology). Take a mid-stream specimen, allowing about 150–200 ml to be voided before the sterile container is placed in the uninterrupted stream. Retract the prepuce in the male, and wash the glans with saline. In women a tampon should be inserted, the labia parted and the periurethral area cleaned with sterile saline. Catheter specimens are much more liable to contamination and should be performed only if the patient is unable to co-operate.

Cystitis

Inflammation of the urinary bladder is commoner in women than in men. It may be due to trauma (e.g. after intercourse—honeymoon cystitis) or bacterial infection. Repeated attacks should always lead to investigation of the urinary tract. Most cases of cystitis are due to ascending infection.

Predisposing factors

Urinary stasis as a result of:
- Pregnancy
- Prostatic obstruction
- Stricture of the urethra or external meatus
- Diverticulum of the bladder
- Spinal injury
- Bladder calculus, foreign body, or tumour
- Immunosuppression

Infecting organisms

E. coli is the commonest, but *Proteus mirabilis*, *Pseudomonas*, *Klebsiella* spp., *Staphylococcus*, and *Streptococcus* are frequently implicated. Less commonly the causal agent is the tubercle bacillus. *Schistosoma haematobium* is the causal organism in many tropical countries (p. 504).

Clinical features

- There is pain on micturition with frequency. Sometimes pus or blood is passed.
- Examination reveals tenderness suprapubically or on pelvic or rectal examination.

Diagnosis

- Midstream urine specimen + microscopy. Look for pus cells or a colony count greater than $10^5/mm^3$. This implies significant infection.
- Ask for bacterial sensitivity.

Treatment

- Give oral analgesia.
- Use amoxycillin 250 mg 8 hourly orally until sensitivities are available.
- Further investigation is indicated for patients who fail to respond or suffer recurrent attacks.

Indications for cystoscopy ± intravenous pyelogram

- Haematuria
- Recurrent attacks

Pyelonephritis

This is infection of the renal pelvis, calyces, and parenchyma which can originate from haematogenous or ascending routes. It can be acute or chronic.

Associations

Female sex, ureteric reflux (35 per cent of patients). Urinary tract abnormalities. Childhood, pregnancy, recent intercourse (honeymoon cystitis and pyelonephritis), menopause.

Causal organisms

Coliforms, usually *Streptococcus faecalis*, *Pseudomonas pyocyanea*, *Proteus*, and *Klebsiella*.

Clinical features

Acute Prodromal symptoms include headache and malaise. The onset is often sudden with rigors, vomiting, pain in the hypochondria or renal angles, and painful micturition (often described as 'scalding' or 'burning'). Uraemia can develop in cases associated with obstruction. Severe cases develop abdominal rigidity and hyperpyrexia (temperature greater than 40 °C).

Diagnosis
- Is by clinical features but may not always be obvious. Intravenous urogram may show limited excretion on the affected side with occasional mild extravasation.
- Radioisotope renography may show scattered uptake on the affected side.
- Midstream urine specimen (+ microbiology + sensitivity)

Treatment
- Bed rest and increased oral fluids
- Alkalinization of acid urine with potassium citrate mixture 10 ml by mouth 6-hourly
- Analgesia. Narcotics, e.g. morphine, should be used if pain is severe.
- Broad-spectrum antibiotics until sensitivities are available (amoxycillin 250 mg 8-hourly by mouth, or co-trimoxazole 2 tablets by mouth, twice daily).

Chronic This may develop *ab initio* or after multiple acute attacks, and is a cause of death from uraemia.

Pathology Females:males 3:1. The lesions are asymmetrical and involve predominantly the renal tubules which become progressively atrophic.

Clinical features Intermittent pyrexia, progressive anaemia, hypertension (40 per cent), lumbar pain, frequent painful micturition, general malaise.

Diagnosis
- *Midstream urine specimen*: Pyuria is common. White blood cells predominate (the count varies). Bacteriology usually reveals *E. coli*, *Strep. faecalis*, *Proteus*, or *Pseudomonas*.
- *Intravenous urogram*: Poor definition and 'rimming' (narrowing) of the renal cortex are important features.
- *Cysto-urethroscopy*: Features include chronic urethritis, trigonitis, stricture, prostatic hypertrophy, posterior urethral valves. Ureteric catheterization is carried out for bacteriology and urinalysis to estimate renal concentrating ability on each side.
- *Renal function tests*: Plasma creatinine, urea, and electrolytes

265

Aims of treatment
- Eradicate predisposing factors, e.g. correct urinary tract abnormalities
- Eradicate local infection. Give long-term mandelic acid in the form of ammonium mandelate 3 g orally 6-hourly + ammonium chloride 2 g orally 6-hourly.
- Eradicate distant foci of infection as above.
- When unilateral and causing hypertension, nephrectomy may be curative provided the remaining kidney is functioning.
- Patients with advanced bilateral disease may need dialysis or renal transplantation.

Hydronephrosis

This is dilatation of the kidney and its pelvis due to obstruction of urinary outflow. It is usually non-infected and may affect one or both sides. When the ureter becomes dilated due to distal obstruction the term *hydroureter* is applied.

Causes

- *Unilateral*: abnormal or aberrant vessels, calculus, abnormalities of the pelvi-ureteric junction, congenital stenosis, inflammation, ureteric invasion by carcinoma, retroperitoneal fibrosis.
- *Bilateral*: Usually due to urethral obstruction in the form of congenital valves or urethral stenosis. Any of the above causes can also occur bilaterally. Pregnancy causes bilateral ureteric and renal pelvis dilatation.

Pathology

The effects of obstruction vary, depending on whether it is intermittent or permanent. The pelvicalyceal system becomes progressively dilated and the kidney parenchyma becomes stretched over it until it becomes a mere rim. The surface of the kidney takes on a lobulated appearance.

There is irregular atrophy of the kidney substance depending on the degree to which the blood supply is impaired. The tubules are reduced but the glomeruli are relatively spared.

Clinical features

Usually the onset is insidious with unilateral or bilateral loin pain which is described as 'aching' and may be more apparent to the patient after a heavy fluid load. There may be episodes of renal colic associated with a loin swelling which reduces after passing a large volume of urine (Dietl's crisis). When there is chronic bilateral hydronephrosis due to lower urinary tract obstruction uraemic symptoms and signs may be apparent—tremor of the hands, hiccup, dry brown tongue, ammoniacal halitosis, thirst, anorexia, and hissing expiration develop terminating in renal failure. Clinically, unilateral or bilateral masses may be felt. When there is renal failure the kidneys may not be palpable.

Investigations

- Intravenous urogram: In early cases this will show renal pelvic dilatation followed by clubbing of the minor calyces and dilatation and broadening of the major calyces.
- US scan: May confirm a cystic loin lesion which may be temporarily decompressed with a needle passed percutaneously.

Treatment

Is aimed at eradicating the primary cause by freeing the ureters of obstruction or dealing with congenital valves. Reconstruction of the hydronephrotic kidney can then be carried out (see pyeloplasty).

Renal calculi

The prevalence of renal stones in the community is 3.8 per cent. Men and women are equally affected.

Types

- Metabolic related to excessive excretion of calcium, oxalate, urate, and cystine
- Infection stones (25–30 per cent) related to urea-splitting organisms, e.g. *Proteus mirabilis*, *E. coli*. 50–60 per cent contain calcium, except pure urate (5–15 per cent) or cystine stones (2 per cent) and are therefore radio-opaque, but the calcium involved in the stone may not be the prime constituent. Therefore, analysis of a removed or spontaneously passed calculus is essential to determine whether an abnormal metabolic process is involved.

Clinical presentation

Classically there is renal or ureteric colic. Pain radiates from the affected kidney or ureter to the groin followed by passage of the stone or portions of it.

Other patients have haematuria, recurrent urinary tract infection, abdominal pain or infection, and septicaemia due to pyonephrosis.

Diagnosis

- Plain abdominal X-ray (kidneys, ureters, and bladder—KUB)
- Intravenous urogram to localize the stone
- Retrograde pyelography to localize an obstructing ureteric calculus

Management problems

Single ureteric stone 80 per cent pass spontaneously but this depends on the site and size. Use the following rules to decide if a stone will pass:

- < 6 mm in diameter usually pass spontaneously
- 4–6 mm: 50 per cent will pass. If in the upper half of the ureter they may lead to obstruction.
- > 6 mm are unlikely to pass and need to be removed

Renal calculi May present with a variety of symptoms and signs. They should be removed to prevent progressive destruction of renal tissue. Indications include pain, obstruction, recurrent infection, haematuria, and deteriorating renal function. Complex staghorn calculi should be removed.

Operative procedures

In almost all patients a flank incision is used.

Stones in the ureter
- Ureterolithotomy

Renal stones

- Pyelolithotomy for pelvic or solitary intrarenal calculi
- Extended pyelolithotomy in which a bloodless plane is developed in the renal sinus up to the division of the major calyces allowing more complex calculi to be removed
- Multiple nephrolithotomy + extended pyelolithotomy for staghorn calculi. Stone fragments can be removed from the peripheral parts of the kidney by the use of multiple radial incisions into the kidney substance.

New techniques

These now supplant open lithotomy whenever possible.

269

- Percutaneous nephrolithotomy involves the insertion of a nephroscope into the kidney via a previously dilated tract. The stone can then be removed, ultrasonically disintegrated and removed by suction or disrupted into small fragments by electrohydraulic lithotripsy.
- Rigid ureteroscopy is used for ureteric calculi. A rigid ureteroscope is passed via the dilated ureteric orifice, the stone visualized and removed.
- Extracorporeal shock wave lithotripsy involves the use of generated shock waves guided by three-dimensional X-rays to fragment the calculus into minute particles which the patient passes in the urine.

Recurrence rates

50 per cent will have a further stone in 5 years.

Prevention of recurrence

- Prevent and treat urinary tract infection.
- Diet containing less than 700 mg calcium, low in oxalate and purines, high in bran
- High fluid intake, more than 2 l clear fluid/day
- In some patients bendrofluazide or allopurinol may be prescribed to reduce calcium excretion and urate formation respectively.

References

Manberger, M. and Stackle, W. (1983). New developments in endoscopic surgery for ureteric calculi. *British Journal of Urology Supplement*, **55**, 30–40.

Vallensiek, E. W. (1982). Urolithiasis: a continuing problem. *Urological Research*, **10**, 159.

Epididymo-orchitis

Inflammation of the epididymis and testis may be acute or chronic.

Causes
- *E. coli*, *Streptococcus*, *Proteus*, *Gonococcus*
- Following urinary tract instrumentation or prostatic operations
- Associated with mumps or Bornholm disease (viral myalgia)
- TB, brucellosis

Clinical features

In young males the differentiation of epididymo-orchitis from torsion is almost clinically impossible.

Pain, swelling, and redness of the testis associated with scrotal oedema and desquamation of the skin. Resolution of the swelling can take 2 months.

Investigations
- Midstream urine specimen and bacteriology
- Three early morning cultures in Loewenstein–Jensen medium (for TB)

Treatment
- Bed rest
- Scrotal support (if the patient finds it more comfortable)
- Analgesia
- Broad-spectrum antibiotics until urine culture results are available (e.g. tetracycline 500 mg orally 6-hourly)
- Specific antibiotic therapy following bacteriological confirmation (e.g. procaine penicillin with probenecid (for Gonorrhea 2.4 g IM as a single dose) but the incidence of penicillin-resistant *Gonococcus* is rising fast

Chronic epididymo-orchitis is due to TB in more than 90 per cent of cases. In the remainder of patients it follows acute attacks and may be related to urethral stricture.

Treatment

Tuberculous
- Treat the primary source with antituberculous drugs.
- If there is no improvement after 2 months carry out epididymo-orchidectomy.

Non-tuberculous
- Long-term specific antibiotic therapy
- Epididymo-orchidectomy if the patient remains no better after 2 months

271

Benign prostatic hypertrophy

Essence

Benign nodular hyperplasia probably affects the prostates of most men over 40 years but only about 10 per cent will present with problems of urinary outflow obstruction. The size of the gland does not correlate with the degree of obstruction. Small prostates may lead to urinary retention whereas larger glands may not. The condition increases in frequency with advancing age.

Pathology

The glandular elements of the prostate are involved probably under the influence of androgens. Adenomatous hyperplasia affects the submucous glands and lateral lobes of the prostate compressing the urethra into a slit and the outer surface of the gland into a pseudo-capsule. There may also be an associated hypertrophy of the circular muscle of the bladder neck. Such a gland becomes bulky, displacing the seminal vesicles.

When hypertrophy affects the middle lobe it protrudes into the urethra to varying degrees.

Effects

Urethra The urethral canal becomes compressed into a slit. That portion above the verumontanum becomes elongated.

Bladder There is muscular hypertrophy to overcome the obstruction, stasis of urine, urinary infection, and calculus formation. With time the bladder muscle becomes atonic.

Ureters and kidneys The ureteric orifices dilate leading to free reflux of urine and ascending pyelonephritis. As obstruction progresses, ureteric dilatation, hydronephrosis, and renal parenchymal damage develop.

Clinical features

Nocturia is a significant symptom. The patient also complains of having to empty his bladder more often than usual. He may also notice delay in starting the stream of urine or that the flow is weak. Acute retention of urine may develop and cause him to seek early treatment. More often as the bladder muscle becomes atonic the volume of residual urine increases and chronic retention develops. The patient may then find that he leaks urine involuntary on coughing, sitting up, etc. He can no longer control the flow and dribbles constantly. This is termed 'chronic retention with overflow'. Such patients have protuberant lower abdomens and smell strongly of stale urine. Other features include pain from cystitis or chronic hydronephrosis and blood at the beginning and end of micturition due to rupture of a prostatic vein.

Clinical examination

Examine the urogenital tract. Look for enlargement of the kidneys and evidence of renal insufficiency (e.g. uraemia—dry tongue, etc.). Inspect the lower abdomen, palpate the bladder. Distinguish acute from chronic retention by the presence of pain and a palpable bladder in the former. Is there a urethral meatal stricture? Are the testes and epididymis normal? Rectal examination usually reveals a smoothly enlarged prostate. It is of no value in estimating the severity of outflow obstruction since the size of the gland is immaterial. However, it is of value in detecting possible prostatic cancer when the gland may feel hard and craggy. If there is a craggy prostate or a firm nodule, histological confirmation should be established at transurethral resection which also relieves the patients urinary symptoms. Routine transrectal Trucut or Franzen needle biopsies may cause septicaemia.

273

Investigations

- Examination of urine for organisms and malignant cells
- Urea and electrolytes; haemoglobin
- Intravenous urography. The postvoiding film shows a thickened bladder wall with marked trabeculation and a large volume of residual urine.

Treatment

Medical Prazosin hydrochloride (alpha-adreno receptor blocking drug) may buy time until prostatectomy by relaxing the smooth muscle at the bladder neck. (Initially 500 μg orally for 3–7 days. Give first dose at bedtime to avoid hypotensive collapse. Maintenance is 2 mg twice daily.)

Surgical Prostatectomy. Indications: prostatism, acute retention unrelieved by catheterization followed by removal of the catheter; chronic retention and hydronephrosis, complications (stones, bladder diverticula, etc.) (see transurethral resection, p. 830).

Varicocoele

There is dilatation and varicosity of the testicular veins, more commonly affecting the left side, due to reflux from the renal vein. Obstruction of the testicular vein may also occur due to compression from the testicular artery which arches over it on occasions. Rarely obstruction can be caused by growth of a renal tumour along the renal vein.

Clinical features

Varicocoeles often occur transiently in healthy adolescents. Often there are no symptoms. In some patients as the scrotum elongates pain is felt. It is described as dragging and is due to the weight of the unsupported testis pulling on the cord.

Examination

With the patient standing the dilated veins are obvious and the scrotum hangs lower on the affected side (Bow sign). When the patient lies down or the scrotum is elevated the veins tend to empty and discomfort is diminished. The affected testis may be smaller than its neighbour. In cases of renal tumour the veins do not empty on elevation of the scrotum.

Associations

Varicocoele is associated with infertility which is often the reason why the patient attends his doctor.

Treatment

- Support with a jockstrap
- Ligation of the testicular vein above the inguinal ligament
- Individual ligation of dilated cremasteric veins or veins of the pampiniform plexus

Complications

- Testicular infarction
- Infection

Prognosis for infertile patients

Although it has been claimed that ligation restores fertility there is unfortunately no good statistical evidence to prove this. However, like the use of hormone preparations the procedure varicoloelectomy is currently popular in the treatment of infertile men.

Paraphimosis

The retracted foreskin is trapped behind the glans especially when the penis is erect. This can lead to gangrene of the glans, so reduction is urgent. After catheterizing a male patient, ensure that the foreskin is properly brought forward. Forgetting this can cause a paraphimosis.

Reduction

May be achieved by a dorsal slit in the prepuce (messy!) with circumcision later or by reduction of the paraphimosis and later circumcision (tidy).

Method

Place two gauze swabs either side of the oedmatous, slippery foreskin and put the penis on the stretch. This will allow the glans to decompress and it can then be reduced by gentle pressure through the constriction ring. The patient will experience immediate relief. Circumcision can be performed later at leisure with a better cosmetic result.

N.B. Be prepared to keep the penis extended for up to 15 minutes.

277

Priapism

This is painful, sustained erection of the penis. It may develop after intercourse or sports cycling in healthy males. More commonly it is associated with idiopathic thrombosis of veins in the prostatic plexus. Less often there is an association with blood dyscrasias or sickle cell disease. Only the corpora cavernosa are affected. The glans and corpus spongiosum remain soft.

Investigations

- Take an accurate history.
- Exclude blood dyscrasias with FBC.

Treatment

Non-operative methods using spinal anaesthesia, ganglion blockers, anticoagulants, saline washouts of the corpora with instillation of noradrenaline are uniformly unsuccessful.

Surgical methods A window may be created between the flaccid glans and corpus spongiosum and the turgid corpora cavernosa. This may be done by thrusting a fine knife or Trucut biopsy needle through the glans into the corpora cavernosa. If this fails surgical anastomosis between the corpus spongiosum and the cavernosum of one side may succeed. The saphenous vein may also be anastomosed to the corpora cavernosa on each side (if a unilateral anastomosis fails to decompress the penis).

Complications

Impotence is common, but is less likely if the patient is operated upon in the first 24 hours after onset. Those who remain impotent may be improved by a prosthetic penile implant.

Peyronie's disease

This is a benign condition affecting the penis of obscure cause and unknown incidence. There is discomfort, pain, and penile deformity on erection which causes difficulty with sexual intercourse.

Cause

Unknown.

Clinical features

Usually occurs in men in their 50s. Symptoms are often present for some time before the patient seeks advice. When the deformity occurs a lump is palpable on the shaft of the penis. The natural history is of gradual resolution over months or years, the discomfort goes, the plaque disappears, then the deformity decreases.

Pathology

There is early periureteric inflammation followed by fibrosis into the corpus cavernosum making excision difficult. Dupuytren's contracture is often present.

Management

Reassure the patient that he does not have cancer. The following treatments are used:
- Mineral waters (suggested by Peyronie). Steroids (systemically and into the plaque). X-rays to the plaque; ultrasound; potassium paraminobenzoic acid (Potaba, a B vitamin); Vitamin E; Excision of the plaque with grafting of the resultant defect.
- Vitamin E 200 mg three times daily, Potaba 15 g daily in divided doses and excision are the most commonly used treatments.

Problems

Potaba has a high incidence of side-effects (diarrhoea, fevers, rashes, vomiting) and 3 g sachets or 50 mg tablets make compliance difficult
- Excision is associated with impotence which may require an implant.

Results

When the condition is stable, ask whether the deformity prevents coitus. If it does, carry out a Nesbitt's procedure.

Nesbitt's procedure Via a circumcision incision dissect the skin back to the base of the penis and apply a tourniquet. Insert a venflon into a corpus cavernosum, inflate it to produce an artificial erection and show the site and extent of the deformity. Use serial relaxing incisions (circular sutured longitudinally) on the short side and shortening incisions (ovals excised and sutured) on the long side to straighten the penis. Close the skin. Insert a small silicone catheter and bandage the penis firmly. Remove the tourniquet. Remove the bandage and catheter at 36 hours.

Implants

There are various types: the 'bendy toy'; those with self-contained inflation; those with a separate inflation reservoir. The cost and failure rates both increase with complexity.

Complications

- Infection
- Erosion of the glans penis } remove the device
- Valve failure preventing inflation or decompression—revision
- Cost (who pays?)

281

Prognosis

5-year complication-free rate for implants is approximately 50 per cent.

Urethral stricture

Causes

'Valves', infection, trauma (including iatrogenic).

Symptoms

Infection, poor stream, straining, post-void dribbling.

Results

Stones, renal failure.

Diagnosis

Ascending and descending urethrography, urethroscopy.

Management

Strictures may not present for 20 years after the original injury. Any treatment must be evaluated with this in mind.

The valves may be destroyed by rupture with a Fogarty balloon or diathermy. The stricture may be dilated, cut, or repaired.

Dilatation

Pass dilators until a normal size is achieved (male: 28 Ch., female: 32 Ch.). Repeat 3-monthly to yearly.

Cutting

Endoscopic cutting may be carried out with either a 'hot' or 'cold' knife. The stricture may be excised with a Colling's knife (the hot resectoscope) or with a urethrotome (cold). Failures occur due to inadequate resection and the fact that the stricture tends to recur. Repeat procedures may be needed.

Repair

Urethroplasty.

Complications

Each time a stricture is dilated there is the risk of bacteriogenic shock (30 per cent), bleeding, false passage and fistula formation.

Urethroplasty

The formation of a neourethra. There are many methods, depending on site, length, and duration of the stricture. The approach is usually on the ventral surface of penis or perineum. Complex posterior urethral strictures may best be approached by a transpubic route with less risk to continence and potency. The procedure may be 'one-stage' or 'two-stage'.

One-stage Laying open of the stricture and insertion of a free skin patch (usually scrotum) or a pedicle skin patch with immediate closure (e.g. Turner–Warwick).

Two-stage The stricture is laid open and a skin flap inlayed with later closure by a skin patch or tubularization of free skin and closure over the top (e.g. Blandy). Between stages the skin is depilated (hair-free skin inrolled) and checks are made to ensure that normal-calibre urethra is present anteriorly and posteriorly. (Prestenotic pressure and poststenotic turbulence may result in an apparently normal calibre urethra which later stenoses.) The technique chosen must be safe, maintain continence and potency, and stand the test of time.

Short stricture of anterior urethra

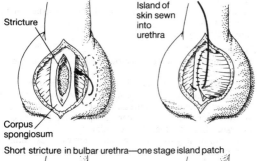

Short stricture in bulbar urethra—one stage island patch

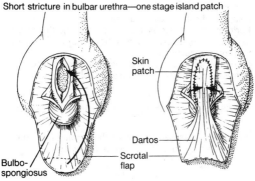

Second-stage urethroplasty

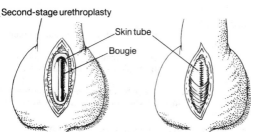

Urethral stricture

The ureter

Ureteric stricture

Causes Malignant disease (commonest), iatrogenic (second commonest), post radiotherapy, inflammatory (TB, endometriosis), periureteric fibrosis, e.g. retroperitoneal fibrosis.

Presentation There may be no symptoms. The stricture may be an accidental finding. The patient may also present with recurrent infections or stone disease.

Investigations
- Intravenous urography will demonstrate obstruction or a dilated ureter.
- Ureterography may be helpful.

Management
- Short strictures can usually be excised and the ends re-anastomosed.
- Long strictures may need interposition with appendix or bowel. The ureter may also be mobilized across the midline to be joined to the opposite ureter (uretero-ureterostomy).
- At the lower end the ureter can be rejoined to the bladder by the Psoas hitch or Boari flap techniques.
- Postradiotherapy strictures require much wider excision than is apparent. This is because associated arteritis may delay healing so healthy tissue must be found.
- Periureteric fibrosis is treated by freeing the ureter and placing it more laterally, sometimes wrapping it in omentum to prevent recurrence.
- Malignant strictures may be treated by nephro-ureterectomy or less commonly by resection and ileal interposition.

Ureteric fistula

Causes Surgery, radiotherapy, penetrating injury, blunt injury (rare).

Presentation Leakage of urine via the vagina or an abdominal drain usually around the fifth postoperative day.

Investigations
- Intravenous urography may demonstrate the leak.
- Ureterography

Management
- If noted at operation carry out an immediate repair after spatulating the ends.
- Later leaks will sometimes permit the passage of a ureteric stent and heal without problems (this is unlikely after radiotherapy).
- If impassable carry out a percutaneous nephrosotomy and re-explore after 3 months when the reaction has settled. Repair as above.

Anastomosis Should be with plain catgut. A self-retaining ureteric stent diminishes leakage and aids healing. Insert a tube drain for 5 days. Remove the stent at 6 weeks. Carry out intravenous urography to check the result at 3 months.

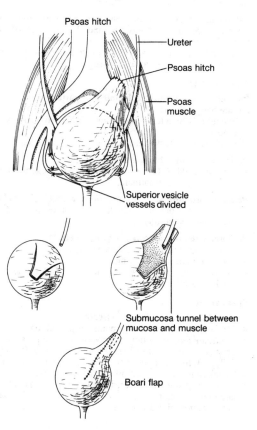

Psoas hitch and Boari flap

Male infertility

Investigations

History Problems with coitus? (premature ejaculation, etc). Does the patient shave daily? Any exposure to mumps? Has there been testicular injury? General health: any history of asthma, coeliac disease, mucoviscidosis? Drugs, e.g. Salazopyrin, nitrofurantoin, antimitotics affect spermatogenesis. Smoking reduces sperm quality. Ask about occupation. Does the patient work in a hot atmosphere?

Examination Check hair patterns (male or female distribution?), the size and consistency of the testicles. Note the presence of hernia or varicocoele, the type and design of underwear (e.g. nylon briefs, boxer shorts, etc.).

Laboratory tests: Semen analysis Measure the volume, viscosity, sperm count motility, and the presence of normal forms. The normal sperm count is 50–150 million with 50 per cent remaining mobile at 1 hour postejaculation. Abnormally shaped sperm may be caused by drugs.

Management

Patients with azospermia, female hair patterns, female body habitus
- Exclude Klinefelter's syndrome (XXY) or gonadal dysgenesis.
- Measure FSH, testosterone, chromosome analysis.

Patients with male hair pattern and habitus
- Measure FSH. A low level suggests spermatogenic arrest for which there is no treatment.

Patients with normal testes, azospermia, and high FSH levels
- Suggest obstruction with the possibility of epididymovasostomy as therapy.

Patients with normal testes and oligospermia
- Advise the patient to wear loose pants.
- Advise him to stop smoking.
- Consider ligation of a varicocoele (if present). Remember the results are still equivocal.

Treatment
- *Klinefelter's, raised FSH*: No treatment
- *Mild oligospermia*: Change pants, stop smoking
- *Moderate oligospermia*: Ligate varicocoele if present. Use of drugs, e.g. clomiphene (25 mg/day) mesterolone (50 mg twice daily). Both are anti-oestrogen. Treatment is continuous. Check semen every 3 months to assess progress.
- *Severe oligospermia*: Consider the use of gamete intra-fallopian transfer or *in vitro* fertilization for the patient's partner after joint discussion.

Abdominal hernias

A hernia is the protrusion of a viscus, or part of a viscus, from a situation where it should be to one where it should not be through a defect in the cavity wall in which it is contained. The protrusion occurs through a normal or abnormal aperture in the abdominal wall. Approximately 1 in every 100 people has a hernia at some time. The common sites are the inguinal (70 per cent), femoral (20 per cent), or umbilical (10 per cent) orifices. In addition herniation may occur in the supraumbilical region or through the linea alba. Incisional hernias are iatrogenic and of particular importance in relation to surgical incisions. There are also many rare types of hernias (p. 298). Hiatus hernia is discussed on p. 150.

Aetiology

Congenital An abdominal wall weakness may be aggravated by increased intra-abdominal pressure as in coughing, straining, or making an excessive muscular effort, e.g. lifting a heavy weight.

Acquired These are more common with increasing age when there is weakness and fatty infiltration of muscle. They may also be related to muscular weakness or damage post appendicectomy.

Anatomy Abdominal hernias are enclosed in a *sac*, usually the peritoneum. However in a sliding hernia (*en glissade*) the posterior wall of the sac is formed by the caecum on the right and the sigmoid colon on the left. The sac may be covered with the tissues through which it protrudes or uncovered, lying subcutaneously. The *neck* of the sac is that portion at the level of the orifice in the abdominal wall. The contents of the sac may include bowel (an enterocoele) or omentum (omentocoele), often both.

Classification

Reducibility A hernia is said to be *reducible* if it can be returned to the abdominal cavity. It is *irreducible* when it cannot.

Obstructed hernia This is an irreducible hernia in which the bowel is obstructed but viable. It is worth while considering gentle reduction by giving analgesia, elevating the foot of the bed, and 20 minutes later attempting to milk the contents back. If this succeeds, operative repair may be carried out 24–48 hours later once the oedema has settled. If it fails, emergency repair is required to prevent strangulation.

Strangulation Strangulated hernias need emergency surgery. They occur when the bowel or omentum in an irreducible hernia becomes ischaemic, due to obstruction of its blood supply, and subseqently gangrenous. Strangulation is commonest at the neck of the sac, the point of the greatest constriction. Gangrene and perforation lead to discharge of the intestinal contents into the peritoneal cavity and peritonitis.

One form of strangulation common in femoral hernias occurs when only part of the circumference of the bowel becomes enclosed in the hernial sac—Richter's hernia.

Clinical features

There is sudden onset of pain over the hernia. This then becomes generalized and is associated with vomiting. The patient will often be able to confirm that the once reducible hernia became stuck and larger. Clinically there is a tense, tender irreducible hernia which transmits no cough impulse.

Treatment

Emergency surgery to prevent gangrene, perforation, and endotoxic shock.

Inguinal hernia

This is the most common hernia and occurs most frequently in men, when it is related to a patent processus vaginalis extending for a variable distance along the inguinal canal and into the scrotum. When bowel or omentum fill the sac the hernia descends along the path of the processus, is **indirect**, and may reach the upper pole of the testis. The sac carries with it all the coverings of the spermatic cord. In women an indirect hernia lies alongside the round ligament and appears at the labia.

A **direct** hernia is more often acquired. The sac passes directly through the abdominal wall, medial to the internal ring and the inferior epigastric vessels. It is distinct from the cord and its coverings.

Sliding hernia (hernia *en glissade*)

In this the posterior wall of the sac includes peritoneum and caecum on the right side and peritoneum and sigmoid colon on the left side. A portion of bladder may descend on either side.

Clinical features

Some patients have no symptoms. Others describe an ache or dragging sensation especially towards the end of the day. Some can relate the onset of pain or swelling to a specific activity (e.g. lifting). Persistent severe pain is suggestive of strangulation.

Points in the examination of scrotal swellings

- Examine the patient standing. There may be no swelling. If there is an indirect hernia it appears at the external ring or extends into the scrotum. A direct hernia appears as a diffuse bulge in the medial part of the inguinal canal.
- Ask the patient to cough. Does a swelling now appear? Decide if it is direct or indirect (see below). Is there a palpable impulse?
- Feel the hernial orifices. Use the right hand for the left side and vice versa with the patient standing facing you. Pass the little finger upwards along the spermatic cord until you feel the external ring. Does it just admit the little finger? (normal). As the patient to cough. Is there a palpable expansile impulse? If so the diagnosis is confirmed. Distinguish an inguinal hernia from a femoral. The former is above and medial to the pubic tubercle, an important landmark in examination.
- Can you 'get above' an obvious scrotal lump by grasping it between finger and thumb. If yes, it does not descend from the inguinal canal and is probably a hydrocoele.
- Is the swelling reducible? Ask the patient to lie down and gently milk the hernia back into the abdomen. The patient usually does this for you. Do not reduce a hernia if you suspect obstruction or strangulation.

- Is it an indirect or direct hernia? Invaginate the scrotum as above. Is there a defect in the posterior wall of the inguinal canal (direct). When the patient coughs does the impulse hit the pulp of the finger (direct) or its tip (indirect). Can the hernia be controlled by pressure over the deep ring? (indirect).
- Transilluminate the swelling. Hydrocoeles are often translucent. Remember inguinal hernia in children may also transilluminate. *Do not use needle aspiration to diagnose scrotal swellings.*

Treatment

291

Uncomplicated elective herniotomy (children), herniotomy and repair (adults) (pp. 714, 780). Truss if operation is contraindicated. Use them rarely. If you must prescribe one make sure it works.

Strangulated emergency surgical exploration, resection of bowel if indicated, and repair.

Advice

Instruct the patient not to drive for 10 days or lift heavy objects for 2 weeks.

Recurrence rate

Less than 1 per cent at five years (Devlin, 1982).

References

Devlin, H. B. (1982). Hernia. In *Recent advances in surgery*, Vol. 11 (ed. R. C. G. Russell), pp. 209–23. Churchill Livingstone, Edinburgh.
Welsh, C. I. and Hopton, D. (1980). Advice about driving after herniorraphy. *British Medical Journal*, **1**, 1133–4.

Umbilical and epigastric hernias

Umbilical hernias

Occur in children due to incomplete closure of the umbilical orifice. They close spontaneously and only require repair if they persist beyond 2 years.

- Repair may be effected by a circumferential subcutaneous suture. If the sac is large, formal exploration is indicated.

Paraumbilical hernias are acquired. *In children*, usually girls, the sac is supraumbilical. As it enlarges it hangs down and may be an embarrassment. Strangulation is not common.

- Repair consists of excision of the sac with one or two nylon sutures inserted into the defect.

In adults the sac passes through the linea alba above or below the umbilicus. It may become very large and hang down below the pubis. It contains omentum and loops of bowel. Strangulation occurs in 20 per cent of cases and the mortality rate is high.

- Repair is indicated by formally identifying and reducing or excising the sac followed by repair of the defect.
- Recurrence is uncommon.

Epigastric hernias

Herniations of fat through the linea alba. They occur at any age and appear between the xiphisternum and umbilicus. They may be tender and troublesome. In patients with dyspepsia the chance finding of an epigastric hernia should not preclude investigation with endoscopy or barium studies.

- Repair is indicated. The herniation is exposed, the fat excised and the defect closed with two or three interrupted non-absorbable sutures.

Incisional hernia

Herniation may complicate any abdominal incision. Midline vertical incisions below the umbilicus are most commonly affected. Predisposing factors include obesity, wound infection, pulmonary disease, poor wound healing due to vitamin deficiencies, malignancy or steroid therapy, postoperative distension, age over 60 years, burst abdomen, incorrect suture materials, and poor operative technique. 'Never judge the surgeon until you see him close the wound' (Lord Moynihan).

Incidence

Around 10 per cent of wounds develop incisional hernias. 90 per cent of these have been infected.

Clinical features

Incisional herniation can occur up to 5 years after the operation. When the patient sits up or coughs, the swelling protrudes through the wound. As time passes the hernia enlarges, and peristaltic bowel movements may be seen through a paper-thin skin. Strangulation is uncommon because the neck of the sac is wide. If obstruction occurs, it is usually due to intestinal adhesions.

Treatment

- If the hernia is small or the patient unfit for surgery a corset may be worn.
- Surgical repair may be effected by several methods (see p. 784). Most involve the excision of the old scar and the sac, followed by the identification of normal aponeurosis. The edges may be approximated side to side, by overlap or by the interposition of a mesh or Teflon patch if the defect is large.

Recurrence rate

10–20 per cent.

Review appointment

This is indicated only to check that the wound has healed. Thereafter follow-up is for the primary disease alone.

Femoral hernia

N.B. Femoral hernias are particularly liable to obstruct or strangulate. Always examine the groins of patients who are vomiting, especially obese old women.

A femoral hernia lies in the most medial compartment of the femoral canal in the upper part of the thigh. It is four times as common in women but still not so frequent as inguinal hernia. It may occur spontaneously or follow inguinal hernia repair on the same side. Most are painless and irreducible.

Clinical features

It appears below and lateral to the pubic tubercle (cf. inguinal hernia which is above and medial). It may follow a J-shaped path to turn upwards and backwards so that it lies over the inguinal ligament. The absence of a cough impulse over the inguinal ring distinguishes a femoral from an inguinal hernia.

Differential diagnosis

Inguinal hernia, lymph node, saphena varix (disappears on lying, coughing produces a palpable saphenous thrill distal to the femoral canal), lipoma, psoas abscess, femoral artery aneurysm (pulsatile).

Treatment

Surgical treatment is advocated for all patients. Three approaches are described (see p. 782). In strangulation an incision over the hernia and a further abdominal incision may be necessary, preferably combined in a one-incision high approach.

Recurrence rate

About 1.5 per cent recur.

Unusual hernias

Obturator hernia

The sac passes through the obturator canal. Men:women 1:6. Age over 60 years. The diagnosis is not usually made until strangulation occurs.

Clinical features A sac is rarely palpable because it is overlain by pectineus muscle. It may become apparent if the hip is flexed, abducted, and externally rotated. When strangulated pain is referred along the inner aspect of the thigh to the knee via the geniculate branch of the obturator nerve. Rarely is this sign seen in non-strangulated hernias. Rectal or vaginal examination should be done if there is suspicion, when the the hernia may be felt as a tender mass at the obturator foramen.

Treatment An obturator hernia may only be recognized at laparotomy for small-bowel obstruction. Identify the sac and reduce it by stretching or incising the obturator fascia with the patient in Trendelenburg position. Resect non-viable bowel if indicated. Avoid contamination of the peritoneal cavity with fluid from the sac. Excise the sac. Suture the broad ligament over the defect in the obturator fascia to prevent recurrence.

Lumbar hernia

Herniation usually occurs through the inferior lumbar triangle bounded by the iliac crest, external oblique and latissimus dorsi. Incisional hernias may follow nephrectomy.

Differential diagnosis Tuberculous abscess (cold abscess), lipoma.

Treatment Small hernias are easy to repair. Large defects can be closed by prosthetic patches or fascial flaps.

Interparietal hernia

The sac passes between the layers of the abdominal wall. It leads to intestinal obstruction and may communicate with the sac of a femoral or inguinal hernia.

Treatment Surgical reduction or resection and repair.

Spigelian hernia

A variant of interparietal hernia. The sac protrudes adjacent to the rectus sheath, below the umbilicus at the level of the arcuate line.

Treatment Split the external oblique aponeurosis. Deal with the sac and its contents. Repair the defect in the transversus, internal and external oblique muscles.

Perineal hernia

This is herniation of the abdominal contents through the pelvic floor. It may occur spontaneously as in rectal prolapse or post-operatively after abdomino-perineal excision of the rectum or pelvic exenteration (a patch is sutured across the pelvic floor to prevent herniation).

Treatment is of rectal prolapse. Other causes require laparotomy, reduction of the sac, and repair of the pelvic defect by patch prosthesis.

Internal hernias

May result from iatrogenic causes such as defects left in the mesentery or in relation to adhesions. Rare spontaneous hernias may also occur into the foramen of Winslow and the paraduodenal fossa. They are causes of intestinal obstruction.

Diaphragmatic hernia may be congenital through the parasternal foramina of Morgagni or the pleuroperitoneal canal of Bochdalek. Acquired hernias are hiatus hernias through the oesophageal hiatus. Traumatic hernias result from blunt abdominal trauma, surgery, gunshot, or stab wounds. They often appear late. Carry out chest X-ray on the trauma patient regularly. Inspect the diaphragm carefully at laparotomy.

299

8 Paediatric surgery

Lumps

Lumps are common. Remember how to examine them with the help of the mnemonic: Should The Children Find Lumps. (recommended by MacMahon, 1984).

Key points in examination (of all lumps)

S site	T tenderness	C colour	F fluctuation	L lymphatic
size	temperature	contour	filling-emptying	drainage
shape	transillumination	consistency	flow-bruit	lumps
shift	thrill	cough impulse		elsewhere

Additional points

Although the site of a lump is important, additional factors should be established. This will involve a careful review of its duration as well as the clinical signs and may necessitate additional investigations like X-rays. Clinically it is often possible to determine whether a lump is deep or superficial, whether it lies subcutaneously, subfascially, intramuscularly, or intra-abdominally. The site of abdominal masses can often be determined by body movements, e.g. the lump becomes fixed when the patient sits up (lump lies in the anterior abdominal wall) or moves with respiration or coughing (usually intra-abdominal). Establish too whether the lump interferes with normal functions, e.g. swallowing, speech, appetite, defecation.

Common sites for lumps in children

The most common and important lumps are to be found in the neck and abdomen.

Neck lumps may be lateral or midline.

Lateral:
- lymph nodes
- branchial sinuses and cysts
- cystic hygroma
- sternomastoid tumour
- haemangioma
- lymphangioma
- submandibular gland
- parotid gland
- neoplasm

Midline:
- lymph nodes
- submental nodes
- thyroglossal cyst
- thyroid swelling
- dermoid cyst

Abdominal lumps Mass in right upper quadrant:
- liver (biliary atresia, metastases, portal hypertension)
- choledochal cyst
- pyloric tumour (pyloric stenosis)
- solid tumours (nephroblastoma, neuroblastoma, non-Hodgkin's lymphoma, hepatoblastoma)

Mass in left upper quadrant:
- spleen
- solid tumour

Mass in lower abdomen or pelvis:
- pelvic rhabdomyosarcoma
- ovarian tumour
- intussusception (sausage-shaped, concave towards umbilicus)

Reference

MacMahon, R. A. (1984). An approach to lumps and swellings: swellings in the neck. In *An Aid to Paediatric Surgery*, pp. 17–21. Churchill Livingstone, Edinburgh.

The child with a large head

If there is suspicion that the head is enlarging rapidly or is becoming asymmetrical, measurements must be repeated regularly and centile variations compared to the normal growth curve of the infant head.

Causes of a large head

- *Hydrocephalus* is the commonest.
- *Craniostenosis* due to premature or abnormal fusion of cranial sutures leads to abnormalities of shape.

Hydrocephalus

In this condition there is an imbalance between the production and absorption of cerebrospinal fluid (CSF) leading to increased pressure in the cerebral ventricular system which becomes progressively dilated.

Clinical features

- Abnormal enlargement of the head. This is the earliest sign.
- Crack-pot sign on percussion (MacEwen's sign): The raised intracranial pressure leads to widening of the anterior fontanelle, separation of the sutures, and thinned bones. All contribute to the high-pitched percussion note.
- Frontal bossing and prominent scalp veins
- Eyeballs are rolled down—'setting-sun' eyes.
- Transillumination of the skull is usually diffuse, indicating general dilatation. Local areas of lucency may indicate cysts or subdural collections.

If hydrocephalus occurs after the sutures have fused the circumference of the skull may not change much but signs of increased intracranial tension develop rapidly and in older children may lead to headache, vomiting, and papilloedema. Slowly progressive hydrocephalus may lead to delayed milestones in mental development.

Causes

The imbalance between the production and absorption of CSF may be due to excessive production of CSF, obstruction along the CSF pathway, or impaired absorption of CSF into the veins. The commonest is due to diminished absorption due to a block, congenital or acquired, somewhere in the CSF pathway. The three common congenital causes are: stenosis of the aqueduct of Sylvius, stenosis or occlusion of the foramina of Lushka and Magendie, and basal arachnoid adhesions around the foramina of the fourth ventricle including the Arnold Chiari malformation. Acquired causes include meningitis, trauma, and intra-uterine infection (rubella, toxoplasmosis, herpes, congenital syphilis, and cytomegalovirus). Tumours, usually of the posterior cranial fossa and pineal area, account for 3–5 per cent of cases.

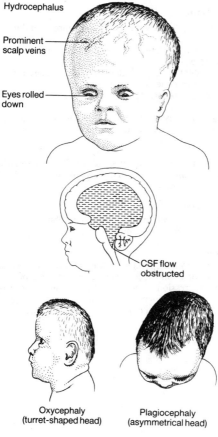

Hydrocephalus

Prominent
scalp veins

Eyes rolled
down

CSF flow
obstructed

Oxycephaly
(turret-shaped head)

Plagiocephaly
(asymmetrical head)

The child with a large head

The child with a large head (*cont.*)

Types

- *Communicating*: Ventricles communicate with the subarachnoid space. Therefore ventricles, basal cisterns and cerebral sulci are all dilated.
- *Non-communicating*: the obstruction lies in the ventricular system and leads to ventricular dilatation alone.

Progression

Progressive ventricular dilatation, oedema, demyelination of white matter, thinning of cerebral cortex, cortical atrophy, pressure on cranial nerves.

Investigations

- Straight X-ray shows widening of the sutures and thinning of the vault.
- CT scan demonstrates ventricular dilatation or intracranial tumour or cyst.
- CSF examination is indicated in possible acquired cases. Samples are obtained by ventricular tap.
- CSF pressure recording is indicated in possible acquired cases.
- Angiography may be helpful when there is an intracranial bruit or possible sagittal sinus thrombosis.

Treatment

Specific for infections and tuberculosis. Ventriculo-atrial or ventriculo-peritoneal shunt. (Drainage of CSF to the right atrium or peritoneum.)

Complications

Shunt blockage, infection, sudden decompression may lead to subdural haematoma or midbrain herniation.

Prognosis

75 per cent do well if operated on within the first 10 weeks.

Children with craniostenosis need decompression of the skull to prevent progression of mental retardation and optic atrophy, as well as an increasingly unacceptable cosmetic appearance which makes them victim to taunts at school and play.

Infections in childhood

Infection is due to micro-organisms and may be localized or generalized.

Presentation

Localized The affected area is hot, swollen, and tender. Limitation of function is common, e.g. limp if a limb is affected.

Generalized Vague malaise, loss of appetite, sweats, chills, and rigors, thirst, tachycardia, tachypnoea, febrile with hot clammy skin. Hypotension and oliguria. May progress to febrile convulsions or present as fits.

Diagnosis

History, clinical examination to determine onset, type and site of infection. Take throat swabs if indicated. Examine pus and urine for culture, organisms, and sensitivity. Take blood for WBC and U&E. Are X-rays indicated? E.g. in chest infections, in cases of suspected osteomyelitis. (Bone scan is more accurate but not always immediately available.)

Deal with specific infections promptly
- *Ingrowing toenails*: Treat by wedge excision initially. Definitive procedures can be performed later if necessary.
- *Perianal abscess*: Cruciate skin incision and drainage of pus. Fistula formation is a risk. Review the patient.
- *Adenitis*: Common at all ages and in many sites e.g. neck, axilla, groins, mesentery and retroperitoneum. Ask the laboratory for help to exclude bacterial or protozoal infections, zoonoses, TB, viral and fungal infections.
- *Acute osteomyelitis*: see p. 454.

Peritonitis (see p. 186).

Postoperative sepsis

This ranges from wound infection which may be treated by drainage to septicaemia which may occur postoperatively due to Gram-negative infection resulting from instrumentation, immunodeficiency, malignant disease, or resistant organisms. The child becomes clinically shocked due to associated endotoxin release. Disseminated intravascular coagulation is a complication and the mortality is high.

Treatment

Involves ventilation, fluid volume replacement, cardiovascular support with inotropic agents, e.g. IV dopamine 10 μg/kg/min, antibiotics or steroids, and specific surgery (if indicated).

309

The child with a naevus

Parental anxieties about the cosmetic aspects of naevi are usually the reason for presentation to the general practitioner or hospital. Serious complications are uncommon but may occur in extensive lesions. Malignancy in pigmented naevi in childhood is virtually unknown.

Naevi may be vascular or pigmented.

Vascular naevi

Haemangiomas are commonest. They usually involve skin or mucous membrane and are the result of hamartomatous malformations of blood vessels. In some there may be arteriovenous communications. If large these can lead to pathological shunting with cardiac hypertrophy and failure.

Haemangiomas are described according to the vessels which predominate, e.g. arterial, venous, capillary, or combinations.

Arterial angiomas (uncommon) Sometimes seen in the scalp. Treatment is total ligation where possible.

Capillary angiomas
- *Strawberry naevus*: Confined to the skin and often unnoticed at birth begins to expand after 1–2 weeks until 3–4 months of age. Regression usually begins during the second year and most resolve by the age of 5 or 6 years.

    Treatment. Conservative until resolution occurs. Surgery is required for complications such as pressure on vital structures or repeated haemorrhage.
- *Port wine stain* (salmon patch, flame naevus, according to its colour): These angiomas change little during life and may present cosmetic problems on the face.

    Treatment. Surgical excision if possible with grafting. Extensive lesions are best camouflaged by make-up.

Venous angiomas These consist of blood lakes, often large, and submucous, subcutaneous, or intramuscular. Excision is usually impractical.

Composite naevi May involve a limb or large area. Surgery is indicated for rapid growth or heart failure.

Pigmented naevi

Junctional naevus This consists of increased melanocytes in the junctional layer of the skin. It is black or brown, flat and common. In childhood the risk of malignant change is extremely low but after puberty it may develop malignant potential (to melanoma).

Compound naevus In this lesion naevus cells bud from the junctional layer to the dermis and the lesion becomes palpable. In later years junctional activity regresses and the lesion becomes intradermal.

⁀sion of pigmented naevi is indicated only for cosmetic

The child with a mediastinal mass

Presentation may be due to symptoms caused by compression of vital structures. Some mediastinal masses are asymptomatic and detected as an incidental finding on chest X-ray. Although investigations are valuable, the final diagnosis is usually established at laparotomy.

Clinical features

Some patients are asymptomatic. Features of respiratory infection with coughing, wheezing, and dyspnoea are common. Facial oedema and engorgement of the veins of the neck and arms may be caused by superior vena caval compression. Involvement of the spinal cord and sympathetic chain may lead to neurological symptoms like progressive paraplegia. Horner's syndrome, myoclonic jerks, ataxia and abnormal eye movements ('dancing eyes'). Patients with neuroblastoma may also present with GI symptoms, usually diarrhoea suggesting possible ectopic hormone secretion by the tumour. Symptomatic dilatation of the urinary system has also been described.

Investigations

- *Chest X-ray* (lateral and PA): FBC, ESR, U&E
- *CT scan*: Bone scan for metastases or leukaemic infiltration
- *Bone marrow aspiration* (lymphomas, metastatic neuroblastomas)
- *Examination of the urine* for 3-methoxy 4-hydroxy mandelic acid CMHMA or VMA in a 24-hour sample (neural crest tumours)
- *Gastrograffin swallow* to determine site of compression, oesophagoscopy, bronchoscopy, myelography if indicated

Sites, types, and treatment

There are three mediastinal compartments. The common sites of each lesion are indicated in the table.

Compartment	Lesion	Treatment
Anterior	Teratoma	Surgery
	Thyoma	Chemo-radiotherapy
Middle	Bronchogenic cysts	Surgery
	Oesophageal duplication	Surgical biopsy + combined chemo-radio-therapy
Posterior	Neuroblastoma Ganglioneuroblastoma ('dumbell' tumour on chest X-ray)	Surgical resection + radiotherapy (survival rate is inversely proportional to age. Those under two years old at diagnosis do better. 70 per cent survival rate).

Reference

McLatchie, G. R. and Young, D. (1980). Presenting features of thoracic neuroblastoma. *Archives of Disease in Childhood*, **5**, 958–62.

Oesophageal atresia

This is a congenital anomaly in which there is complete interruption of the oesophageal lumen. The commonest type is of a blind-ending upper oesophageal pouch with a tracheo-oesophageal fistula to the lower pouch. Other variations also occur (Vogt's classification).

Other anomalies

Congenital cardiac, intestinal (atresias, anorectal abnormalities), and skeletal abnormalities (hemivertebra, absent radius) are often present and should be sought. There is a very strong association with maternal hydramnios. All babies with such a history should be investigated for atresia by passing a fine rubber catheter into the gullet to confirm its patency (see below).

Clinical features

Vomiting of all feeds, persistent drooling of saliva, choking, cyanotic attacks (especially when feeding), regurgitation of froth and mucus. Regurgitation of gastric juices or aspiration of saliva or milk from the upper pouch may cause aspiration pneumonia.

Diagnosis

Keep the diagnosis in mind. Pass a size 10 (French gauge) soft rubber catheter into the oesophagus through the mouth. It cannot be advanced more than 10–12 cm from the gums in cases of atresia.

Investigations

Although the diagnosis is almost certain following failure to pass a catheter, chest and abdominal X-rays should be taken to establish the presence of a tracheo-oesophageal fistula (when air is noted in the stomach and small intestine) and also the state of the heart and lungs.

Treatment

The patient should be rehydrated and have surgical correction carried out within 12 hours of diagnosis if possible. Nurse the patient prone with aspiration of the upper pouch. Maintain body temperature before and during surgery. Complete correction of the abnormality is achieved by end-to-end anastomosis through a right thoracotomy in the bed of the fifth rib. The azygos vein is doubly ligated, divided, and the mediastinal pleura incised. The proximal dilated pouch is then mobilized and an end-to-end anastomosis established.

Complications

Leaks and chest infection. The postoperative care of these neonates requires specialized nursing care, so affected children should be transferred to specialist centres.

Prognosis

More than 90 per cent survive.

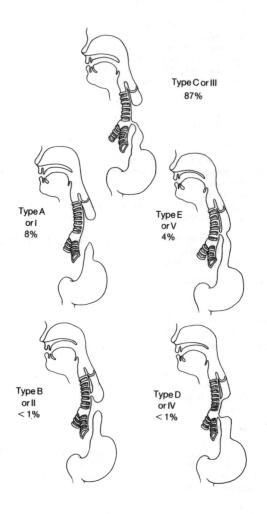

Type C or III
87%

Type A
or I
8%

Type E
or V
4%

Type B
or II
<1%

Type D
or IV
<1%

The neonate with vomiting

Vomiting is the commonest symptom in children, especially in the neonatal period. Vomiting is said to be 'significant' when bile-stained, blood-stained, projectile, persistent, or associated with failure to thrive.

Causes

- Neonatal intestinal obstruction
- Sphincteric disorders
  - gastro-oesophageal reflux
  - congenital hypertrophic pyloric stenosis (see p. 324)

N.B. In neonates, bile in the vomit suggests intestinal obstruction. This cause must be excluded or confirmed.

Neonatal intestinal obstruction can be caused by:

- Luminal problems, e.g. meconium ileus, meconium plug syndrome
- Stenosis or atresia of the wall of the bowel
- Extrinsic compression, e.g. peritoneal bands
- Neurogenic cause, e.g. Hirschprung's disease

Effects of vomiting

Symptoms have often been present for some weeks. This leads to muscle hypertrophy of the bowel wall above the obstruction, with dilatation of the bowel proximal to the obstruction. Distally the bowel is empty.

- Dehydration results from loss of fluids and electrolytes.
- Metabolic acidosis/alkalosis result from acid–base disturbances.
- Respiratory difficulties result from diaphragmatic splinting by the distended gut.
- Perforation may result from distension.

Points in examination

General Temperature, skin of the limbs, state of peripheral circulation, state of fontanelle (depressed = dehydration), volume and frequency of vomits may indicate extent of dehydration and metabolic disturbance.

Abdomen Look for distension, visible peristalsis, palapable mass, bowel loops.

Examine umbilicus, scrotum (don't miss a hernia, which is often forgotten as a possible cause of vomiting), and the rectum using the little finger. Is there meconium? Is the rectum empty? Is the anus imperforate?

Investigations

- U&E, FBC, acid–base balance
- Cultures for assessment of possible sepsis (exclude neonatal infections and the adrenogenital syndrome)
- Radiology—plain erect abdominal X-ray. Contrast studies as indicated.

Treatment

Depends on the diagnosis. Laparotomy is frequently indicated.

The neonate with jaundice

Newborn babies usually develop 'physiological' jaundice in the first 2–5 days which lasts for 7–14 days. This is due to haemolysis of fetal red blood cells. Unconjugated bilirubin is released which is not excreted by the immature liver.

Causes of jaundice

There are traditionally medical and surgical causes.

Medical (prehepatic, hepatic jaundice)
- Haemolytic disease: Rhesus or ABO incompatibility. The stools are normal in colour.
- Infection: Septicaemia, neonatal hepatitis, congenital syphilis, toxoplasmosis
- Galactosaemia
- Abnormalities of the bile transport

Surgical (obstructive jaundice)
- Atresia of the bile ducts
- Choledochal cyst

N.B. The haemolytic diseases, infections, and galactosaemia can be identified by appropriate investigation. Differentiation between neonatal hepatitis and atresia of the extrahepatic bile ducts can be difficult because the age of onset and progress of the two conditions can be similar initially.

Surgical causes only are considered here.

Atresia of the bile ducts

Clinical features Jaundice may appear after the first days of life or be delayed several weeks. It may fluctuate but does not progress. The stools are putty-coloured and look normal. The infant is generally in good health but develops hepatosplenomegaly and ascites. Death from liver failure occurs within 2 years.

Investigations The differential diagnosis lies between hepatitis and atresia. The compromise is to investigate jaundice of more than 4 weeks duration by liver biopsy which may confirm changes of hepatitis or intrahepatic biliary atresia. Surgical exploration, open liver biopsy, and operative cholangiography should be resorted to if initial biopsy is unhelpful. If there is free flow to the duodenum and the duct system is well outlined, the jaundice is probably caused by hepatitis.

Types of atresia
- Extrahepatic
  —correctable: short lengths of extrahepatic ductal stenosis
  —incorrectable: extensive extrahepatic ductal stenosis from portahepatis to duodenum
- Intrahepatic

Procedures
- *Correctable lesions* are treated by choledochojejunostomy Roux-en-Y. The divided jejunum is sutured to the proximal patent ductal system bypassing the area of atresia. Intestinal continuity is restored by Roux-en-Y anastomosis.

- *Incorrectable lesions* may be treated by hepatico-porto enterostomy in which the extrahepatic atretic ducts are excised to expose patent ducts in the porta hepatis and a jejunal Roux loop is anastomosed.

Prognosis Good in correctable lesions. 20 per cent survival in incorrectable.

Choledochal cyst

Localized areas of dilatation of the bile ducts may be intra- or extrahepatic. They are more common in female infants and are a cause of biliary obstruction. They may present less commonly in later life when they are said to be premalignant. Their cause is unknown. Diagnosis is by ultrasound. Treatment is excision with Roux-en-Y choledochojejunostomy.

Less common causes of jaundice

Inspissated bile syndrome is associated with haemolytic disease. The ducts are irrigated with saline introduced through the gall bladder.

319

Intestinal obstruction in infants and children

Features

Vomiting, abdominal pain, constipation, abdominal distension (associated with low small-bowel and large-bowel lesions).

Causes

Oesophagus—atresia

The commonest form is of a blind-ending oesophageal pouch and a tracheo-oesophageal fistula involving the lower oesophagus (p. 314).

Stomach Congenital hypertrophic pyloric stenosis (p. 324).

Duodenum Atresia, stenosis, duodenal septum, malrotation with bands or volvulus, and annular pancreas. Vomiting is early and profuse, leading to dehydration and electrolyte upset. Distension is absent. Constipation is present but not prominent. Treatment involves resuscitation and surgical correction.

Jejunum and ileum Atresia, stenosis, and septum are uncommon. Examine the groin: *incarcerated inguinal hernia* is common. *Meconium ileus* occurs in 10–15 per cent of infants with fibrocystic disease. Meconium becomes desiccated in the distal ileum with a mass of soft meconium above accompanied by distended proximal bowel containing fluid levels and air. *Meconium peritonitis* results from perforation of the gut either *in utero* or postnatally. There is distension and vomiting. Treatment of these types of obstruction is surgical. Bands or hernias are dealt with at laparotomy. Atresia is treated by end-to-end anastomosis. Meconium ileus may be managed by gastrograffin enema or laparotomy and ileostomy. The chemical peritonitis of leaked meconium is treated by peritoneal toilet and repair of the perforation or resection with anastomosis of the affected gut. *Duplications* can occur anywhere in the GI tract. They may be short closed cysts or long tubular communicating segments. They can cause obstruction by pressure or volvulus and may bleed or perforate. Treatment is surgical resection.

Colon *Hirschsprung's disease* is the commonest cause of intestinal obstruction in the neonate.

The functional obstruction is caused by an aganglionic segment which fails to propagate peristalsis (p. 328).

Investigations

- Electrolyte and acid–base status with replacement correction as required
- Plain X-ray of abdomen (double bubble in duodenal obstruction) (fluid levels, distension), chest X-ray
- Arrested passage of blunt oesophageal catheter (oesophageal atresia)
- Barium studies to show malrotations or stenosis
- Rectal biopsy

In all cases of suspected intestinal obstruction, full physical examination is mandatory. Do not forget to look for hernias.

Anorectal malformations

If unrecognized these can lead to the symptoms and signs of
large-bowel obstruction. Fistulous connections are common.
Treatment depends on the level of the lesion and may involve
simple dilatation to colostomy and abdomino-perineal pull-
through of the large bowel (modified Duhamel procedure).
Patients with low lesions usually become continent; 75 per cent
of those with high lesions have some degree of control.

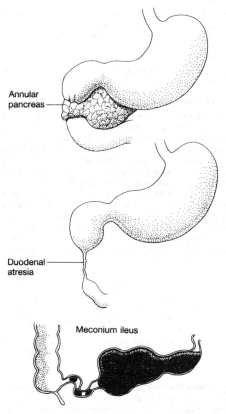

Annular
pancreas

Duodenal
atresia

Meconium ileus

Intestinal obstruction in children

Acute abdominal emergencies in childhood

There are three cardinal symptoms: pain, vomiting, and diarrhoea.

Pain

If abdominal pain lasts longer than 3 hours it should be regarded as evidence of intra-abdominal mischief until proven otherwise.

Vomiting

Especially if persistent or painless should raise the suspicion of small-bowel obstruction. Painless vomiting may imply high small-bowel obstruction which is not accompanied by abdominal distension.

Diarrhoea

If this lasts longer than 24 hours it should be regarded as significant. Appendicitis and intussusception must be excluded before accepting a diagnosis of gastroenteritis.

Common causes

Acute appendicitis, intussusception, intestinal obstruction, mesenteric adenitis, strangulated inguinal hernia.

Management

- Establish whether pain, vomiting, diarrhoea is still present, becoming worse, or subsiding. If these symptoms persist, investigation must continue until a cause is found.
- Clinical examination of the abdomen may elicit localized or general tenderness which when accompanied by localized reflex abdominal rigidity are among the most reliable signs of peritoneal inflammation. Probability should then be used to ascertain the likely cause, e.g. localized signs of underlying peritoneal irritation in the right iliac fossa is probably acute appendicitis. In intussusception and other forms of intestinal obstruction rigidity localized to the obstructed gut may imply impending perforation, gangrene, or secondary peritonitis.
- If there is a mass, exclude faeces as a cause but remember the commonest mass in the right iliac fossa in childhood is an appendiceal abscess. A sausage-shaped mass may be an intussusception. Most occur in children under 1 year old.
- Generalized rigidity suggests general peritonitis. Gross abdominal distension and visible peristalsis suggest advanced intestinal obstruction. These are late signs. Try to make your diagnosis early.
- Examine the chest, ears, throat, neck, skin. Many conditions present with abdominal pain. Remember Apley's rule: 'the further pain is from the umbilicus the more likely it is significant'.
- Do examine the rectum. Occasionally the apex of an intussusception can be felt, or there may be deep tenderness due to a pelvic appendix. Bimanual examination in the dorsal position may permit the detection of ovarian masses in girls.

Moral

In children with abdominal pain, repeated observation and examination will allow a trend to be identified. When there is doubt about the diagnosis it is better to look and see than wait and see.

Congenital hypertrophic pyloric stenosis of infancy

There is hypertrophy of the pyloric sphincter which leads to pyloric obstruction and subsequent projectile vomiting and dehydration. Three children in 1000 are affected.

Clinical features

The child is usually a first-born male aged 3–6 weeks. After initial 'possetting' of infancy, the following features develop:
• Projectile vomiting
• A palpable 'tumour' with visible peristalsis
• Intermittent upper abdominal distension
• The infant is hungry, appears alert initially, and may feed immediately after vomiting only to vomit again.

Complications

Dehydration, loss of weight, alkalosis (due to vomiting hydrochloric acid), constipation.

Diagnosis

Palpable 'tumour' at the pylorus, best felt with a warm hand from the left side of the infant, below the liver, during a 'test' feed. Other features include visible peristalsis and projectile bile-free vomit.

Treatment

• Correct dehydration by IV electrolytes
• Ramstedt's pyloromyotomy (laparotomy and longitudinal division of the pylorus permitting the *intact* mucosa to point through)

Intussusception in childhood

In this peculiar variant of normal peristalsis, one segment of the gut 'telescopes' to be engulfed by its distal segment.

Incidence

0.3 per cent of admissions to childrens' hospitals. Most occur between 6 and 9 months. 5 boys:1 girl.

Aetiology

In infants it may be related to swelling or inflammation of Peyer's patches in the lower ileum due to dietary change resulting in changed intestinal flora or due to upper respiratory tract infection. In older children there may be an abnormality of the bowel which becomes the leading point, e.g. polyp, Meckel's diverticulum, cyst.

Pathology

The outer tube is the *intussuscipiens*, the inner and middle tubes form the *intussusceptum*.

Ileo-colic intussusception is commonest, but ileoileal and colocolic can occur. The blood supply of the intussuscepted bowel may become impaired, leading to the so-called 'red-currant jelly' stool and later strangulation in advanced cases.

Clinical features

Sudden onset abdominal pain occurring in spasms. The child may draw up his legs and vomit afterwards. Attacks last for a few minutes but recur every 10–15 minutes becoming more severe. The child looks pale and listless. When well established, red-currant jelly stools may be passed.

Abdominal examination

There is typically a sausage-shaped lump in the right or left side of the abdomen concave towards the umbilicus. It may harden on palpation.

Rectal examination

Blood-stained mucus on finger, apex of intussusception may be palpable, or may rarely protrude through the anus.

Progression

Untreated, pain will become continuous. Abdominal distension and severe vomiting occur after 24–36 hours. Death results from peritonitis associated with gangrene of the segment.

Diagnosis

- Plain abdominal film may reveal absent caecal shadow.
- Barium or gastrografin enema are the diagnostic tests.

Treatment

- Within the first 24 hours: Hydrostatic reduction with barium or gastrografin enema
- After 24 hours (or failed reduction): Laparotomy with reduction or resection and end-to-end anastomosis if bowel viability is in doubt

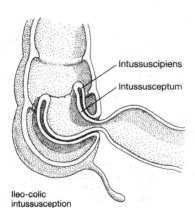

Ileo-colic
intussusception

Ileo-ileal intussusception

Hirschsprung's disease

In this inherited disorder the parasympathetic ganglion cells are absent from the large bowel wall for a varying distance upwards from the anus. When the peristaltic wave of defecation reaches the aganglionic segment it stops. As a result the proximal bowel becomes dilated and hypertrophied. The aganglionic segment remains contracted. In most patients, usually males, a short segment of the sigmoid, rectum, or anal canal is affected—'short' segment disease. Less commonly the affected segment may extend to the ascending colon and even further—'long' segment disease (see below).

Incidence

1 in 5000 live births.

Types

- 'Short' aganglionic segment in sigmoid colon, rectum, or anal canal: 5 boys: 1 girl
- 'Long' aganglionic segment: 1 boy: 1 girl

Diagnosis

This is a cause of neonatal intestinal obstruction. There is delay in the passage of meconium, abdominal distension, and bile-stained vomit. In older children, not diagnosed at birth, chronic constipation is the main feature. Digital rectal examination, usually with the little finger, is useful. The anus and rectum may feel tight and 'unused', as in fact they are.

Investigations

- Anorectal manometry shows characteristic abnormalities in Hirschsprung's disease.
- Barium enema may demonstrate the transition zone between the contracted aganglionic segment (which can also have a normal appearance) and the dilated proximal colon.
- Formal rectal biopsy under anaesthetic gives the definitive diagnosis. Mucosa and submucosa only is required, obtained by punch or suction biopsy. Full-thickness biopsy is easier to interpret and is sometimes necessary (1.5 cm^2 is excised).

Treatment

- Decompression colostomy when the diagnosis is made
- Resection of the aganglionic segment at 6–20 months with restoration of intestinal continuity
- Closure of the colostomy at the time of resection or 2–3 weeks later

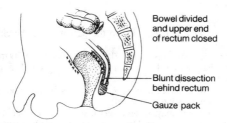

Bowel divided
and upper end
of rectum closed

Blunt dissection
behind rectum

Gauze pack

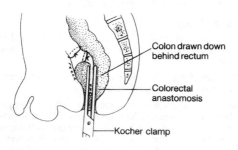

Colon drawn down
behind rectum

Colorectal
anastomosis

Kocher clamp

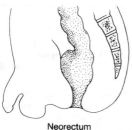

Neorectum

Fissure in ano, prolapse, and polyps in children

Fissure in ano

This occurs almost exclusively in infants and toddlers. Passing a hard stool splits the overstretched mucosa. There then follow episodes of painful defecation with blood often on the surface of the stool. Pain leads to crying on defecation, or may inhibit further defecation. The resulting constipation leads the process to repeat itself.

Examination Often there is little to see in the anus because the fissure heals quickly. When visible it is usually anterior or posterior.

Treatment
- Explain what is happening to the parents (and the child if possible).
- Prevent further constipation (lactulose elixir 2.5 ml by mouth twice daily under 1 year, 5 ml twice daily from 1–5 years, 10 ml twice daily 6–12 years).
- Apply local anaesthetic gel or cream before and after defecation (e.g. lignocaine gel).
- Anal dilatation may be necessary if symptoms persist in spite of the above measures. Do this under anaesthetic. Be gentle.
- Persistent or multiple fissures may suggest underlying chronic disease, e.g. Crohn's disease, TB.

Rectal prolapse

This is common in toddlers less than 2 years old, and is alarming for the parents. In most children the condition settles spontaneously.

Predisposing factors Straining at stool, precipitate defecation, mobile rectum, thin body habitus.

Causes The condition is more common in boys than girls. It usually develops during toilet training, possibly due to parental demands to defecate. This leads to prolonged straining in the absence of constipation. In a few children there is, however, an obvious abnormality, e.g. myelomeningocoele, coeliac disease, cystic fibrosis. In others it may follow anoplasty, operations for ectopia vesicae, or trauma.

Types
- Mucosal: Protrusion of prolapse of the mucosa
- Incomplete: The anterior rectal wall prolapses
- Complete: The full thickness of the rectum protrudes through the anus

Clinical features In most children the prolapse rolls out painlessly during defecation and returns spontaneously, but sometimes manual replacement is necessary. When the mucosa is prolapsed it may become abraded and bleed.

Treatments
- Most settle spontaneously if constipation is relieved
- Persistent prolapse—submucosal injection of 5 per cent phenol in almond oil
- Linear cauterization of the mucosa
- Retrorectal injection of saline to induce fibrosis
- Formal resection of the sigmoid with rectopexy through a transsacral approach

Polyps

Present because of prolapse or rectal bleeding.

Treatment
- Proctoscopy under general anaesthetic with identification of the polyp and transection of its stalk with diathermy
- High polyps may be removed by colonoscopy or at laparotomy

331

Adenomatous polyposis is rarely seen in children. The condition is familial and the polyps premalignant. Consider it when there are three or more polyps, and organize a barium enema.

References

Freeman, N. V. (1984). Rectal prolapse in children. *Journal of the Royal Society of Medicine*, **77**, (Supplement 3), 9–12.

Height, D. W., Hertzler, J. H., Phillipart, A. E., and Benson, C. D. (1982). Linear cauterisation for the treatment of rectal prolapse in infants and children. *Surgery, Gynaecology, and Obstetrics*, **154**, 400–2.

Abdominal malignancies in childhood

The commonest are abdominal neuroblastoma and nephroblastoma.

Abdominal neuroblastoma

This childhood malignancy arises from the foetal neural crest. The commonest site is the adrenal gland, but it may also arise in the mediastinum or posterior abdominal wall. More than 65 per cent present as abdominal mass in a child under two years old.

Diagnosis is based on clinical grounds, plain X-ray (may show calcification). Intravenous urogram often shows a displaced kidney. 24-hour urine collection for HMA and HVA, which are often grossly elevated. Ultrasound is of value in differentiating solid from cystic lesions. CT scan provides accurate detail on size, site, and spread. It can also be used to evaluate the effect of treatment.

Pathology Early metastasis to marrow, skull, long bones, liver, and spinal canal is characteristic. These may lead to varying presenting features like lymph nodes in the neck, paraplegia, proptosis, pain in the thigh, failure to thrive, diarrhoea, etc.

Prognosis relates to age at presentation and site of origin.
- Neonates: 60–70 per cent survive
- < 1 year: 60–70 per cent
- 1–2 years: 20–30 per cent
- > 2 years: 15–20 per cent

Children with abdominal neuroblastomas have a worse prognosis than those with thoracic neuroblastoma.

Treatment Chemotherapy, total body irradiation, autologous bone marrow transplantation, high-dose methotrexate are used in varying combinations. Cisplatin in combination improves the response in advanced disease.

Nephroblastoma (Wilm's tumour)

The peak incidence is around two years of age. 60 per cent of children present with an abdominal mass. 30 per cent have abdominal pain as the main symptoms or in association with a mass or nausea and vomiting. 20 per cent have haematuria.

Associations Genito-urinary abnormalities, aniridia. There may be a genetic basis, as chromosome abnormalities have been noted.

Diagnosis This is made by intravenous urography which shows distortion of the calyceal system.

Prognosis Almost 90 per cent of children are cured with current therapy.

Treatment
- Surgical excision and staging is essential (Stage I = limited to renal capsule, Stage II = invasion of renal vein, Stage III = tumour in regional lymph nodes, Stage IV = distant spread).

- Chemotherapy/radiotherapy in variable schedules. Active chemotherapeutic agents include actinomycin D, vincristine, and Adriamycin in combination. These drugs may act as radiomimetics. Therefore the simultaneous use of radiotherapy may increase toxicity.

References

Shafford, E. A., Rogers, D. W., and Pritchard, J. (1984). Advanced neuroblastoma—improved response using a multi-agent regimen (OPEC) including cis-platinum and VM-26. *Journal of Clinical Oncology*, **2**, 742–8.

Thomas, P. R. M., Lec, J. Y., and Fineberg, B. B. (1984). An analysis of neuroblastoma at a single institution. *Cancer*, **53**, 2079–82.

The child with an abdominal mass

Establish the length of the history and the specific symptoms complained of. Carry out a general physical examination including digital examination of the rectum. Note the site, size, shape, and consistency of the mass. Is there associated hepatomegaly or splenomegaly? Is there evidence of compression, ascites, or portal hypertension? Try to identify the likely pathology according to the age of the child and the duration of symptoms.

Common benign abdominal masses are the *liver* (palpable below the right costal margin even up to the age of 4 years), *faeces* palpable in the colon, the *full bladder*, and *hydronephrosis*.

Malignant masses are commonly *abdominal neuroblastoma* or *nephroblastoma* or *hepatic metastases*. Less commonly *retroperitoneal sarcoma*, *teratoma*. The mass is usually the presenting feature usually first felt by the mother while she baths the child.

Investigations

- Palpate the mass minimally to prevent milking metastases into the lymphatic or blood stream.
- Certain tumours produce metabolites, breakdown products of which can be detected in the urine. A 24-hour urine collection should therefore be carried out. The urine should also be tested for blood. These metabolites (markers) can also be used to measure response to treatment.
- X-rays: plain abdominal film, IVU, CT scan (to identify size, site, and possible spread)
- Ultrasonography is non-invasive and valuable in differentiating between solid and cystic lesions and in assessing the vena cava (for involvement).
- Full blood examination will diagnose most leukaemias and indicate disorders associated with metastases, but specific changes are not usually seen with solid tumours.
- Other investigations, e.g. skeletal survey, biopsy, marrow sample, chest X-ray, as dictated by the results of the above.

Ectopia vesicae and epispadias

Ectopia vesicae is more common in boys than girls, and is part of a pathological anatomical complex in which the bladder forms the anterior wall of a ventral hernia usually associated with rectal prolapse and a waddling gait due to separation of the pubic symphysis.

Incidence

1 in 10 000–15 000 live births. 3 boys : 1 girl.

Abnormalities in the male child

Epispadias, deficiencies of corpora cavenosa, gap in rectus muscles occupied by bladder mucosa, exposed verumontanum, inguinal hernias, undescended testes.

Abnormalities in the female child

Septate or duplex vagina, divided clitoris, anteriorly placed anal opening, rectal prolapse, ventral hernia.

In both sexes the condition is expressed with varying severity, but abnormalities of the upper urinary tract are common and urinary incontinence is a feature.

Aims of treatment

- To deal with the obvious defects
- To deal with associated abnormalities
- To prevent upper urinary tract damage from back pressure
- To achieve urinary continence (often an impossible ideal)

Treatment

Less serious abnormalities may simply involve repair of the epispadias and bladder neck.

Major abnormalities involve urinary diversion with staged reconstruction of the numerous defects.

Epispadias

In this condition the urethral opening is on the dorsal surface of the penis which is short and angulated dorsally. There are usually associated bladder sphincter abnormalities.

Treatment

Involves urethroplasty and procedures to deal with incontinence by urinary diversion and subsequent bladder neck repair (as in ectopia vesicae).

337

Hypospadias

This congenital disorder of the penis occurs in 1 in 300 live male births. The urethral meatus fails to reach the glans penis and commonly emerges in the corona or subglandular areas. Less commonly the meatus is mid-penile; rarely penoscrotal or perianal.

Pathological anatomy

There is an absence of corpus spongiosum between the glans of the penis and the abnormally sited urethral opening. It is substituted for by a sheet of fibrous tissue (chordee) extending from the base of the glans to the urethral opening which causes angulation of the glans on the shaft so that the glans appears spatulate with an exaggerated ventral curvature. The prepuce is deficient ventrally.

Clinical features

- Ventral bowing of the penis. Cosmetically unacceptable
- Urine is sprayed between the legs on micturition or drips back along the penile shaft. Older children squat to urinate.
- Uncorrected the chordee could interfere with intercourse
- Associated urinary tract abnormalities are often present and should be looked for.

Aims of treatment

- To restore a urethra of adequate length and calibre allowing normal micturition
- To restore a normal-looking penis before school age
- To permit normal sexual intercourse

Treatment

Glandular hypospadias Meatoplasty is often an adequate treatment and may even be required.

Coronal hypospadias At a single procedure a new urethra is constituted from local flaps.

Proximal forms of hypospadias These forms usually require a two-stage procedure. In the first the chordee is excised and new skin introduced to the defect to allow the penis to straighten. At the second procedure the new distal urethra is constructed from preputial skin, usually about 6 months later.

There are more than 100 described procedures. In every case the ideal is to complete the repair before the boy goes to school.

Complication

Urethral fistula occurs in less than 10 per cent, but further surgery is required.

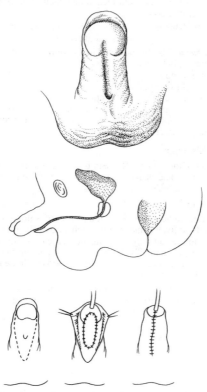

Crabtree urethroplasty for urethral reconstruction

Hypospadias

Disorders of the prepuce (see also p. 276)

Phimosis

This is stenosis of the preputial orifice. As a result, the space between the prepuce and the glans cannot be adequately cleaned. The accumulation of smegma and stagnant urine from difficult micturition leads to recurrent attacks of balanitis which causes further fibrosis and stricturing. In the adult, sexual intercourse may be difficult or painful and lead to paraphimosis.

Treatment

Manual retraction of the foreskin This procedure may be performed at the outpatient clinic using local anaesthetic gel or under general anaesthetic in theatre. The foreskin is retracted, and adhesions between the gland and prepuce manually broken down. This is a successful technique which often obviates the need for circumcision.

Dorsal slit This may suffice in adults who have difficult intercourse or recurrent balanitis, but it is untidy.

Circumcision May be indicated in the following conditions:
- Recurrent balanitis
- Phimosis and paraphimosis
- Repair of hypospadias
- Following a dorsal slit of prepuce

The commonest request for circumcision is on religious grounds. Compliance with the request is a decision for the individual surgeon.

Paraphimosis

In this condition a tight foreskin has been retracted but cannot be returned. The glans becomes constricted and the prepuce engorged and oedematous.

Treatment

Manual reduction After injection of 1 ml of isotonic saline with 150 units of hyaluronidase into each side of the prepuce, wait 10–15 minutes until the swelling has decreased and reduce the paraphimosis manually by placing thumbs on the glans to stabilize it and pulling the prepuce forwards with the index and middle fingers of each hand.

Dorsal slit In this technique the constricting band is incised to allow reduction. Circumcision is often performed at a later date once oedema has settled.

Haematuria in children

Confirm the presence of blood in the urine (stix test). Centrifuge a sample and do a red blood cell count.

History

Ask about recent sore throat or other streptococcal infection (?glomerulonephritis). Fever, dysuria, frequency, malaise suggest infection. Ask about injury, exercise (haemoglobinuria), and the family.

Causes

Diagnosis apparent

- *Palpable renal mass*: ?neoplasm
- *Bleeding disorders*: clotting screen, bleeding time
- *Glomerulonephritis*
- *Urinary tract infection*: malaise, dysuria, abdominal pain, positive MSSU
- *Meatal ulcer*: patient screams with pain on micturition

Diagnosis apparent after investigation

- *Straight X-ray*: urinary calculi
- *Intravenous urogram*: renal and ureteric abnormalities
- *Retrograde pyelogram*: neoplasms of urinary tract
- *MCU*: (micturating cysto-urethrogram) vesico-ureteric reflux, bladder diverticulum, urethral polyps
- *Endoscopy*: diverticulum, neoplasm, vascular disorders
- *US scan*: renal cysts, hydronephrosis, hydrometer
- *CT scan*: neoplasm (site and extent)
- *Arteriography*: tumour circulation, vascular anomalies

Examination

Check the genitals, abdomen, loins, throat, and ears.

Investigations

- *Urine*: Midstream urine specimen, or catheter specimen. Request examination for granual or proteinuria. Sterile pyuria associated with haematuria suggests atypical infection, possible tuberculosis.
- *Radiology*: Plain films, IVU, MCU, US scan, CT scan, aortography if indicated
- *Endoscopy*: Cystoscopy, urethroscopy
- *Renal biopsy*

The child with wetting

Wetting is defined as the passage of urine at times other than during deliberate micturition. The most common cause is involuntary micturition in bed at night or in varying frequency during the day (involuntary micturition = enuresis).

Causes of wetting

Functional
- Enuresis (commonest)

Organic
- *Neurogenic*: Myelomeningocoele, sacral agenesis, spinal cord—abnormalities, infection, tumour, trauma
- *Urinary tract infection*: Pyelonephritis, cystitis, urethritis
- *Structural anomalies*: Urethral valves, urethral diverticulum, phimosis, ectopia vesicae, ectopic ureter, epispadias

History

Organic causes are uncommon but must be excluded before a diagnosis of enuresis can be made. Identify the patterns of wetting, as shown in the table.

Organic cause	Neurogenic and sphincter trauma	Urinary tract infect	Structural anomalies
Features:	Uncontrollable wetting (day and night)	Episodes of wetting Intervening control Respond to antibiotics	Ectopic ureter—constantly wet but also voids urine normally (girls only) Valves, diverticula —flow is fine or slow, achieved by straining—dribbling and infection may occur
	Expressible bladder (normal bladders are not)		Phimosis. Stenosis of prepuce leads to ballooning on micturition with dribbling after until it is empty

Functional wetting produces an extremely variable picture with unpredictable, irregular wetting but there are *usually periods of normal or near normal micturition*.

Investigations

Eliminate overt abnormalities (see above). Carry out intravenous pyelography, MCU, and endoscopy to identify a latent organic cause. Examine the urine for evidence of infection, protein, sugar.

Treatment

Organic causes Specific antibiotic therapy, surgery where indicated.

Functional causes
- Reassurance that spontaneous cure is expected
- Bladder distension training for daytime enuresis with delayed micturition
- Bell and pad for nocturnal enuresis. When the child wets the pad a bell rings to waken the child who then gets up to pass urine. Often wakens the rest of the family while the child sleeps on. Try for up to 6 weeks only. Worthwhile in children over 6 years old.

The child with an empty scrotum

The child's mother is usually a good indicator of whether there are testes in the scrotum. Undescended testes are of two types, retractile or maldescended.

Retractile testes

Prepubertal boys have retractile testes. They can be drawn up from the scrotum into the superficial inguinal pouch by contraction of the cremaster muscle. Test this by stroking the inner aspect of the boy's thigh during examination. The testis on the same side is drawn up into the groin.

Clinical examination A retractile testis can be brought down into the bottom of the scrotum by applying firm pressure along the inguinal canal. The child should be warm and relaxed. The procedure is facilitated by applying baby oil or talcum powder to the skin. If it cannot be brought down the testis is maldescended.

Maldescended testis

This occurs in 2 per cent of boys. The abnormality may be unilateral (more common) or bilateral, and the right testis is more frequently affected. Examine the scrotum, superficial inguinal pouch, inguinal canal, upper thigh, perineum, and abdomen. If the testis cannot be palpated it may be absent. CT scan is useful to detect an intra-abdominal testis. Maldescended testes may be arrested in the normal line of descent or ectopic.

Arrested The testis is often small and abnormal with a short spermatic cord. An associated inguinal hernia is often present. Most are felt close to the pubic tubercle, some in the inguinal canal. Rarely the testis is the abdomen.

Ectopic The testis is usually normal with a normal spermatic cord but has diverged from the usual pathway during its descent. Most lie in the superficial inguinal pouch, but if impalpable there should be sought for in other areas (see above).

Complications

Trauma, torsion, and tumour are more common in undescended testes. As a result of trauma, testicular atrophy may occur. Infertility occurs in bilateral cases.

Treatment

Orchidopexy reduces the risks of complications and can restore fertility.

Torsion of the testis

This is rotation of the testis with resultant interference of its blood supply. It does not occur in a normal fully descended testis and usually some abnormality is present such as inversion of the testis, high investment of the tunica vaginalis with a horizontal lie, separation of the testis from the epididymis. It may be intravaginal (commoner) or extravaginal.

Clinical features

The highest incidence is between 10 and 25 years, with infancy the next most common time.

In the older child there is usually sudden, severe pain in the groin or lower abdomen accompanied by vomiting. Its onset may follow heavy straining or lifting, bicycle riding, etc. It is therefore mandatory that the testes be examined in young boys with abdominal pain. Neonatal torsion is easy to miss but should be suspected if the scrotum appears swollen or bruised. Appropriate action (exploration) will prevent subsequent atrophy of the affected testis and may even influence subsequent fertility (see below).

Differential diagnosis

- Acute epididymo-orchitis (exclude mumps and urethritis)
- Torsion of a testicular or epididymal appendix
- Idiopathic scrotal oedema

Treatment

If torsion is suspected or diagnosed, urgent exploration of the scrotum through a transverse incision is indicated. The torsion should be identified and corrected and the affected testis fixed in the scrotum by anchoring it at its upper and lower poles with fine nylon sutures through the tunica. Since torsion may also occur on the other side, the contralateral testis should be fixed in similar manner.

An obviously infarcted testis must be removed. If there is doubt about viability, wrap the untwisted torted testis in a warm moist pack while the other side is fixed. If there then appears to be hope of viability return the testis and fix it. Observe the patient carefully postoperatively for signs of inflammation or infection. Arrange outpatient follow-up to assess the testis. If a non-viable testis is left in the scrotum it may lead to the development of testicular antibodies and infertility. If an atrophic testis has to be removed give the patient the option of an implant. The loss may affect his sexuality and body image (cf. mastectomy in women).

Infantile hydrocoele

A hydrocoele is a fluid-filled sac lying in front of the testis. It results from a patent processus vaginalis communicating with the abdominal cavity. Hydrocoeles are fairly common in neonates and tend to resolve spontaneously. If not, there may be an associated inguinal hernia.

Clinical features

The swelling
- Can be 'got above'
- Transilluminates
- Is often small in the morning and increases in size by evening, when the child may complain of dragging pain

Treatment

Most disappear by the time the child is a year old. If a hydrocoele persists or appears for the first time around the age of two years, surgery is usually necessary. This involves a procedure similar to herniotomy with excision of the outer layer of the tunica vaginalis.

Inguinal hernia in children

Inguinal hernias are common. The patent processus vaginalis permits descent of intrabdominal or pelvic contents into the sac. Most occur in boys, most are indirect, and the sac usually contains small bowel or omentum. 10 per cent occur in girls, when they may be bilateral. The sac often contains an ovary which can be felt.

History

The parents may report a lump 'coming and going' in the groin. Establish exactly what they mean. Where is the lump? Does it get bigger during the day? Does it increase with coughing or straining? Does it go away on gentle pressure or on lying down?

Examination

The older child may be examined standing when the lump can be seen and felt. If the lump descends into the scrotum it is not possible to get above it (see hydrocoele). In babies and young children the history alone is enough but the cord of the affected side may feel thickened (a positive finding). In any child with abdominal pain and vomiting hernia must be considered, with examination of the abdomen, groin, scrotum, or labia.

Differential diagnosis

Hydrocoele, undescended testis, direct or femoral hernias (uncommon), inguinal lymph nodes.

Complications

- Incarceration: The bowel becomes stuck and the hernia will not reduce.
- Strangulation: The blood supply is cut off. Strangulated bowel becomes gangrenous and may perforate. Strangulation is most common in the first six months.
- Testicular infarction: Due to damage to its blood supply. A non-viable or subsequently atrophic testis should be removed.
- Peritonitis: Can follow reduction *en masse* of a strangulated hernia. When attempting to reduce a hernia manually, make sure that you can feel the testis and cord afterwards. Do not attempt this if there is a history of an irreducible lump in the groin for several hours.

Treatment

Elective
- Herniotomy of the sac at the internal ring (removal of the sac)

Emergency
- An irreducible hernia may reduce spontaneously if the child is sedated and the foot of the bed elevated. If successful, herniotomy can be performed electively in a few days. If unsuccessful within a few hours, surgery is necessary on the day of admission.

- Strangulated or suspected strangulated hernias should be operated upon as soon as possible after resuscitation.

    N.B. Almost one third of infants under 10 months with an inguinal hernia will present with strangulation.

The child who limps

Gait is the manner of walking or running which, when normal, involves rhythmic movement of the body. Limping implies an uneven step—an abnormality of gait. The commonest orthopaedic causes of limp in the UK are irritable hip and Perthes' disease (5–10 year olds). Slipped upper femoral epiphysis (10–15 year olds) is less common. Congenital dislocation of the hip should be detected at birth. Keep it in mind, however, when faced with a limping toddler.

History

Ask specifically how long the child has been limping, whether it is getting worse, and whether there is pain. Details of difficult delivery, birth injury, or prematurity may point to a cerebral cause.

Physical examination

Examine the child in his underwear. Watch his gait. Learn to recognize the *pattern* of the limp. Ask the child to hop on alternate legs. If this can be done easily, the limp is probably of little significance. Observe *spinal movements* by asking the child to touch his toes, bend from side to side, etc. Measure *leg length* from the anterior superior iliac spines to lateral malleolus. Are both equal? Test the *hips for instability*: ask the child to stand on one leg. Normally, the buttock on the opposite side will rise a little; if it falls, this implies instability of the weight-bearing hip—*Trendelenburg's sign*. Look for *wasting of the buttocks*. This implies hip joint or nerve lesion. Test *muscle power* on the affected side, and compare it to the other. Flex one thigh on to the abdomen. Does the other hip stay flat on the couch or can it be pushed flat? If not, there is a fixed flexion deformity (Thomas' test).

Common gait abnormalities

Neurological (birth injury, prematurity) Affected children may be *spastic* with a tip-toe gait and scissoring (the legs cross when walking) or *ataxic* with loss of balance and poor co-ordination—*stamping gait*.

Trauma, infection, overuse, crack fracture of tibia or fibula, osteomyelitis, septic arthritis, ?irritable hip, ?osteochondritis The child tries to spare the affected limb or joint by adopting an *antalgic* or *hopping* gait.

Hip instability (Perthes' disease, congenital dislocation) The gait is waddling, with considerable lateral movements of the spine and buttocks.

Leg length disparity The pelvis is tilted when standing and the longer limb bent at the knee when walking.

Neurosis/psychiatric illness Bizarre gaits may be related to psychiatric illness or hysteria. No pattern is identified. Always exclude organic diseases before making the diagnosis.

Investigations

- X-rays (of all the joints of the affected limb, if necessary)
- FBC. Blood culture (in suspected infection)
- Aspiration of the affected joint. Microscopy and culture of fluid.

Treatment

Is directed against specific causes (pp. 356–67).

The child with a fracture

Sprains are uncommon in children. If there is pain, local tenderness, swelling accompanied by loss of function (pseudo-paresis in young children), always exclude a fracture in the first instance. Deformity, crepitus, or abnormal mobility are absolute indications for radiological investigation.

Nerve injuries may be associated with fractures, especially around the elbow. The prognosis is good if the injury is recognized early, so test nerve function distally (motor and sensory).

Arterial ischaemia should be excluded by feeling the pulses in the injured limb. Repeat if necessary. This is a recognized complication of supracondylar fracture of the humerus, but can occur equally in other sites. Damage to the main artery produces spasm followed by ischaemia and necrosis especially of the flexor muscles of the forearm. This leads to contracture (Volkmann's ischaemic contracture). About 12 hours ensue before changes are irreversible.

Radiology

- Ensure X-rays are taken in at least two planes. Include the associated joints. X-ray the other limb for comparison.
- Know the sites of epiphyses and the normal maturation of the child skeleton, especially around the elbow.
- Look for greenstick fractures which are easily missed. The bone cracks or fractures on the side opposite the deforming force, like a greenstick (try bending a young willow branch). Check that the joint above and below the fracture are intact (e.g. Monteggia fracture).
- Look for the three Cs: comminuted, compound, or complex (associated with soft tissue injury) fractures.

Reduction

Restoration of alignment is the aim. Rotational deformity should be avoided, but minimal angulation is acceptable because the bone remodels. Reduction should be done under general anaesthetic, and will involve one or more of the following: traction, manipulation, or open reduction.

Immobilization and retention of the reduced fracture can be achieved by external fixation in the form of continued traction, plaster of Paris cast, or cast bracing. Internal fixation may be indicated in specific fractures (e.g. the three Cs, radial head, femoral neck fractures). Methods include pins, plates, screws, or intramedullary nails.

After-care and rehabilitation

- Examine the limb after the application of a plaster cast to ensure normal digital movements. Split the cast if excessive swelling around the fracture site is anticipated. If the child is going home, give the parents a printed set of instructions warning them to report marked swelling, blueness, severe pain, paralysis, or numbness of fingers. Examine the limb at the hospital the next day.

- If there is pain the plaster is either too tight or causing localized pressure. Either split it or cut a window to inspect the area. In cases of severe pain, suspect arterial ischaemia.
- Carry out a check X-ray if there is suspicion of poor reduction and also at about a week after injury to check the position. Remanipulation or wedging of the plaster may be necessary.
- Non-union is rare in children. Physiotherapy is usually not needed. Most child fracture patients return to normal.

Congenital dislocation of the hip

This serious disorder results in an unstable hip joint which dislocates spontaneously. It occurs in approximately 12 in 1000 live births, and can be diagnosed at birth. In 1 in 4 cases, dislocation is bilateral. Early detection emphasizes the importance of neonatal screening; if recognized and treated most hips will develop normally.

Aetiology

Heredity The condition is commoner in girls than boys (4:1) and there is a higher incidence in the daughters of mothers who had the condition. It may be related to maternal transfer of relaxin across the placenta, which leads to joint laxity in the female foetal pelvis.

Breech presentation This position *in utero* over the last 2 months of pregnancy and at delivery may lead to dislocation.

Clinical features

The gluteal folds are asymmetrical. The perineal gap is abnormal and the affected side moves poorly by comparison to the other side. In older children a limp becomes apparent.

Early diagnosis

This should be in the first days of life, when treatment is almost always successful.

Ortolani's test The naked infant is placed supine on an examination table. The knees and hips are flexed with the thumbs overlying the font of the hip joints. As the hips are abducted there is a palpable forward jerk. This is a positive test.

X-ray of the pelvis shows a small femoral head lying outside Perkin's lines.

Late diagnosis

Such children present when there is *delay in walking* or *abnormal gait*. Bilateral dislocation leads to a characteristic waddle.

Features
- Short leg on affected side, asymmetrical adductor creases, limited abduction especially in flexion. Positive Trendelenburg's sign.
- X-ray shows a small capital nucleus on the affected side. The acetabulum is shallow, and the femoral head is displaced both outwards and upwards.

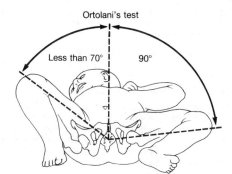

Ortolani's test

Less than 70°   90°

359

Normally each hip should be abducted to 90°
Abduction less than 70° is abnormal

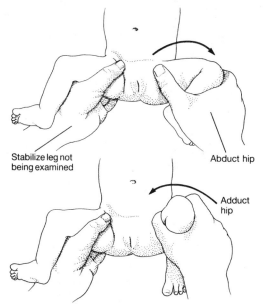

Stabilize leg not
being examined

Abduct hip

Adduct
hip

A 'clunk' heard on abduction and adduction confirms
congenital dislocation of the hip

Congenital dislocation of the hip

Congenital dislocation of the hip (*cont.*)

Treatment

Infants Reduce the dislocation and maintain reduction with a frog plaster inspected every 6 weeks and changed 3-monthly. Checking X-rays at 6-monthly intervals will indicate reformation of the acetabulum. The legs may gradually be brought into lesser degrees of abduction.

6–12 months Traction may be required to obtain reduction. An abduction hip splint or hip spica plaster is then applied for 3–6 months. Open reduction may be required.

1–2 years Traction and open reduction.

Older children Traction, manipulation, and open reduction. Attempts to relocate the head involve dividing the capsule, adductor tendon, and psoas tendon. This may permit development of the acetabulum. If the acetabulum fails to develop reconstruction by osteotomy techniques (Chiari's operation, Salter's osteotomy) may be indicated.

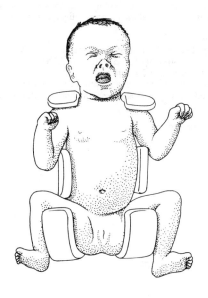

Slipped upper femoral epiphysis

This common condition most often presents in adolescent boys who are above average height, moderately obese, and with small external genitalia. It can also occur in boys or girls aged 10–15 years who are tall and thin.

Types

Acute or chronic.

Clinical features

The epiphysis slips downwards and backwards. The leg becomes externally rotated. There is often an associated synovitis.

In the *acute slip* the features are similar to a fracture of the neck of the femur. The child cannot walk, the leg is externally rotated. There may be severe groin to knee pain, and passive movement is painful.

In the *chronic slip* there is a history of intermittent or increasing pain for several months. The child develops an antalgic limp. On examination there is shortening, irritability of the hip, and fixed external rotation.

Radiology

Anteroposterior and lateral views must be taken. There is widening and irregularity of the epiphyseal growth plate. If a line is drawn along the upper border of the femoral neck it fails to intersect the femoral head. On the lateral view it can be seen that the axis of the neck and the epiphysis diverge from each other.

Treatment

- *Acute*: The femoral head is reduced under anaesthesia and the reduction maintained with pins. There should be no weight-bearing for 3 months.
- *Chronic*: The femoral head is pinned as it is and the resultant deformity accepted. Osteotomy techniques may improve the position.

Complications

Early onset osteoarthrosis and avascular necrosis of the femoral head are real risks.

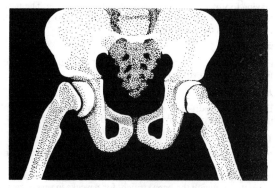

Slipped right upper femoral epiphysis

Slipped upper femoral epiphysis (adolescent coxa vava)

The osteochondritides

These conditions result in avascular necrosis of epiphyseal bone, similar to Perthes' disease but less common. In general they are troublesome rather than serious. They are also usually self-limiting.

Features

Most lead to aching and muscle spasm. Radiographic changes occur in the affected epiphysis, with varying degrees of density change and fragmentation. Resolution results in restoration of radiological appearance although the shape may be altered.
Epiphyses commonly affected are:
- Lateral condyle of the humerus (Panner's disease)
- Carpal lunate (Kienboch)
- Capitellum of the radius (Pannett)
- Head of a metatarsal (Freiberg)
- Tarsal navicular (Kohler's)
- Patella (Larsen–Johanssen)
- Vertebral epiphyseal plates (Scheuermann's)
- Vertebral body (Calve's)
- Femoral capital epiphysis (Perthe's)

Traction apophysitis

An apophysis is a traction epiphysis which may undergo partial avulsion with avascular change followed by subsequent repair. These changes can be the result of trauma, overuse or rapid growth. The commonest affects the apophysis of the tibial tubercle into which the patellar ligament inserts. (Osgood–Schlatter's disease).

Clinical features

The patient is usually an adolescent who presents with aching, swelling, or pain in the knee often related to physical activity, e.g. playing football. There may also be pain on kneeling. The tibial tuberosity is enlarged and tender. There is also pain in the tuberosity on resisted extension of the knee.

Radiology

Fragmentation of the tibial tuberosity with a degree of distraction from the growth plate.

Action

Explain the condition to patient and parents. Reassure them of its self-limiting nature and that return to full activity is usual.

Treatment

- Restrict physical activity, especially kicking, jumping, running.
- Symptoms can be alleviated by immobilizing the limb in a plaster cast with the knee straight for 8 weeks in severe cases.

Other sites of apophysitis

Sever's disease of the calcaneal apophysis. The affected child is usually a boy who has pain over the site of insertion of the Achilles tendon. Treatment is by limiting activity and inserting a heel raise in the child's shoe.

Perthes' disease

This is a form of osteochondritis juvenilis which affects the capital femoral epiphysis. The changes are believed to result from one or more episodes of ischaemia of the femoral head which ultimately lead to avascular necrosis. The cause is unknown, but the epiphyseal changes can lead to permanent deformity and osteoarthrosis in adult life. Children between 3 and 12 years are affected, boys four times as often as girls. In around 15 per cent of cases the disease is bilateral.

Clinical features

The presentation is variable, but pain and limping are common. The pain may be located in the groin or referred to the thigh or knee, and is aggravated by activity. The child is otherwise well.

Clinical examination

ESR, WBC are normal. There may be tenderness in the affected hip, wasting of the buttock or thigh, and a decreased range of movements, especially abduction and internal rotation. Trendelenburg's sign is positive in well-established cases.

Radiological features

Early changes are an increased joint space and capsular bulging. When avascular necrosis occurs, the epiphysis looks dense. This is followed by fragmentation when vacuoles and pseudocysts appear in the epiphysis and broadened femoral neck. The femoral head then collapses if weight-bearing continues and there is reduction in the neck shaft angle. Although reconstitution and regeneration occur over the next 1–3 years, the deformity is permanent.

Classification

Four categories are recognized (Catterall 1971).
- *Group 1*: Only the anterior part of the epiphysis is involved. There is no collapse and no sequestrum.
- *Group 2*: The affected area is greater (< 50 per cent) with sequestrum formation. However, the head remains spherical and complete healing occurs.
- *Group 3*: Only a small part (< 25 per cent) of the peiphysis is unaffected. The centrally placed sequestrum collapses with normal medial and lateral segments. The neck is broadened and there are extensive metaphyseal changes.
- *Group 4*: The whole epiphysis becomes a sequestrum. There is total collapse and extensive metaphyseal changes. This group has the poorest prognosis.

Treatment

Suspect this condition in 3–12 year olds. Carry out X-rays and bone scan.
- *Group 1*: Aspiration or decompression of the joint capsule and restricted weight-bearing during the active phase are required. A weight-relieving caliper or crutches can be prescribed for older children. Most do well.

- *Groups 2 and 3*: Patients should restrict activity and may need bed-rest with the hip abducted so that the head is contained within the acetabulum until reconstitution occurs. Older children (> 6 years) can be mobile with crutches. If the disease progresses, operative treatment is indicated to maintain the head in the acetabulum. 75 per cent of Group 2 do well, but only 25 per cent of Group 3.
- *Group 4*: Patients have advanced disease, with poor results from treatment. Most efforts are directed against the symptoms of osteoarthrosis.

Reference

Catterall, A. (1971). The natural history of Perthe's disease. *Journal of Bone and Joint Surgery*, **53B**, 37–53.

367

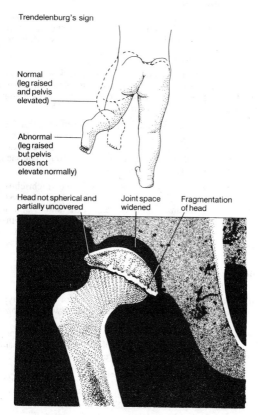

Trendelenburg's sign

Normal (leg raised and pelvis elevated)

Abnormal (leg raised but pelvis does not elevate normally)

Head not spherical and partially uncovered

Joint space widened

Fragmentation of head

Spina bifida

In this congenital condition there is a defect in the vertebral arches with or without abnormalities of the spinal cord and its investing membranes. Many infants are stillborn. 50 per cent die in the first year of life.

Incidence

3 in 1000 births.

Aetiology

Unknown: genetic and environmental factors are involved. The risk is increased in:
- future siblings of an affected child
- children of an affected parent
- consanguinity
- infants conceived in spring

Antenatal detection

Estimation of alpha-fetoprotein in the blood of pregnant women indicates whether amniocentesis should be undertaken. Women with positive tests may wish to have their pregnancy terminated.

Spina bifida cystica

The child is born with a cystic swelling on the back which may be a meningocoele (less than 5 per cent) or myelomeningocoele (greater than 95 per cent). About 60 per cent of myelomeningocoeles occur in the lumbosacral region.

Meningocoele is a simple meningeal sac which contains CSF and only occasionally neural tissue. The sac is covered by skin which is usually intact but may be ulcerated. Mild hydrocephalus may be present. Death is rare in this group. The sac is excised and closed at around 3 months.

Myelomeningocoele may be open (myelocele, rachischisis) with an exposed spinal cord or closed when there is a cystic swelling in the midline which contains CSF and dysplastic neural tissue. Neurological signs are always present. Other malformations: hydrocephalus (80 per cent of cases), renal tract (50 per cent), orthopaedic (60 per cent) are commonly associated.

Clinical features

- *Sac is midline*: Variable in size, usually lumbosacral
- *Motor loss*: Falccid paralysis of lower motor neurone type depending on the level of the lesion. Mixed lower and upper motor neurone lesions also occur.
- *Sensory loss*: May not correspond to motor loss but predisposes to pressure ulceration of the feet, sacrum, and perineum exacerbated by faecal or urinary soiling.

- *Neurogenic bladder*: Sphincters and parasympathetic outflow from $S_{2,3,4}$ are almost always affected. There are varying degrees of bladder dysfunction. Training is impossible. If the bladder is inert it may be emptied by manual pressure. If the bladder is uncoordinated with an external sphincter which does not relax, back pressure can lead to chronic renal failure.
- *Bowel dysfunction*: Faecal incontinence, rectal prolapse, and impaction of faeces are features of myelomeningocoele.

Aim of treatment

A child who can walk, is dry, and can attend school. About half of the affected children achieve this ideal.

Contra-indications to treatment

Surgery is avoided in children who have severe associated abnormalities and complications like severe infection, although this approach is ethically difficult to justify.

Surgery

The child should be transferred to a paediatric centre as soon as the diagnosis is made. The sac is repaired within 24 hours, but this may be postponed if there is skin overlying the defect.

Future management

This is multidisciplinary, involving the treatment of hydrocephalus, bladder training, the management of faecal incontinence, psychological support, and training for employment.

Spina bifida occulta

Affects 10 per cent of adults. The common sites are the fifth lumbar and first sacral vertebrae. Although the vertebral arches are defective, there is skin cover and no neural sac. In most people it is an incidental finding on X-ray. A small proportion develop neurological signs during adolescence. The site of the arch defect may be apparent on the skin as a dimple or hairy patch.

9 Plastic surgery

Cosmetic skin closure

Warn your patients that the procedure may leave a scar, especially when operating on exposed sites like the face. Reassure them that you will try to make this as cosmetic as possible.

Closure of skin wounds

Wounds can be closed with sutures, skin clips, or Steristrip tapes. These permit wound healing and should be removed when the tensile strength of the wound is sufficient to maintain closure.

Ways to avoid suture marks

- Remove sutures within 7 days for cosmetic closure (especially the face and breast). Some sutures may have to be left longer—up to 14 days—and these will leave permanent marks on the skin.
- Avoid tension. The suture should approximate the skin edges only.
- Subcutaneous sutures in the long axis of the wound permit less tension on the skin.
- Evert the skin edges. When the wound heals, gradual flattening will occur. Inverted skin edges will produce a depressed scar.
- Obliterate the dead space of a wound either by buried sutures or by suction drainage. Infection is commoner with buried absorbable than non-absorbable sutures.
- Suture marks are common on the sternum and trunk. Warn the patient. Fortunately the face, eyelids, and mucous membranes usually heal well without excessive scarring or suture marks, provided the basic principles are adhered to.

Cleft lip and palate

These constitute the commonest congenital abnormalities of the head and neck, with an incidence between 1 in 500 (cleft lip) and 1 in 1000 (cleft palate) live births. If one child is affected, the risk is increased in subsequent children. If the parents themselves are affected, the risk is three times that of the normal population. Associated abnormalities include tongue tie, retroposition of the mandible, and collections of fluid in the middle ear.

Associated factors

Drugs like antimitotics, phenytoin, used during pregnancy may be implicated.

Classification

Cleft lip may be unilateral and incomplete or bilateral and complete. If the alveolus is involved, the cleft is median.

Cleft palate

- Unilateral abnormalities usually affect the soft palate and posterior third of the hard palate.
- Bilateral clefts affect the soft and hard palate.
- Beware the submucous cleft in which the palate appears intact. Oronasal escape will be noticed during phonation.

Diagnosis

Inspect the lips and palate at birth.

Management

Explain the defect, its treatment, and results to the parents. Advise them that counsel is available from CLAPA (Cleft Lip and Palate Association). Tel. 01 405 9200 ext 256 (headquarters).

Feeding

- Palate intact: No problem with sucking or swallowing
- Palate cleft: Cannot suck. Swallowing is normal. In these children feeds can be fed by delivering the milk over the back of the tongue into the pharynx by spoon, pipette, or 'squeeze' bottle. This has a spatulate feeding spout which makes delivery of milk easier.

Aims of surgery

- To achieve an intact lip, alveolus, and palate
- To correct cosmetic deformity
- To permit normal speech and dentition

Preoperative preparation

Orthodontic assessment for the provision of a dental plate may be necessary in children with bilateral complete cleft lip and palate.

- *Cleft lip*: 'Rule of 10'. The child should: weigh 10 lb (4.5 kg), be 10 weeks old, have a haemoglobin of 10 g. The lip, nose, and anterior palate are repaired at this stage.

- *Cleft palate*: This is repaired at 6–12 months.
- *Secondary procedures*: May be required to improve the appearance of the nose and lip and to improve speech (pharyngoplasty = palatal lengthening).

Prognosis

1 in 5 children need speech therapy in the pre-school period. Further cosmetic procedures may be required after adolescence for those who have had severe defects.

Lacerations and fractures of the hand

Take an accurate history, then look, feel, and move the affected or related parts, specifically testing all muscles/nerve functions, etc. Skin injuries may appear trivial, but deep structures may be injured and missed if examination is inadequate.

History

When, where, how? Recent injury? Chronic? Occupation and hobbies?

Look

Is the posture normal compared to the uninjured hand? Where is the laceration? What structures underlie it? Think specifically of nerves, tendons, and vessels. Is the hand/finger viable? Is there skin loss?

Feel

Ulnar and radial pulses. Is the digit/hand cold? Is pin-prick sensation present distal to the injury?

Move

Test the integrity of the flexor tendons: superficialis test, profundus test. Is the range of joint movement normal?

Fractures

Minimal or undisplaced but stable Rest for a few days. Reduce swelling by elevation (on a pillow) at night. Mobilize early.

Unstable fractures Most heal with reduction and splintage with the metacarpophalangeal joint flexed and interphalangeal joints in extension. Use Zimmer splints, plaster of Paris front slab, or strap to the neighbouring finger. Check the fingertips for malrotation.

Some unstable fractures require internal fixation.

Flexor tendon injuries

The superficial and profundus tendons may be injured together or individually.

Treat both by early surgical repair imediately after injury. Use a tourniquet and a large enough incision to bring both ends together. Repair the flexor tendon sheath at the same time.

Postoperatively, use dynamic splints—a plaster back slab with the wrist and fingers flexed on tight elastic bands. Maintain for 6 weeks, then actively rehabilitate.

Nerve injuries

Primary surgical repair gives the best results. Age of the patient is related to prognosis. The young do well, the elderly badly. Use magnification and avoid tension. Immobilize for 4 weeks. Many months of re-education are required.

Complication

Neuroma This is a mass of nerve tissue growing from the proximal end of the injured nerve which has failed to find the distal sheath. Symptoms vary, but localized pain or generalized stiffness with swelling, trophic skin changes, and hypersensitivity may result. Treatment involves rehabilitation, electrical counterstimulation, and sympathetic blocks as indicated. Surgery is rarely of value.

Trauma to the hand

Oedema is glue. Try to reduce swelling in the hand, and encourage early mobilization when possible.

Foreign bodies

These are common injuries which may present as an acute or chronic infection. Two types of foreign bodies are described: irritant and non-irritant.

Treatment

Non-irritant foreign bodies can be observed, e.g. broken needles, glass, steel, etc. Fibrous tissue tends to encapsulate the object, which can be left if it is small and causing no trouble or excised otherwise.

Irritant foreign bodies, especially wooden splinters or thorns, should be removed by exploration to prevent abscess formation.

X-rays should be taken in two planes when the foreign body is radio-opaque, and skin markers placed to facilitate access. Exploration must be done under good light and with a bloodless field.

Nail injuries

Subungual haematoma Common and painful. Evacuate the haematoma by painlessly trephining the nail with the end of a heated paper-clip.

Avulsion of nail Provided there is no damage to the nail bed normal regrowth will occur. Prevent adhesions by replacing the avulsed nail in its bed accurately. It acts as a splint.

If there are lacerations they should be accurately approximated. If tissue loss occurs, use partial-thickness skin grafts.

Late deformities after injury are difficult to treat. It is best to remove the nail and its bed and provide an artificial nail.

Pulp injuries

Can be lacerations, amputations or crush. Avoid primary suture, use Steristrips instead.

Assess skin loss and viability. Is there exposed bone? Treat conservatively in *children*, except for the most severe. Simply apply longitudinal Steristrip and splint the finger-tip.

In adults the repair depends on the injury. Partial-thickness grafting or local flaps may maintain length. In many patients terminalization by amputation of the distal phalanx will permit rapid healing.

Joint injuries

The patient presents with a painful, swollen finger joint. Check for stability and exclude fracture. Reduce dislocations. Open reduction is indicated for collateral ligament rupture or volar plate disruption. Early mobilization prevents post-traumatic stiffness.

Mallet finger

This deformity results from disruption of the extensor mechanism to the distal phalanx with or without bony avulsion. When there is no fracture, splint the injury for 6 weeks. Open reduction is indicated when large bone fragments are avulsed.

Infections of the hand

Paronychia (whitlow)

The causative organism is usually *Staphylococcus aureus*. Chronic candidal infection may be present. This needs scrapings for diagnosis, as the treatment is different.

Features Pus accumulates around the nail between cuticle and nail matrix.

Treatment
- *Acute*: Surgical drainage and elevation of the hand. Take bacteriology swabs.
- *Chronic*: Take swabs and scrapings. Specific antifungal agents are usually applied topically. They include clotrimazole (Canesten), miconazole (Daktarin), econazole (Pevaryl), benzoic acid ointment (Whitfield's), amphotericin (Fungilin) and nystatin (Nystan, Tinaderm-M). If unsuccessful, systemic therapy or excision of the nail may be indicated.

Felon (pulp infection)

Features The condition is very painful and can lead to digital vessel thrombosis distally with avascular necrosis of the phalanx.

Treatment Surgical drainage of pus is usually adequate.

Web space infection

There are three web spaces: palmar, thenar, and hypothenar. All may become acutely infected, usually by *Staphylococcus aureus*.

Features Swollen flexed fingers, pain, and swelling over affected space.

Treatment Surgical drainage. Specific antistaphylococcal antibiotics, e.g. flucloxacillin 250–500 mg 6-hourly by mouth.

Bites (human and animal)

Human bites are often sustained in street or pub brawls. They can produce unpleasant mixed infections. Commonly a punch to the opponent's face will cause his teeth to become embedded in the hand of the aggressor, leading to a laceration over the dorsum of the hand. Animal bites (commonly due to dogs) occur during defensive movements and may therefore be more extensive.

Features There are teeth marks (human or animal) on the hand. Dog bites may present as severe injuries.

Treatment
- All bites where skin has been penetrated should be explored. Retained tooth fragments and foreign bodies must be removed.
- Take bacteriology swabs for aerobic and anaerobic culture.
- Carry out adequate wound excision. Primary suture should be avoided.

- Give appropriate antibiotic and tetanus prophylaxis.
- Splint and raise the hand postoperatively.
- Delay the repair of damaged structures under antistaphylo-coccal and antianaerobic antibiotics (e.g. flucloxacillin and metronidazole).

Dupuytren's contracture

This genetically determined condition affects the palmar or plantar fascia, and is characterized by the formation of nodules and contractures. Thought to originate in Celtic races, it has been disseminated worldwide by their migration.

Incidence

It is commonest in men over 65 years. In Scotland, 25 per cent of males over 65 years are affected.

Clinical features

Classically nodular formation in the palmar fascia is accompanied by contracture which causes flexion deformities at the metacarpophalangeal and subsequently the proximal interphalangeal joints. The little finger is most often affected, with the ring finger next. The condition is painless but the affected fingers become caught on clothing, etc. Contracture of the plantar fascia may also occur but it is less common.

Treatment

If the patient cannot place his hand flat on a table, surgery is indicated. The metacarpophalangeal joint can usually be satisfactorily mobilized, but the proximal interphalangeal joint may never regain full movement.

Four procedures are described:

- *Fasciotomy*: Blind or open division of fascial bands. A useful procedure for the frail elderly patient with hand disease.
- *Fasciectomy*: Limited excision of fascial bands and their extension into the fingers. The wound may be left open, or closed with sutures or by skin graft.
- *Open palm technique* (McCash): Through a transverse incision the fascia is excised and the hand splinted with the fingers extended and the wound open. The wound heals in 4–6 weeks.
- *Fasiectomy + skin grafting*: This is indicated for recurrent disease or in patients in whom recurrence is predicted. The skin defect is closed with a full-thickness graft under which recurrence is uncommon.

Prognosis

50 per cent of patients will have further problems over 5 years.

Complications of surgery

20 per cent of patients develop problems: commonly infection, haematoma, and skin necrosis, less commonly digital nerve damage (which needs specialist care) or loss of a finger.

10 Endocrine surgery

Multiple endocrine neoplasia

Multiple endocrine neoplasia (MEN) is a disorder inherited as an autosomal dominant. The cells involved, irrespective of site, are called APUD cells (Amine Precursor Uptake Decarboxylase cells), and are of neuro-ectodermal origin. The condition should always be considered in patients with endocrine syndromes. They are also thought to be responsible for the ectopic hormone secretion of certain tumours, e.g. ACTH or serotonin by bronchial cancers. All are rare. Three main groups are described.

Men I

This involves the parathyroids, pancreatic islets, pituitary, thyroid, and adrenal cortex in variable degrees. Thus patients commonly present with peptic ulceration. Other presentations include acromegaly (human growth hormone, HGH) or watery diarrhoea (vasoactive inhibitory peptide, VIP secretion).

Diagnosis is made by confirming excess gastrin or parathyroid hormone (PTH) secretion.

Treatment is surgical. The whole family should be screened.

Men II

Triad of parathyroid hyperplasia, phaeochromocytoma, and medullary carcinoma of thyroid. Two subgroups are recognized:

Men IIa Medullary carcinoma predominates. In 30 per cent of patients there is either a phaeochromocytoma or parathyroid hyperplasia or both.

Men IIb Medullary carcinoma and phaeochromocytoma predominate. The patient also has neurological abnormalities—multiple neuromas affecting neuronal surfaces and the gut. The appearance is of nodular skin with a marfanoid body habitus.

Diagnosis elevated plasma calcitonin with marked elevation of plasma calcitonin after injection of IV Peptarlon ICI. Exclude and treat phaeochromocytoma first (see p. 400).

Treatment Thyroid medullary carcinoma is treated by total thyroidectomy. Doxorubicin for metastatic disease. Monitor response by plasma calcitonin levels.

Carcinoid syndrome

This is due to excessive serotonin secretion by an argentaffin (carcinoid). Common sites are the appendix and small intestine.

Symptoms Occur only in advanced disease with liver metastases. Intermittent attacks of flushing, diarrhoea, and bronchospasm are characteristic. Blood pressure variations, cardiac valvular lesions, peripheral oedema and cutaneous telangiectasia are also features.

Diagnosis suspected by increased urinary excretion of 5-hydroxy indole acetic acid, a metabolite of serotonin.

Treatment Excision of tumour in small bowel, mesentery, and appendix. Excision of localized hepatic metastases (hepatic lobectomy).

Prognosis variable.

Asymptomatic carcinoid tumours These are the commonest tumours found in the appendix usually coincidentally after appendicectomy. They may also occur in the small bowel. Provided there is no spread and no evidence of the carcinoid syndrome local excision or appendicectomy is adequate as the tumour may eventually bleed or lead to obstruction as it grows.

Anterior pituitary disorders

The anterior pituitary develops from the epithelium of Rathke's pouch, part of the embryological pharnyx. It is connected by a portal venous system to the hypothalamus which permits the transfer of hormonal messengers from the hypothalamus to stimulate or inhibit the endocrine tissue of the pituitary.

Hypothalamic messengers

Thyroid stimulating hormone releasing factor, luteinizing hormone and follicle stimulating hormone releasing factor, prolactin inhibiting factor, growth hormone releasing and inhibiting factors.

Anterior pituitary hormones

ACTH (adrenocorticotrophic hormone), TSH, gonadotrophins, HGH (human growth hormone), prolactin.

Disorders of the anterior pituitary

Most are due to adenomas which may result from hypothalamic overstimulation. As they expand they may compress the optic chiasma leading to visual defects. Spontaneous haemorrhage into an adenoma can lead to hypopituitarism.

Diagnosis skull X-ray, CT scan. Three syndromes are amenable to surgery.

Acromegaly

Due to excess HGH secreted by an acidophilic adenoma, rarely a chromophobe adenoma.

Effects before closure of epiphysis there is proportional bone growth resulting in pituitary gigantism. After closure of epiphysis there is periosteal overgrowth, cortical thickening, protrusion of the jaw (prognathism). It is said that Goliath may have suffered from acromegaly. David was able to approach closely because of the giant's poor peripheral vision (limited by overgrowth of the lateral aspects of his orbits) and so kill him with an accurately directed stone from his sling. The soft tissues are hypertrophied. The hands and feet enlarge. The skin is thickened with acne and hypertrichosis is common. There is myasthenia, glycosuria, hypertension in 20 per cent of patients and galactorrhea in some women.

Diagnosis Characteristic appearance. Skull X-ray shows thickened frontal sinuses, expansion of sellaturcica. Fasting HGH levels greater than 10 μg/ml.

Treatment Supervoltage irradiation, radioactive implant, transphenoidal excision of adenoma.

Cushing's disease

This is due to a functional basophilic or chromophobe adenoma which results in pituitary ACTH excess.

Features Moon face, plethoric appearance, truncal obesity with dorsal cervical fat pads—'buffalo-hump'. Myasthenia, muscle atrophy. The skin is atrophic with purple striae. Wounds heal poorly.

Hypertension, renal calculi, glucose intolerance, psychiatric features are common.

Diagnosis Characteristic appearance. Plasma cortisol elevated with loss of diurnal variation. Synacthen test: Tetracosactrin 250 μg IV after breakfast. Plasma cortisol levels are measured at 30 min, 60 min, and 5 hours later. An exaggerated increase is seen in pituitary disease.

Treatment surgical excision of adenoma.

Hyperprolactinaemia

This causes galactorrhea and amenorrhoea. Usually due to a chromphobe adenoma (Forbes–Albright syndrome). It may occur after pregnancy (Chiari–Frommel syndrome).

Diagnosis Basal levels of serum prolactin are high. Response to thyrotrophin releasing hormone (TRH) and metoclopramide is reduced. Skull X-ray and CT scan.

Treatment Surgical for small tumours. Bromocriptine to inhibit prolactin release (large tumours).

389

Thyrotoxicosis

There is overproduction of T_3 and T_4. These are responsible for the clinical features of the condition.

Types	Pathophysiology
Graves' disease (antibodies against TSH receptors → diffuse hyperplasia (75 per cent)	Antibodies against TSH etc. here. Autoimmune types may rarely present with elevated T_4 levels
Toxic multinodular goitre (20–25 per cent)	Several nodules in a non-toxic goitre begin to function independent of TSH stimulation
Solitary toxic adenoma (1 per cent)	Functions autonomously

Clinical features

8 females : 1 male. Peak age 30–50 years. There is intolerance of warm atmospheres, sweating, increased appetite, weight loss, frequent defecation, oligomenorrhoea, tremor, emotional lability, mental irritability.

Signs

Tachycardia (even when asleep), palpitations, atrial fibrillation, cardiac failure in older patients, fine tremor, increased tendon reflexes, exophthalmos, lid lag (as the patient follows the examiner's finger downwards the upper lid lags behind the eye), lid retraction, general hyperactivity. Palpable thrill occasionally over thyroid. Bruit to auscultation. Pretibial myxoedema.

Diagnosis
- Often clinically evident
- Elevated T_3 and T_4 with low TSH
- Thyroid radioisotope scan will confirm a 'hot' toxic nodule in a normal or suppressed gland.

Treatment

Antithyroid drugs Give carbimazole (Neo-Mercazole) in an initial daily dose of 30–60 mg by mouth until the patient is euthyroid. Then reduce gradually to a maintenance dose of 5–15 mg daily. It is not uncommon for hypothyroidism to develop and this should be avoided especially in pregnant patients when fetal goitre may result.

Prognosis 70 per cent relapse within 2 years.

Radioactive iodine (^{131}I) This should be reserved for those over 40 years unless there are obvious cosmetic or pressure effects when surgery will be indicated.

Prognosis 80 per cent are hypothyroid after 10–15 years. Monitoring is lifelong. Thyroxine is indicated for hypothyroidism.

Surgery The treatment of choice for those under 40. Prepare the patient with β-adrenergic blocking drugs like propranolol 10–40 mg 3–4 times daily by mouth to abolish increased sympathetic activity. Antithyroid drugs may also be given as above. Iodine should be given for 2 weeks before partial thyroidectomy with carbimazole in a dose of 0.1–0.3 ml 8-hourly by mouth (aqueous iodine solution). This will reduce the vascularity of the thyroid.

N.B. Propanolol in a dose of 40 mg 6 hourly by mouth, is indicated at any age to control symptoms in the first month of treatment.

Operations

- Graves' disease
- Toxic multinodular goitre   bilateral subtotal thyroidectomy
- Solitary toxic adenoma: total lobectomy

Results

95 per cent thyrotoxic patients are cured by surgery. 5 per cent recur.

391

Complications of surgery

(see operative surgery p. 832).

Hyperparathyroidism

This is caused by an increased secretion of parathyroid hormone (PTH).

Primary hyperparathyroidism

Is an unstimulated PTH excess by an adenoma inappropriate to the serum calcium level (which is not low). Prevalence is 1:800 commoner in women. Occurs between the ages of 20–60 years.

Clinical features relate to the effects of hypercalcaemia. The commonest presentation is the detection of asymptomatic hypercalcaemia on routine biochemical screening. 50 per cent of patients suffer from bone disease (osteitis fibrosa cystica, cysts, pseudotumours), renal stones, peptic ulcer or pancreatitis or psychiatric symptoms: 'bones, stones, abdominal groans, and psychic moans'.

Risk factors neck irradiation, hyperthyroidism.

Associations Pancreatic islet cell tumour and pituitary adenoma (multiple endocrine neoplasia (MEN type 1), Medullary carcinoma of thyroid and phaeochromocytoma (MEN type 11a) (see p. 386). Duodenal ulcer, pancreatitis, hypertension.

Investigations Clinical investigation is usually negative except in patients with severe bone disorders.

Biochemical
- Elevated serum calcium, reduced serum phosphates. Fasting patient. Do not use a tourniquet when taking blood.
- Increased excretion of calcium in urine (24-hour specimen)
- Increased alkaline phosphatase in bone disorders
- Elevation of serum PTH. Collect blood in plastic syringe after overnight fast. Transport in a heparinized ice-cooled tube.

Other X-rays of chest, skull, pelvis, hands.

Treatment Surgery if patient has symptoms. Preoperative localization of the tumour can be made by:
- Selective venous sampling
- Neck ultrasound
- CT scanning (useful for glands in mediastinum)
- Thallium–technetium scan may locate up to 95 per cent of adenomas

All four glands must be identified, studied, biopsied if necessary, and the adenoma(s) removed. 10 per cent will be missed. Give calcium and vitamin D supplements postoperatively to prevent tetany.

Secondary hyperparathyroidism

The stimulus is chronic hypocalcaemia from chronic renal failure or malabsorption syndromes. All four glands are hyperplastic. Treat low calcium with oral calcium. Reduce plasma phosphate with oral aluminium hydroxide.

Tertiary hyperparathyroidism

Is a consequence of secondary hyperparathyroidism. Autonomy occurs usually due to an adenoma. Treat as for primary hyperparathyroidism.

Myasthenia gravis

This affects 1 in 30 000 adults. There is chronic weakness, fatiguability, and paralysis of voluntary muscle related to failure of transmission of motor impulse at the myoneural junction due to too few functional acetylcholine receptors.

Associations

Thyrotoxicosis, Hashimoto's thyroiditis, rheumatoid arthritis, systemic lupus erythematosus, carcinoma of lung, thymitis, thymoma.

Clinical features

The commonest feature is weakness of the extraocular muscles in young adults. There is ptosis, diplopia, and squint. Muscles of deglutition, the neck, and upper limbs are next affected. The disease is chronically progressive. Untreated, patients will die of respiratory paralysis.

Physical fatigue, emotional upset, and medical interventions such as enemas or drugs exacerbate symptoms. Pregnancy may lead to a temporary remission followed by an exacerbation.

Confirmation

Tensilon test Give 10 mg IM or 2 mg IV followed by 8 mg IV after 30 seconds. If the symptoms improve dramatically, but briefly, the diagnosis is confirmed. Neostigmine (2 mg) + atropine (0.65 mg) may also be used but the response is much slower.

Treatment

Medical 80 per cent will be controlled. Give neostigmine (Mestinon, Prostigmin) tablets (15 mg) orally, timed so that the patient can eat meals. Give doses of 15–30 mg to a total daily dose of 5–20 tablets according to needs.

Side-effects include diarrhoea, vomiting, and abdominal cramps. Pyridostigmine (0.3–1.2 g/day orally) causes less diarrhoea and vomiting.

Surgical Thymectomy. 30 per cent of patients remit. Up to 60 per cent improve. Best results are in women with a short history who do not respond to medical measures.

Cushing's syndrome

Cushing's syndrome is caused by chronic and inappropriate secretion of cortisol (glucocorticoid).

Causes
- Functioning tumour of adrenal cortex (20 per cent)
- Ectopic ACTH production (small cell carcinoma of lung)
- Cushing's disease due to excess ACTH from pituitary basophil or chromophobe adenoma (see p. 388)
- Alcoholism

Clinical features

Commoner in young women. Centripetal obesity, buffalo-hump, moon face, myopathy, myasthenia, skin atrophy, acne, poor wound healing, skin striae, osteoporosis, water retention.

Investigations
- Plasma cortisol. The diurnal variation and low-dose dexamethasone suppression or cortisol are lost. In the low-dose test dexamethasone (which suppresses ACTH production under normal circumstances) is given 0.5 mg IM 6-hourly for 48 hours. Urinary and plasma cortisol are measured on the second and third days respectively. Normal values are less than 170 nmol/24 hours and less than 170 nmol/l. In Cushing's syndrome suppression is absent and the values remain high.
- Insulin-induced hypoglycaemia fails to cause an elevation in plasma cortisol.
- ACTH absent in plasma in adrenal tumours.
- ACTH is high in pituitary causes. Dexamethasone suppression of urinary cortisol occurs.
- Chest X-ray, abdominal X-ray, intravenous urography, skull X-ray, CT scan

Treatment
Adrenal tumour
- Unilateral adrenalectomy. Cortisone is given postoperatively and gradually reduced until remaining adrenal resumes function.

Pituitary tumour
- Microsurgical removal of adenoma
- Bilateral adrenalectomy but permanent replacement therapy is required. The pituitary adenoma may cause visual problems.
- Internal or external irradiation of pituitary

Prognosis

Death in 5 years if no treatment. Good results for benign disease. Poor prognosis in carcinomas.

Hyperaldosteronism

Primary hyperaldosteronism is often due to a benign functional adenoma of the adrenal cortex (Conn's syndrome). Other causes include adrenal and ovarian carcinoma and bilateral adrenocortical hyperplasia.

Features

Middle-aged females who develop hypertension, hypokalaemia, sodium retention, and alkalosis, all the result of excess aldosterone production independent of the renin angiotensin system.

Diagnosis

Prove:
- Hypokalaemia. Take repeated samples of venous blood without a tourniquet.
- Elevated plasma aldosterone by radioimmunoassay (no diuretics)
- Reduction of hypertension with spiranolactone (aldosterone antagonist)
- Absent plasma renin
- Presence of adenoma by CT scan or selective venous sampling for aldosterone levels to identify the side

Treatment

Unilateral adrenalectomy.

Preoperative care

Give spironolactone 100 mg three times daily orally for 1 month preoperatively.

Secondary hyperaldosteronism

Due to excess renin production which in turn activates the adrenal zona glomerulosa to produce excess aldosterone. It is caused by renal artery stenosis, cardiac disease, hepatic cirrhosis, accelerated hypertension and diuretics.

Treatment is that of the underlying cause.

399

Phaeochromocytoma

This is a benign tumour of the adrenal medulla (90 per cent) or extra adrenal paraganglionic (10 per cent) tissue. Adrenal medullary tumours secrete large amounts of adrenaline and noradrenaline. Extra-adrenal tumours secrete noradrenaline. The tumours may be multiple.

Associations

Renal artery stenosis, medullary carcinoma of thyroid, neurofibromatosis, duodenal ulceration (MEN type II) (see p. 386).

Symptoms and signs

Headache, palpitations, sweating, pallor, nausea and vomiting and dyspnoea occur for periods lasting around 15–20 minutes. There is paroxysmal hypertension associated with weight loss and hyperlgycaemia.

Differential diagnosis

Hyperthyroidism (thyroid swelling rarely occurs in phaeochromocytoma), hypocalcaemia, carcinoid syndrome, paroxysmal atrial tachycardia, acute anxiety syndromes.

Investigations

- Vanillylmandelic acid (VMA) estimations on 3×24-hour urine specimens. The level is elevated in phaeochromocytoma (N = 7 mg/24 hours). 24 hour meta-adrenaline secretion can also be used
- Radiology. Abdominal X-ray, intravenous urography arteriography, CT scan detect more than 80 per cent
- Venous sampling detects more than 95 per cent

Treatment

Surgical excision. Severe postoperative hypotension is a real risk due to a contracted vascular bed, the result of catecholamine excess. Prevent this by controlling hypertension with phenoxybenzamine and propanolol for 10–14 days preoperatively. Measure the plasma volume 4 days before surgery. Do not use atropine for premedication. Use alternative induction agents to thiopentone. Tumours may be multiple so both glands may have to be explored.

Acute attack

Patient typically says 'I am going to die'. There is pallor, palpitation, pulsating headache.

Treatment

- Phentolamine 5–10 mg IV
- Atropine for tachycardia greater than 120/min—1 mg IV
- Propranolol 1 mg IV slowly. Do not give first as may precipitate cardiac failure.
- Control is then maintained by oral propranolol 20 mg 8-hourly.

Prognosis

Elective treatment gives good results. Patients may live a normal life. Acute attack can be a cause of sudden death after trauma or surgery. Keep the possibility in mind.

11 Orthopaedic surgery

Examination of a joint

Develop an ordered system to avoid missing vital clues.

Ask the patient where the problem is. Can he point to it specifically? Does he indicate a vague area of discomfort? Can he move the joint actively?

Look at the affected joint. Is there swelling? Is the skin broken? Is there obvious deformity? Compare it to the other side. Is there muscle wasting? Is there a lump? Does the joint look normal?

Feel Compare sides. Examine the good side first, then the injured joint. It the skin normal, cold, moist, etc? Can the patient feel you touch him? Does it feel normal? Look at the patient's face for grimacing. Is there tenderness on palpation? If so, is it localized? The diagnosis may be immediately apparent. Is there fluid or an effusion? Test for fluctuation by alternatively pressing the skin with the examining fingers on either side of the joint. Can abnormal movement be demonstrated by stressing the ligaments?

Move

Test active and passive movement. Ask the patient to move the joint. Note whether the movement looks normal or if there is difficulty, discomfort or pain. What range of movement can he move the joint through?

Passive movements are carried out by the examiner. Move the joint through its anatomical range first. Are the movements full? Is there pain? Make a note of the range and compare it to the good side. Now move the joint outwith its normal anatomical range to stress its ligaments. Do this gently. Is there excessive or abnormal movement? Is there pain?

Test for muscle power. Use the same system always. State clearly what you are going to do. Lift the limb, tell the patient to hold the position, then to keep the position while you push against the limb. Do this for different muscle groups. Examine the good side first and compare it with the other.

Now ask the patient to perform specific activities like walking, hopping, skipping for the lower limb or reaching for objects, lifting weights, etc. for the upper limb. Note how well each task is performed and decide whether function is abnormal.

Examination of limbs and trunk

Develop a system. Compare arm with arm and leg with leg. Make allowances for the patient's 'handedness'.

Follow the order: muscles and joints, power, co-ordination, reflexes, sensation.

Muscles and joints

Look for deformity, asymmetry (often due to muscle wasting), twitching (fasciculation). Move each joint through a full range of active and passive movements. Learn to recognize normal 'feel' and that due to flaccidity, increased tone, or spasm. Is the tone the same throughout the whole movement? Is there rigidity (resistance throughout the whole movement) or spasticity (only initially)? Is there limitation of movement due to joint disease or muscle contracture? Are there involuntary movements, e.g. twitch, tremor? Are they regular or modified by movement or relaxation of the limb?

Power

Test the patient's ability to carry out active movements independently and against resistance. Test individual muscle groups, the intactness of their nerve supply and their spinal segment, e.g. biceps and brachialis, musculocutaneous nerve, C_6.

Co-ordination

Ask the patient to touch his nose with his index finger with his eyes first open then shut or to apply the contralateral heel to his shin passing it downwards to the foot then to alternately pronate and supinate his forearms as quickly as possible. Note whether these movements are performed smoothly or jerkily. Disorders of co-ordination may occur generally in e.g. demyleninating diseases or locally following injury.

Reflexes

Test the following with a tendon hammer:

Upper limbs
- Biceps (C_6)
- Triceps (C_7)
- Brachioradialis (C_6)

Lower limbs
- Knee (L_3, L_4)
- Ankle (S_1)

Plantar responses the normal is flexor. Extensor is abnormal (Babinski's sign).

Grade the reflexes 0 (absent), 1 (diminished), 2 (normal), 3 (increased), 4 (clonic).

Sensation

Test:
- Pinprick sensation, touch, and temperature. Map out any abnormalities found.
- Vibration sense (tuning fork 128 c/s applied over bony prominences)
- Proprioception: the patient shuts his eyes. Move joints up, down, to neutral, etc. Ask the patient to say which and what position.
- Deep pain: Pressure on the Achilles tendon is normally painful

Fracture healing

Three stages of fracture healing are described:
- Fracture haematoma and granulation tissue
- Callus formation
- Ossification

When a bone is broken, a variable amount of haematoma forms between the bone ends and extends into the surrounding soft tissue. The size of the haematoma depends on the degree of violence causing the fracture and may continue to expand for 12–36 hours, leading to complications because of pressure.

The haematoma then becomes permeated with capillary loops and fibroblasts to form granulation tissue. Osteoblasts from the elevated periosteum at the fracture site proliferate into the mass to form callus. This subsequently becomes partly calcified to form woven bone. How much callus is formed depends on local factors like the type of fracture, the proximity of the bone ends, movement of the bone ends, and the amount of haematoma.

New blood vessels migrate into the callus leading to calcification and subsequent ossification as a result of osteoblastic activity. The first radiological evidence of healing occurs with the onset of calcification which usually takes 3–4 weeks. There then follows a period of consolidation and the deposition of lamellar bone. The final stages are of bone remodelling. The swelling around the fracture site decreases and bony trabeculae can be seen on X-ray crossing the fracture site. Remodelling is most marked in infants and children.

Factors adversely affecting fracture healing

- Displacement of the bone ends
- Periosteal interposition in the fracture site
- Inadequate mobilization
- Infection
- Disturbances of ossification, e.g. osteomalacia, osteoporosis, metastatic tumour
- Site of fracture, e.g. femoral neck, distal third of tibia, carpal scaphoid, mid-humerus. Non-union and delayed union are common at these sites.

Delayed union If movement occurs at the fracture site, cartilage will tend to be formed instead of bone. If the blood supply is adequate the fracture will heal especially if firm fixation is achieved.

Non-union This occurs when the blood supply is poor at the fracture site, e.g. femoral neck fractures, waist of carpal scaphoid. Healing takes place by the formation of fibrous tissue. Bone necrosis may occur.

Fixation of fractures

There are three methods: internal fixation, external fixation, and external support from a cast or traction.

Internal fixation
Indications
- Intra-articular fractures. Osteoarthritis is common unless anatomical reduction is achieved. Fractures of the ankle, knee, and elbow are commonly treated by anatomical reduction and fixation.
- Repair of vessels or nerves. Fracture stability is essential for good results following neurovascular damage with surgical repair.
- Patients with multiple injuries to facilitate nursing care and patient mobility
- Elderly patients tolerate immobilization badly, e.g. fractures of the femoral neck.
- Long-bone fractures, e.g. radius and ulna in adults, femur. The aim is to achieve anatomical reduction in the forearm and early mobility with reduction of time in hospital.
- Failure of conservative treatment
- Pathological fractures

Methods used Plates and screws, intramedullary nails, tension banding, Kirschner wires.

Complications Osteomyelitis, nerve and vascular injury, fracture of the implant, fracture through a screw hole once the screw is removed.

External fixation

The main indications are compound fractures, especially of the tibia. Soft tissue loss can be managed by skin grafts, rotation, cross-leg flaps or free vascularized flap techniques. When the skin defect has healed the limb may then be immobilized in a plaster or brace.

N.B. External fixation is of value in major pelvic ring disruptions associated with urinary tract injuries and in the management of unstable neck injuries.

Complications Bone infection related to the insertion of transosseous pins, pressure necrosis of the skin because the device is too close to the skin.

Cast bracing

This method has the advantage of producing early fracture healing whilst the joints of the affected limb are kept mobile. Muscle shortening is also prevented. Fractures commonly treated in this manner are closed tibial fractures and femoral shaft fractures. The patient with a tibial fracture is first immobilized in a plaster of Paris cast for about 2 weeks until swelling subsides. Treatment is then with a functional brace or Sarmiento plaster. Most simple fractures will unite in 3 months.

The patient with a femoral shaft fracture may be treated conservatively for about 1 month on skeletal traction to achieve alignment. After this a brace may be applied, weight-bearing encouraged, and mobility reacquired.

N.B. The technique can be used for fractures of the lower and mid femoral shaft. A waist band must be added to prevent varus deformities of upper femoral shaft fractures.

Some upper limb fractures (ulna, humerus) may also be treated by functional cast bracing.

Reference

Newman, J. H. (1986). New developments in fracture management. In *Recent advances in surgery*, **12**, (ed. R. C. G. Russell), pp. 199–213. Churchill Livingstone, Edinburgh.

The skeletal X-ray

This is the most important investigation in making a diagnosis in disease or injury of bone.

Radiological features

Diffuse disease

Osteoporosis is the commonest form of metabolic bone disease. It affects middle-aged and elderly women, predisposing to fractures of the distal radius, femoral neck, and vertebral bodies. Localized osteoporosis follows disease. The cortices are thin with reduced medullary trabeculae. The bone looks normal but there is just too little of it.

Osteomalacia There is reduced mineralization of osteoid. The trabeculae are blurred. Symmetrical transverse or oblique cortical defects appear (Looser's zones, pseudofractures). In children changes are most marked at the metaphysis (rickets).

Hyperparathyroidism There is bone resorption. Look for it in the phalanges of the hands in the subperiosteal cortex. Note generalized cortical striations. Carry out parathormone levels and iliac crest biopsy.

Diffuse increased density Think of neoplasms, fluorosis sarcoidosis, uraemia osteodystrophy, bone dysplasias (osteopetrosis—Albers–Schonberg disease).

Abnormalities of bone modelling

Developmental disorders e.g. osteochondrodysplasias. The changes are often present from birth. Look for abnormalities of eye, ears, heart. Carry out biochemical and genetic studies.

Localized abnormalities may occur in congenital disorders, e.g. enchondromatosis (Ollier's disease), fibrous dysplasia, neurofibromatosis or acquired disorders, e.g. Paget's disease.

Solitary lesions (sepsis, primary tumours, and metastases are causes). Location and age are important, e.g. an epiphyseal lesion in a child may be a chondroblastoma. A lesion in the subarticular region in a young adult may be a giant cell tumour, in an older patient a metastasis.

Guidelines for evaluating the skeletal X-ray

- Note the history, race, occupation, pastimes, age, sex, and recent laboratory results.
- Clearly state the site to be X-rayed on the request form.
- Take two projections, preferably perpendicular to each other.
- Examine the films in a viewing box. Have a bright light and magnifying glass for fine detail.
- Note the character of the lesion. Is it destructive/sclerotic?
- Look for cortical/medullary changes, periosteal reaction, deformity, soft tissue swelling and fracture.
- Supplement radiological findings with biochemical investigations, bone scanning, and biopsy if indicated.

Reference

Stoker, D. J. (1988). Interpreting the skeletal X-ray. *British Journal of Hospital Medicine*, **39**, 143–52.

413

Osteoarthrosis (osteoarthritis)

This is the process of joint degeneration secondary to disease of articular cartilage. It is limited to the joint only. There is no systemic upset. It can involve central and peripheral synovial joints, and is the most common form of musculoskeletal disease.

Types

Primary osteoarthrosis affects especially the following joints: distal interphalangeal, first carpometacarpal, the hips, knees, and apophyseal joints of the spine. Women are more commonly affected. Heredity may be involved.

Secondary osteoarthrosis affects previously damaged joints and is more common in weight-bearing joints. Both sexes are equally affected. Local causes are fractures, acquired or congenital deformities, joint injury, diabetic neuropathy, and avascular necrosis.

Clinical features

Pain, stiffness, deformity. Most patients complain of aching pain and stiffness of the affected joint which becomes steadily worse throughout the day. Pain in major joints often disrupts sleep. The onset of primary generalized osteoarthrosis may be acute. The involved joints may be hot and painful and the ESR raised. Look for Heberden's nodes at the distal interphalangeal joints and Bouchard's nodes at the proximal interphalangeal joints.

X-ray features

There is loss of joint space, subchondral bone sclerosis, and cyst formation (especially in the hip). Osteophytes are seen at the joint margins. Physical symptoms do not correlate with the severity of radiological change.

Treatment

Relieve pain, improve mobility, correct deformity.

Medical treatment
- Pain relief and increased mobility are effected by non-steroidal anti-inflammatory drugs. Beware of GI side-effects, especially in the elderly.
- Radiant heat in the form of infrared light or a hot-water bottle can be effective. Short-wave diathermy penetrates more deeply.
- Other measures include weight loss, walking sticks, raised heel, self-raising chair to facilitate standing up, household aids, e.g. stair lift, muscle-strengthening exercises.

Surgical treatment is indicated for pain relief and improvement of joint function. Techniques include osteotomy, joint excision, arthrodesis, and joint replacement.

Carpal tunnel syndrome

This results from compression of the median nerve under the flexor retinaculum of the wrist. It is commoner in women than men (6:1) and rare in children. It is often bilateral, but when unilateral may affect the dominant hand.

Age incidence

30–60 years.

Aetiology

Compression of the tunnel walls
- Trauma (Colles' fracture, vibrating machinery)
- Rheumatoid arthritis
- Subluxation of the wrist
- Acromegaly

Compression within the tunnel
- Fluid retention of pregnancy
- Myxoedema
- Benign tumours
- Chronic proliferative synovitis

Changes in the median nerve
- Diabetes mellitus
- Peripheral neuropathies
  Hypertrophy of the flexor retinaculum is common.

Clinical features

Numbness, 'pins and needles' over the radial 3.5 fingers. Pain at night, on flexion of the wrist, and after repetitive movement. There may be weakness and wasting of the thenar muscles in 25 per cent of patients, with complaints of weakness in the hand when holding a book or writing.

Clinical examination
- Decreased sensation over thumb and index
- Weakness of thumb abduction
- Delayed conduction time in the median nerve at the wrist (Tinel sign)
- Exclude cervical spondylosis or other cervical abnormality (cervical ribs, tumour, cysts) by plain cervical X-ray.

Treatment

Mild symptoms (60 per cent of patients)
- Splintage
- Corticosteroids
- Diuretics
- Restricted activity

Severe or persistent symptoms (40 per cent)
- Surgical decompression under regional or general anaesthetic
- Use a tourniquet

- Protect the motor branch to the thenar muscles by making the incision on the ulnar aspect of the thenar crease.
- Divide the carpal ligament under direct vision to completely decompress the nerve.

Regional anaesthesia may be accomplished by axillary nerve block when local anaesthetic is infiltrated around the cords of the brachial plexus in the axilla taking care not to damage the axillary artery or vein. It may also be accomplished by regional intravenous anaesthesia. Both techniques should be observed and undertaken initially under strict supervision.

Complications

- Persistent weakness for several months (warn the patient)
- Pain in the scar (due to neuromata)
- Recurrence (due to incomplete division of flexor retinaculum or fibrosis in tunnel)
- 10 per cent of patients have little or no improvement

Fractures and dislocations of the metacarpals and phalanges

Thumb

Causes Blows to the point of the thumb. Forced opposition of the thumb.

Fractures Transverse at base of first metacarpal.

Treatment Reduce under traction. Apply a plaster of Paris splint with the distal phalanx free for a month.

Bennett's fracture dislocation There is an oblique fracture of the metacarpal base which involves the articular surface. A small fragment of bone remains in position and the metacarpal shaft subluxates proximally.

Treatment Easily reduced with traction on the thumb but difficult to hold. Methods include the application of plaster of Paris splint with padding over the fracture, continuous skin traction with the thumb abducted in a plaster of Paris splint. If the X-ray still shows displacement internal fixation is indicated and is often the method of choice. Treatment is for 3 weeks.

Treat other dislocations by reduction and strapping for 3 weeks to a neighbouring finger.

Metacarpal fractures

Cause Usually due to punching someone or something.

X-ray Neck fractures usually show forward angulation. Transverse fractures of the shaft are often displaced or angled. Spiral fractures may overlap, but the alignment is usually satisfactory.

Treatment
- *Single fractures*: Strap the injured finger to its neighbour. Apply a backslab. Encourage early active mobilization. Reduce only if the patient does not want a 'dropped' knuckle.
- *Multiple fractures*: Reduce and fix internally. Elevate the limb. Ask the physiotherapist to encourage active movements.

Phalanges

Terminal phalanx

Cause Crushing injuries.

Look The fracture is often compound.

Treatment Wound toilet, avoid primary skin closure, apply a pressure dressing. Change at 48 hours to inspect skin. Give tetanus toxoid and broad-spectrum antibiotics on admission.

Mallet finger There is avulsion of the extensor tendon from the terminal phalanx with or without a fragment of bone.

Cause Forced flexion of the terminal phalanx. Occurs often in cricket.

Look The patient cannot actively extend the terminal phalanx.

X-ray A bony fragment may be seen.

Treatment Suture an avulsion fragment back into position. If there is no fragment apply a stack mallet splint for 6 weeks.

Proximal and middle phalanges These are often unstable.

Cause Direct blow, twisting injuries.

Treatment
- *Stable fractures*: strap the affected finger to its neighbour for four weeks. Encourage active movements.
- *Unstable fractures*: require internal fixation with crossed K wires

Immobilization of the hand to minimize post-traumatic stiffness

The interphalangeal joints are immobilized in extension and the metacarpophalangeal joints in flexion.

Injuries to the wrist (carpal/metacarpal bones)

Scaphoid fracture

Is there a history of forced dorsiflexion of the wrist?

Examination
- The wrist often looks normal. There may be fullness and tenderness to light pressure in the anatomical snuff box and over the anterior surface of the scaphoid.
- Examine the hand and arm to exclude other injuries.
- Compare wrist movements. They are present but reduced on the injured side.

X-rays Take AP, lateral, and oblique views. The bone fractures at its waist or proximal pole and may be seen on the oblique view. If there is no obvious fracture but the history is suggestive take further films at 2–3 weeks and treat the patient as if it were a fracture in the meantime.

Treatment Scaphoid plaster extending from below the elbow to the knuckles and around the proximal phalanx of the thumb. The wrist is dorsiflexed. The hand position is as though holding a glass. Remove and re-examine at 8 weeks. If still tender re-apply for a further 4 weeks. Advise active shoulder and elbow exercises immediately.

Complications
- Non-union: treat with bone graft or double-threaded (Herbert) screws
- Osteoarthritis

Other carpal bones

These account for less than 10 per cent of fractures of the wrist. The most common are:

Triquetral a lateral film may show the fragment over the dorsum of the wrist. Treat in a Colles' type plaster of Paris cast for 3 weeks.

Hamate The hook may be fractured. Can occur in racket sports, cricket, or golf. Non-union is common and may be a source of persistent pain. Treat by excision of the hook.

Dislocations

Perilunar or lunar
- In perilunar dislocation the lunate stays where it is and the rest of the carpus dislocates dorsally.
- If the carpus then slips back into place (as it often does) this can cause forward dislocation of the lunate and acute median nerve compression.

Treatment General anaesthetic. Dorsiflex and apply traction to the hand. Push the lunate back into position.

Transcaphoid This is rare. Open reduction and internal fixation is often indicated.

Carpometacarpal/fracture dislocations may be seen on the lateral film. Clinically there is tenderness over the base of the injured metacarpal bone.

Treat by maintaining reduction with Kirschner wire which transfixes the joint.

Injuries of the wrist (distal radius and ulna)

Establish the mechanism. Exclude scaphoid fractures which result from a fall on to the outstretched dorsiflexed hand. Remember to look for other injuries if the mechanism is direct trauma to the radial or ulnar side of the wrist.

Types of injury

Fractures
- Distal radius (Colles', Smith's)
- Separation of lower radial epiphysis
- Ulnar/radial styloid
- Scaphoid
- Other carpal bones

Dislocations
- Perilunar
- Lunar
- Transcaphoid perilunar
- Carpometacarpal
- Fracture dislocation

Distal radius

Colles' fracture Common in middle-aged women who fall on the outstretched hand.

Look: Dinner-fork deformity. All wrist movements are painful.

X-ray: Transverse (usually impacted) fracture of distal radius, tilted backwards and radially. The ulnar styloid may also be fractured.

Treatment:
- *Undisplaced*: Plaster of Paris with wrist slightly dorsiflexed for 3–4 weeks
- *Displaced*: General anaesthetic. Disimpact the fracture by traction and increasing the deformity. Then palmar flex and firmly pronate the wrist. Press the lower radius to maintain reduction. Apply a POP with the wrist flexed, pronated and in ulnar deviation. Check X-ray post reduction and at one week. Remove plaster at 6–8 weeks.

Smith's fracture (reversed Colles') The deformity is the reverse of a Colles' fracture.

X-ray The fracture is often impacted. The distal fragment is displaced forwards and radially.

Treatment General anaesthetic. Reverse the reduction procedure for Colles' fracture. Apply an above-elbow plaster of Paris splint for 6–8 weeks with the fracture held dorsiflexed and in strong supination.

Separated epiphysis (juvenile Colles') The deformity is similar to a Colles' fracture.

Treatment Reduce as for a Colles' fracture. Accept some deformity rather than damage the growth plate. If the growth plate has been crushed, differential radial and ulnar growth may result and excision of part of the ulna may be indicated.

Styloid fractures Reduction is usually not necessary. Apply a Colles'-type plaster of Paris with the wrist in slight dorsiflexion for 3–4 weeks.

Complications

- Mal-union may lead to wrist pain. Relieve by excision of distal end of ulna.
- Late rupture of extensor pollicis longus. Operative repair or re-siting is indicated.
- Post-traumatic stiffness is common. Avoid by good physio-therapy. In Sudek's atrophy the wrist bones show patches of rarefaction. Stiffness persists for months.

Fractures of the radius and ulna

Cause
- Indirect forces from a fall on to the outstretched hand with or without rotation
- Direct trauma usually leads to fractures of a single bone

Types
- Greenstick fractures with angulation (children)
- Transverse or oblique (adults)

Associations
Solitary fractures may be associated with dislocation of the other bone.

Look
There is obvious deformity. The patient cannot move the forearm. Is there apparent shortening of the forearm? This may suggest dislocation. Is the fracture compound?

X-ray
Note the fractures, which are usually at the same level. Check the elbow and wrist for dislocation.

Treatment
Radius and ulna
- Greenstick fractures are usually easy to correct by applying firm pressure.
- Complete fractures are reduced by correcting rotational deformity first, then applying traction and moulding the fracture. Hold the elbow at 90° and apply plaster of Paris from axilla to metacarpal heads. Reduction must be anatomical in adults, but some overlap is acceptable in children. Split the plaster when it has set and elevate the limb. Consolidation takes 6–12 weeks. Use functional bracing after 3 weeks.
- Failed reduction: Proceed to open reduction maintained by a plate and screws.
- Compound fractures: wound excision. The fracture may then be treated by closed means in plaster of Paris or by plating if contamination is minimal. Give tetanus toxoid (or antitetanus serum) and broad-spectrum prophylactic antibiotics which are effective against *Staphylococcus aureus*. (Flucloxacillin 500 mg given IV upon induction of anaesthesia then orally 6-hourly for 48 hours postoperatively).

One bone only
- Minimal displacement: Plaster of Paris as above. Treat as if both bones are broken.
- Open reduction is advocated by many surgeons because of the risk of delayed union if the bone ends do not impinge against each other.

Complications

Ischaemia, mal- and non-union, postimmobilization joint stiffness.

Fracture dislocations

Monteggia fracture dislocation
- Fracture of the upper one third of the ulna
- Forward dislocation of the radial head

Treatment Closed in children. Open reduction in adults. Plaster of Paris from axilla to metacarpal heads. Forearm supinated, elbow at 90°.

Galeazzi fracture dislocation
- Fracture of the lower one third of the radius
- Dislocation of the inferior radio-ulnar joint

Treatment
- Closed reduction + full-length plaster of Paris for 3 months is possible.
- Open reduction with radial plating gives better results.

Fractures around the elbow in children

Cause

Think of epiphyseal injuries as well as fractures. Most are due to a fall on the outstretched hand.

Supracondylar fractures

Types Greenstick, oblique with backward displacement (common). Oblique with forward displacement (uncommon)

Look The child holds the forearm

Feel Check the radial pulse before and after reduction. Test the hand for nerve injury

X-ray Both sides for comparison if you are in doubt.

Treatment

- *Backward displacement*: General anaesthetic. With the elbow slightly flexed apply firm traction for up to 1 minute. Use thumb pressure to engage the distal fragment. Correct tilt or shift. Continue thumb pressure on the distal humerus while the elbow is flexed to 60°. Check the pulse. Keep the elbow flexed with padded elastic strapping and apply a collar and cuff. The fracture unites in about 3 weeks.
- *Forward displacement*: Reverse the above. Apply plaster of Paris back slab.

Difficulties

- *Radial pulse disappears on reduction*. Reduce the degree of elbow flexion. Feel the pulse. If it has not returned, treat the patient in bed and suspend the arm vertically.
- *Displacement cannot be reduced*. Suspend the forearm and hand vertically until swelling settles. Skeletal traction with a pin through the olecranon may be necessary. The fracture may subsequently reduce itself or be reduced by further manipulation.

Complications

- *Ischaemia of the forearm and hand* due to complete or partial obstruction of the brachial artery. Check the radial pulse regularly. If unrecognized, Volkmann's ischaemic contracture can develop.
- *Cubitus valgus*: May require corrective osteotomy (gunstock deformity)
- *Nerve injury*: The median, ulnar or radial nerves may be damaged by bone end or subsequent soft tissue swelling. Check nerve integrity in the hand.

Fracture of the lateral condylar epiphysis

Look and feel There is swelling and tenderness on the outer aspect of the elbow.

X-ray Displacement may be minimal or extensive.

Treatment Open reduction to prevent deformity and traction neuritis of the ulnar nerve. Use a posterior slab from shoulder to wrist with the elbow at a right angle.

Radial neck fractures

Treatment Accept up to 15° of tilt. Try closed reduction first. If this fails, open reduction is indicated. Apply a posterior slab with the elbow at a right angle.

Fractures of the elbow in adults

Humeral condylar fractures

Cause Direct trauma to the point of the elbow. The olecranon is then driven proximally.

Types Medial condylar fracture, intercondylar fracture. Fragments are frequently displaced or fracture comminuted.

Look The elbow is swollen. All movements are painful.

X-ray There is a Y- or T-shaped or medial condylar fracture.

Treatment
- *Medial condyle*: open reduction, screw fixation. Plaster of Paris cast for 5–6 weeks
- *Intercondylar*:
  —Internal fixation (non-comminuted fractures)
  —Collar and cuff after restoration of alignment (comminuted fractures)

Capitulum

Cause A fall on the outstretched hand with the elbow straight.

Look The cubital fossa appears full.

X-ray On lateral view the capitulum is out of line with the radial head, having been sheared off. It sits in front on the lower end of the humerus. There may be an associated radial head fracture.

Treatment
- *Small fragment*: excision
- *Large fragment*:
  —attempt closed reduction and collar and cuff
  —operative reduction if this fails

Olecranon

Cause Direct trauma (comminuted fracture), fall on to hand. Triceps action then fractures the process.

Types Displaced, undisplaced, comminuted.

Look Swelling and bruising behind the elbow. The patient has pain on full extension.

X-ray The type of fracture is usually obvious.

Treatment
- Undisplaced: Immobilize in sling
- Displaced: Internal fixation unless displacement is minor. Aim to restore the extensor mechanism.
- Comminuted: Apply a sling. Treat as a haematoma. Encourage early active movements when pain is relieved.

Head of radius

Cause A fall on the outstretched hand leads to varying degrees of impaction of the radial head against the capitulum depending on the rotation of the forearm.

Types Vertical split, marginal, neck (children), comminuted

Look There may be little swelling but rotation of the forearm is painful.

Feel There is tenderness over the radial head.

X-ray May confirm a vertical split, neck fracture with tilting, marginal fracture with distal displacement.

Treatment
- Vertical split ⎫
- Undisplaced marginal ⎬ Closed reduction. Collar and cuff for 3 weeks. Then active flexion and extension. Rotation is slow to return
- Neck (children): closed reduction
- Displaced or comminuted: Radial head excision ± replacement

Complications of elbow fractures

Joint stiffness, osteoarthrosis, ulnar nerve symptoms.

Dislocations and fracture dislocations of the elbow

Cause

Posterior dislocation results from hyperextension forces after a fall on the outstretched hand.

Anterior dislocation is caused by direct trauma to the back of the forearm: 'side-swipe' injury, e.g. when driving a car with the elbow resting on the open window sill.

Associated fractures

Coronoid process, head of radius, olecranon, medial epicondyle, comminution (side-swipe injury). Exclude these by taking an X-ray.

Treatment

- General anaesthetic. Apply traction to the forearm and gradually increase flexion whilst pressing the olecranon forwards with the thumbs.
- Splint in a split plaster of Paris cast or collar and cuff with the forearm at 90°.
- Treat other fractures conservatively or by open reduction and fixation depending on their site, displacement, and size of fragments.

Complications

- Vascular injury: always check the radial pulse
- Traction neuritis of the ulnar or median nerve
- Myositis ossificans. Forbid passive movements. If X-rays show calcification, rest the elbow in a gutter plaster. X-ray at regular intervals to assess progress.
- Non-reduction. If the diagnosis has been missed, it is still worth while trying to reduce a dislocation up to 6 weeks afterwards. Maintain the elbow in flexion even if complete reduction is not possible. Some useful movement may return.

431

Fractures around the shoulder

Clavicle

Cause Fall on to outstretched hand, fall on to point of shoulder.

Site Commonest is middle third. Fractures at the outer end are associated with coracoid fracture or tears of the coraco-clavicular ligament.

Look The proximal portion is drawn upwards by sternomastoid, the distal end droops due to the weight of the arm.

Treatment
- Middle third: figure-of-eight bandage for 3 weeks. Tighten regularly
- Outer third: triangular sling
- Ligamentous disruption: surgical repair

Complications Poor union, non-union (fix with a small metal plate), deformity, neurovascular damage, neurovascular compression (costoclavicular syndrome). Pneumothorax from bony penetration of the pleura.

Scapula

Cause Direct trauma.

Types Usually comminuted.

Associations Fractured ribs.

Treatment Collar and cuff. Early mobilization to prevent stiffness.

Upper end of humerus

Cause Most are due to indirect forces resulting from a fall on the outstretched hand.

Types Fractures vary from minimal displacement to anatomical or surgical neck, greater and lesser tuberosities, fracture dislocations, and involvement of the articular surface.

Treatment
- Minimal displacement of the anatomical neck—collar and cuff. Gentle mobilization. Beware avascular necrosis. Replace head with a prosthesis.
- Surgical neck: Internal fixation
- Tuberosities: Internal fixation, rotator cuff repair if indicated, prosthetic replacement if blood supply is disrupted
- Fracture dislocations: closed reduction or internal fixation
  N.B. Beware of brachial plexus damage.

Humeral shaft

Causes Direct, indirect trauma, especially road traffic accidents. Fractures may be oblique, spiral, transverse, or comminuted.

Treatment
- Conservative for most. Use a hanging cast; i.e., the wrist is supported by a sling and a U-slab of plaster is applied around the elbow to the humerus. Gravity maintains reduction. Healing takes around 6 weeks.
- Internal fixation is rarely indicated but may be necessary for nerve or vessel injury, pathological fracture, and non-union.

Children

Fractures occur at the surgical neck or through and around the epiphysis at the upper end.

Treatment
- Greenstick of surgical neck: collar and cuff
- Epiphyseal fractures: closed reduction. Remodelling is good. Open reduction is rarely necessary.

Dislocations of the shoulder region

Sternoclavicular joint

Look There is deformity at the medial end of the clavicle. Rarely it dislocates retrosternally, leading to tracheal compression.

Treatment
- Reduction by bracing the shoulder backwards. Figure-of-eight bandage.
- Immediate reduction under general anaesthetic is indicated for retrosternal dislocation. Open reduction may be necessary.

Acromioclavicular joint

Subluxation (tear of joint capsule, superior acromio-clavicular ligament and muscle origins) or dislocation (tear of coraco-clavicular ligament) occur.

Look
- Subluxation: the clavicle is tender and mobile upwards
- Dislocation: the acromion and clavicle are separated

Treatment Accept the deformity. Treat in a sling. Shoulder function is usually good. Open reduction + ligament repair is an option.

Shoulder joint: Anterior dislocations

Cause Fall on to the outstretched hand. Capsule is torn, glenoid labrum avulsed, or postero-lateral aspect of head crushed.

Look The patient supports his arm, which is abducted. The deltoid looks flat or hollow. The arm may appear too long. He cannot move his shoulder.

X-ray The humeral head lies below the glenoid labrum. Exclude associated humeral neck fractures.

Treatment
- *Kocher's manoeuvre*: Flex the elbow under traction. Laterally rotate the humerus, adduct and rotate it medially.
- *Hippocratic method*: Place your foot gently in the axilla, apply traction to the arm, and lever the head into position. Encourage early movement. Avoid lateral rotation and abduction.
  N.B. Either of these procedures can lead to neurovascular damage if there is a humeral neck fracture. Open reduction of the dislocation is safer.

Check Reduction and exclude fracture by X-ray. Shoulder abduction (axillary nerve) to exclude neurovascular injury.

Complications Muscle tears, circumflex nerve injury, associated fracture of the humerus or tuberosities, recurrent dislocation.

Shoulder joint: Posterior dislocations

Cause Forced internal rotation, direct blow to front of shoulder.

Look Coracoid process is prominent. The shoulder is flattened. The arm is held in internal rotation.

Feel The humeral head posteriorly.

X-ray Use an axillary view to confirm the suspicion on AP films.

Treatment General anaesthetic. Rotate the arm outwards under traction. Push the humeral head forwards. Encourage early shoulder movements. Avoid internal rotation.

Complications
- Unreduced dislocation should be treated in the young by open reduction.
- Recurrent dislocation

Shoulder joint: Recurrent dislocations

Cause In most patients the shoulder capsular tear repairs itself. If, however, there has been damage to the glenoid labrum or humeral head at the time of dislocation this is less likely to occur. Inexpert reduction can also be a factor.

Features Dislocation occurs on movement of the arm, especially when raised and abducted.

Treatment Operative stabilization of the shoulder by re-attaching the glenoid labrum or repairing the capsule and suturing it to subscapularis. The coracoid process is resited on the neck of the scapula. The arm is bandaged to the side for 5 weeks, then movements can begin.

Fractures of the ribs and sternum

Injury	Cause	Comment
Single rib fracture	Direct trauma (e.g. fall against furniture)	Confirm by X-ray Treat by oral analgesics, local anaesthetic injection into the painful area (not effective in a haematoma) or intercostal block
Lower 2–3 ribs	May result from bout of severe coughing	Confirm by X-ray Determine the cause of coughing Relieve pain
Sternal fractures/dislocations	Cervical hyperflexion causes the chin to strike the sternum, leading to fracture or dislocation	Exclude associated serious injuries, viz. cervical and thoracic fracture/dislocation Treat the sternal injury conservatively unless there is displacement, when wiring may be indicated.
'Stove-in chest'	Crushing injuries, road traffic accidents	Fractures of the anterior and posterior aspects of the ribs result in a flail segment. Steering wheel compression injuries tend to cause sternal fracture/dislocation and anterior rib fractures. Victims develop respiratory distress with paradoxical respiration: the affected segment is sucked in on inspiration and pushed out on expiration. This leads to hypoxia.

- Associated injuries include cardiac and pulmonary contusions. Suspect them if the first rib is fractured (an indication of severe trauma).
- Cardiac tamponade may be diagnosed by pericardiocentesis (if suspected) and the underlying right atrial tear (by common injury) repaired at emergency thoracotomy.
- Cardiac contusion is difficult to diagnose but should be suspected by ST segment and Q wave abnormalities on ECG. The presence of arrythmias in a trauma victim are a further indicant. Treatment is similar to the patient with myocardial infarction (OHCM, p. 236).
- Pulmonary contusion leads to areas of consolidation. If extensive, thoracotomy may be necessary. Postoperative physiotherapy and flucloxacillin 250 mg orally every 6 hours or ampicillin 250 mg–1 g orally every 6 hours, both for 5–7 days, should be given to prevent and treat secondary infection.

- crossed Garden nails
- 3 AO screws

Displaced/unstable fractures:

Young patient
- Closed reduction + internal fixation
- Check the vascularity of the femoral head with technetium scan at 6 weeks
- Total hip replacement for avascular necrosis

Young people
Treat with bed rest and traction

Elderly patient
- Femoral head replacement
- Primary total hip replacement

Unstable fractures—oblique fracture line, sub-trochanteric extension—may need valgus or medial displacement osteotomy or fixation with AO blade plate or interlocking intra-medullary nail

---

Prognosis

In young people the risks of treatment are minimal. In the elderly many die of intercurrent disease, e.g. deep venous thrombosis, bronchopneumonia, cardiac failure, within a year.

441

Femoral shaft fractures

These are most common in young people, usually the result of massive trauma resulting from road traffic accidents.

Assess the patient

Identify associated injuries:
- Head, chest, abdomen
- Dislocation of the ipsilateral hip, fractured tibia, dislocation or fracture of the knee
- Soft tissue injury: Skin and muscle, vascular injury to femoral or popliteal vessels
- Sciatic nerve injury

Resuscitation and investigation

- Begin IV fluids. Use a plasma expander.
- FBC, U&E, group and/or cross-match (> 6 units)
- Realign the injured limb. Splint to its neighbour or use a Thomas splint.
- Is the fracture compound? Give tetanus toxoid or antitetanus serum (if no previous immunization).
- Give prophylactic antibiotics: Flucloxacillin and benzylpenicillin IV.

Management

Depends on the patient's age, type of fracture, and associated injuries.

Conservative
- Wound excision and irrigation
- Close small wounds primarily. Leave large defects open.
- Insert a Steinman pin through the tibial tubercle at right angles to the bone.
- Apply skeletal traction, with a knee flexion attachment. Manipulate the fracture so that the bone ends are in contact.

N.B. Children's fractures heal quickly. Use gallows traction for babies under 18 months. In children under 6 years use a Thomas splint. In older children use skeletal traction if there is shortening (more than 2 cm). Observe carefully for angulation. Cast bracing using plaster of Paris or plastics permits mobilization after 6–8 weeks in adults.

Surgical

External fixation should be used for compound fractures where there is soft tissue loss. Pins are inserted above and below the fracture site. Reduction and alignment are maintained by an external fixator device.

Internal fixation is commonly achieved by intramedullary nailing. The patient is stabilized in traction and operated on electively, unless there are multiple injuries when the procedure may be part of emergency surgery. The procedure is performed by open or closed techniques. Gentle movements are begun at about 48 hours postoperatively. Weight-bearing may then

463

Bone tumours

The commonest are metastatic from cancers of the breast and prostate. Thyroid and kidney malignancies also show a predilection to spread to bone.

Primary bone tumours are rare. They are important because they often occur in young people, have a poor prognosis, and may involve mutilating surgery.

Benign tumours

Should be excised or removed by curettage if they cause symptoms. Residual bone cavities may be packed with chips of the patient's own bone.

Examples

Osteoid osteoma causes an exquisitely painful area in the long bone of a young person, usually male, which may be relieved by aspirin. Radiologically there is an osteolytic area surrounded by a rim of sclerosis.

Chondroma is a cartilaginous tumour common in the phalanges, metacarpals, or metatarsals. They are classified as en- or ecchondromas depending on their site in the medulla or surface of the bone. Ecchondromas may calcify to form exostoses. If multiple the condition is known as diaphyseal aclasis (inherited dominant disease). Multiple chondromatosis without exostoses is known as Ollier's disease.

Non-ossifying fibroma is a fibrous tissue tumour which usually appears radiologically as an oval cortical defect. It may be a developmental abnormality, not a bone tumour.

Malignant tumours

Investigations

Radiology Key features are poorly defined osteolytic lesions and disruption of the periosteum with local invasion. Skeletal scintigraphy (bone scan) with technetium (^{99}Tc) is important in detecting metastatic disease before it is radiologically visible. Arteriography indicates both the vascularity of a tumour and its relationship to the normal circulation. CT scanning may provide details of the relationship of a tumour to its surrounding tissue and facilitate pre-operative or pre-radiotherapy planning.

Biopsy Open biopsy is a prerequisite to correct treatment. Closed or needle biopsy is less reliable in that it provides less tissue for examination. It leaves a small scar, however, and is advocated for vertebral tumours, the procedure being carried out under image intensification.

Osteosarcoma occurs in the long bones of young people (10–30 years) or in pre-existing Paget's disease. The radiological appearances are of bone destruction, soft tissue invasion, radiating spicules of bone (sunray appearance) and periosteal elevation with subperiosteal new bone formation (Codman's triangle). Spread is by the bloodstream to the lungs. Clinical features consist of bone pain, particularly around the knee which may be misdiagnosed as osteomyelitis.

- Traumatic pneumothorax/haemothorax are recognized on chest X-ray and should be treated by early insertion of an underwater seal drainage system. Re-expansion usually occurs within 48 hours when the drain can be removed. Persistent bleeding warrants thoracotomy.
- Tension pneumothorax results when air leaking from the lungs cannot escape out of the pleural cavity or is sucked in via a wound. The patient develops increasing respiratory distress with absent breath sounds and hyper-resonance on the affected side. Relieve by inserting a wide-bore needle (the air will hiss out) and insert an intercostal drain, cover sucking wounds with a pad.

Treatment of patients with a flail segment
- Tracheostomy and intermittent positive pressure ventilation until the chest is stable (10–20 days)
- Treat associated injuries (above)
- Provide analgesia—intercostal nerve blocks, frequent small doses (2 mg) IV morphine
- Beware the development of acute respiratory distress syndrome (ARDS) (see p. 843)

437

Fractures of the pelvis

Two groups of patients sustain pelvic fractures: those over 60 years in falls at home, those under 60 years in road traffic accidents. The mortality rate varies between 5–20 per cent. Examine pelvic X-rays carefully.

Types of injury

Avulsion fractures Occur outside the pelvic ring. They are usually traction injuries. Rest and pain relief are indicated unless there is marked displacement.

Single ring fractures Always check that the sacroiliac joints are undamaged.

Treatment Fractures of the ilium or pubis are treated with bed rest and analgesia followed by mobilization when pain-free. Acetabular fractures may require internal fixation with skeletal traction if the hip has dislocated centrally.

Pelvic ring disruption Fracture or disruption has occurred at two places. The pelvic ring is unstable. The patient is in pain, cannot stand, may not be able to pass urine, may be in shock.

Treatment
- Treat shock. 40 per cent of patients need blood transfusion.
- Check for damage to the bladder and urethra. Can he pass urine? Yes. Is it clear? Yes. All is well. If the patient cannot pass urine and the bladder is palpable, establish a suprapubic cystostomy to measure urinary output then request intravenous urography (to detect abnormalities of the bladder) or retrograde urethrography (to detect urethral rupture). Bladder rupture is treated by suture and suprapubic cystostomy if intraperitoneal. If extraperitoneal establish suprapubic cystostomy and drain the retropubic space. Urethral injuries require the insertion of a fenestrated catheter to maintain drainage and provide alignment (refer to p. 482).
- Reduce the fracture either with a pelvic sling or, if there is severe ring disruption, by open reduction and plating. Bed rest is required for 6–8 weeks. Weight-bearing should not be permitted for 10–12 weeks.

Complications
- Haemorrhage and shock
- Urogenital injury
- Deep venous thrombosis
- Paralytic ileus
- Mal-union may lead to pregnancy difficulties
- Late osteoarthritis

Femoral neck fractures

These are most common in the elderly and may be due to minor trauma, multiple stress fractures, disturbed bone architecture, and osteoporosis. In the young they are the result of major trauma.

Classification

- *Intracapsular*: subcapital, transcervical
- *Extracapsular*: basal, pertrochanteric

History

The patient is usually an elderly woman who gives a history of minor trauma: 'I tripped on the carpet and couldn't get up'. Pain may also develop spontaneously.

Examination

- Look: There is shortening, adduction, and external rotation due to the action of psoas on the distal fragment.
- Feel: There may be tenderness of the hip on palpation.
- Move: Movements are painful. Weight-bearing may not be possible.

X-ray features

Take AP and lateral films. The fracture is usually obvious. If there is doubt about the X-ray but suspicion of fracture, get a US scan of the hip. This may demonstrate a haematoma. Repeat X-rays over 1–2 weeks may be necessary to show the fracture.

Aim

To achieve stability of the fracture and early mobility of the patient.

Treatment

- Give pain relief—skin traction and analgesics
- Resuscitate if indicated, e.g. correct dehydration, hypothermia, etc.
- Do a chest X-ray, ECG, FBC. Look for and correct cardiac failure, etc.
- Operative treatment

	Intracapsular fractures	Extracapsular fractures
Risks	• Avascular necrosis of the femoral head • Non-union • Prolonged bed rest in the elderly	• Prolonged bed rest in the elderly
Fixation Undisplaced fractures: Fix with:	• hip screw and two-hole plate	Stable fractures—sliding screw and plate

begin, provided the patient's general state is satisfactory. The rates of non-union and infection are low (0.3 per cent and 0.7 per cent).

Complications of femoral shaft fractures
- ARDS due to fat embolism (see p. 843)
- Deep venous thrombosis and pulmonary embolism
- Infection
  —wound excision and prophylactic penicillin prevent gas gangrene
  —tetanus immunization
- Pressure sores: Encourage mobility. Turn the patient regularly

Tibial shaft fractures

The commonest long-bone fractures occur in the tibial shaft after road traffic accidents, during sport (especially riding and skiing), and at work from falls, etc. They may be transverse from direct trauma, oblique, or spiral from indirect trauma. 20 per cent of patients develop non-union, 10 per cent infection.

Surgery should be performed with prophylactic antibiotic cover (flucloxacillin 500 mg IV at induction and 6-hourly for 48 hours postoperatively). The fracture is exposed through a long incision lateral to the tibial crest, reduced, and maintained so with a dynamic compression plate and screws. If bone mass has been lost a primary bone graft may be indicated.

Closure is without tension. Haemostasis and wound drainage must be established. The limb is supported in a back slab.

Classification

- *Closed* (simple)
- *Open* (compound): Minor skin injuries need wound toilet. Major soft tissue loss and associated vascular damage indicates the need for wound excision, reconstruction and secondary closure with external fixation of the fracture.

Methods of treatment

Closed reduction under general anaesthetic is the commonest method. The patient's leg hangs over the edge of the operating table, the ankle is dorsiflexed to a right angle without bowing the fracture site, and a plaster of Paris cast applied with the knee in 5 per cent flexion. Check the position with X-rays while the patient is still asleep. Frequent X-rays should be taken in the next 3 weeks to assess the reduction.

Preliminary traction on the fracture site via a Denham pin through the os calcis for a variable period permits swelling to subside before the application of a plaster of Paris cast.

At about 1 month, a functional cast brace or patellar tendon bearing plaster may be applied to permit mobility of the knee and ankle without losing control of the fracture.

Clinical union occurs between 12 and 16 weeks. Good results are generally obtained with closed reduction, even by relatively inexperienced surgeons, but anatomical reduction is rarely achieved. Ankle and knee stiffness from prolonged immobility can be prevented by functional bracing.

Internal fixation permits anatomical realignment of the fracture with early mobility.

Postoperative care
- Inspect the wound for haematoma at 48 hours. Evacuate if present.
- Start knee exercises after 48 hours.
- Partially weight-bear on crutches at 5 days (20 kg).
- Discharge at 10–14 days.
- Review and X-ray at 6 weeks. If there is no callus (i.e. no movement), increase weight bearing. If callus is present, reduce weight-bearing.

- Most patients walk without support at 3 months. Remove the implant after 18 months.

External fixation This is indicated if there is extensive tissue or bone loss. Pins are inserted on either side of the fracture side and an external fixator applied. This then allows fixation of the fracture and management of the soft tissue injury. A 'uniframe' fixator which penetrates bone only is recommended.

Complications of tibial shaft fracture

- Delayed union
- Non-union has occurred if the fracture remains unhealed after 9 months.
- Infection

Fractures of the ankle

Remember that the ligaments are also injured. Excessive forces of external rotation, abduction, adduction, and compression can cause solitary or multiple malleolar fractures with ligamentous injury and fracture dislocation. Always examine the patient carefully. Use X-rays to assist you, not as an absolute guide.

Diagnosis

- Establish the mechanism of injury.
- Examine the bones and ligaments of the ankle, the tibia, fibula, and upper tibio-fibular joint.
- Distinguish between solitary ligament injuries and those associated with fracture. Pull the heel firmly forwards. Signs of apprehension imply an anterior talofibular ligament strain (anterior drawer sign). Localized tenderness implies combined fracture and ligament injury.

X-rays

- Take AP and lateral films with 10° of internal rotation of the foot. Note the site of fractures and talar shift or tilt.
- Arthrography confirms extravasation of contrast into the peroneal sheath if there is ligamentous disruption.
- Stress views are seldom necessary.

Treatment

Closed Ligamentous injuries may be treated with oral non-steroidal anti-inflammatory drugs, strapping, or plaster of Paris if there is pain with early mobilization in crutches. Physiotherapy is directed towards the re-acquisition of proprioceptive function (e.g. wobble-board).

Some fractures may be successfully manipulated. Correction of rotation and fibular length are important features. Reverse the direction of the injury. Maintain reduction in an above-knee plaster of Paris cast with the foot in supination or pronation as appropriate.

Open reduction This is indicated in failed closed reduction, but is increasingly a first option in treatment. The aim is to re-establish the mortice, maintain reduction, encourage early mobility, and achieve union in the perfect position. This involves the use of screws, plates, or tension band wiring if the fragments are small. The wound is drained and a back slab applied. At 24 hours the slab is removed and active movements are begun. When a full range is achieved, a below-knee partial weight-bearing plaster of Paris cast is applied for about 6 weeks. The implants can be removed after a year.

Fractures of the tarsus and foot

Talus

Cause Fall from a height, road traffic accident. Look for associated injuries.

Sites Head, neck, body, lateral process (of talus).

Clinical features The foot is swollen, deformed, and held rigid. Check the dorsalis pedis pulse.

X-rays AP, lateral, and oblique.

Treatment
- *Undisplaced fractures*: Split plaster of Paris cast reapplied when swelling settles. Remove in 8 weeks.
- *Displaced fractures*: Attempt closed reduction by forced plantar flexion. If it fails, proceed to open reduction.

Complications Skin necrosis. Avascular necrosis.

Calcaneum

Cause Fall from a height. Look for associated spinal injury.

Types Chip, split, crush fractures.

Clinical features The heel is painful and swollen. There may be a bruise on the sole.

X-rays Lateral axial films.

Treatment
- *Chip fractures*: If undisplaced, may be treated with crepe bandaging. If displaced, closed reduction or screw fixation (for large fragments) is indicated.
- *Split and crush fractures*: Most patients are treated by bed rest and elevation of the legs to reduce swelling. After 2–3 days mobility with partial weight-bearing on crutches is begun. This is alternated with periods of elevation to prevent swelling. Most patients take about 3 months, and even then pain and stiffness are common.

Midtarsal fractures and dislocation

If there is no displacement, a walking plaster of Paris cast or elastic strapping is all that is necessary.

Displaced fractures and midtarsal dislocations are reduced and plaster of Paris applied for 6–8 weeks.

Metatarsals and phalanges

Causes Crushing or twisting forces, e.g. being run over by a car. Stress fractures.

Sites Base of first metatarsal, fracture of neck (crush), transverse fractures of shafts, avulsion fracture of the base of the fifth metatarsal. Tarso-metatarsal fracture dislocation.

Special precautions Fractures or fracture dislocations of the first metatarsal base and tarso-metatarsal fracture dislocation are serious injuries which can disrupt the circulation of the foot.

Treatment Identify the type of injury. Check the circulation and reduce fracture dislocations urgently (see above). Apply a split plaster and elevate the limb. Remove after 3 weeks, start exercises and weight bearing at 6 weeks.

Other fractures Once pain and swelling has subsided, encourage walking in normal shoes. If there is considerable pain, a plaster of Paris or elastic strapping may be applied for 3 weeks.

Stress fractures Activities should be reduced and strapping applied. Normal walking is encouraged.

Phalanges
- *Proximal*: Reduce if there is angulation. Use shoes or slippers with the toe cut out.
- *Middle*: Straighten the toe and strap it to its neighbour
- *Terminal*: If severe, elevate the foot. Evacuate haematoma by trephining the nail.

Injury and the spinal X-ray

Take X-rays of the spine if the patient complains of pain there after an accident. Beware of associations of spinal injury with other conditions, viz:

- Bilateral os calcis fractures: Compression fractures of thoracic or lumbar spine
- Facial fractures: Cervical injury
- Head injuries: Cervical injury especially C_1, C_2 fractures. Look carefully at C_1 and C_2 on the skull X-ray
- Dislocation of the sternum: Thoracic spinal injury
- Ankylosing spondylitis: Multiple fractures of cervical and thoracic spine

Hints for reading spinal X-rays after injury

- Develop a mental picture of the normal spinal X-ray. If you feel that a film 'doesn't look right', the chances are it is not. Check especially the lateral masses of the cervical vertebrae, which should be symmetrical. If they are not, request tomograms or 15° off-lateral films to demonstrate the facets.
- Beware the 'empty' vertebra. This may imply damage to the spinous processes.
- Pay attention to the disposition of C_1 and C_2. The lateral masses of C_1 should not overlap the lateral margins of C_2. If there is more than 3 mm space between the front of the odontoid and the back of the anterior arch of C_1, there may be instability.
- Check the trasverse processes. They become displaced when fractured.
- Check the paraspinal soft tissues. A retropharyngeal space greater than 3 mm suggests a haematoma. The paravertebral pleura on the left side should run parallel to the aorta. If it does not, suspect a haematoma.
- If there is an obvious spinal injury, ensure that associated soft tissue injuries, e.g. thoracic viscera retroperitoneal structures, and potential damage to the spinal cord are excluded or prevented before requesting further radiological investigations.

Signs which imply spinal instability

- Complete vertebral dislocation or translocation
- Fractures in a fused spine, especially ankylosing spondylitis
- Signs of movement: Malalignment, avulsion fractures, evidence of prevertebral swelling

Reference

Butt, W. P. (1988). Interpreting the spinal X-ray. British Journal of Hospital Medicine, **40**, 46–54.

Spinal injuries

Establish as accurately as possible whether the injury is stable or unstable. Patients who have an unstable injury may not be paralysed but may become so if they are handled incorrectly. Any movement of the neck may reduce the lumen of the vertebral canal.

Structure of the spine

It helps to think of the spine as three columns: an *anterior* consisting of the vertebral bodies, intervertebral discs, and the longitudinal ligaments, an *intermediate* comprising the facetal joints and ligaments, a *posterior* consisting of the spinous processes and the interspinous ligaments. If there is a triple-column injury, extreme instability can be predicted; whereas if only one is affected, e.g. compression fracture, the spine remains stable.

Stable injuries	Unstable injuries
Transverse process fractures Solitary spinous process fractures Compression fractures	Fracture dislocations of the thoracic or lumbar spine Fractures in a fused spine, e.g. ankylosing spondylitis Burst fractures (due to a fall from a height) Fracture dislocations of the cervical spine Fractures of the atlas and axis
Principles of treatment: • Pain relief • Bed rest	• Avoid tetra- or paraplegia by careful transport (see below) • Reduce dislocation • Maintain reduction by traction or internal fixation • Specialized nursing care is instituted upon the diagnosis of tetraplegia or paraplegia to prevent complications like pressure sores, or chest and urinary infections

452

Spinal shock

There is loss of power, sensation, and reflexes below the level of the lesion. Expect recovery, if it is going to happen, in 24 hours. If there is no return of function by 48 hours the prognosis is poor. Even in irreversible cord damage, the perianal and bulbocavernous reflexes will reappear within a few days. Ankle jerks return in weeks. Knee jerks take months.

Partial injuries

• *Anterior cord syndrome*: Motor loss but sensation is preserved.
• *Posterior cord syndrome*: Sensory loss but power is intact.

- *Brown–Sequard syndrome*: Results from hemisection of the cord. Ipsilateral paralysis and sensory loss below the lesion with contralateral analgesia and thermoanaesthesia a few segments below the lesion.
- *Central cord syndrome*: Due to extension injury of the neck. There is often no fracture on X-ray. The cause is thought to be vascular injury. Incomplete tetraparesis, upper limbs are worse affected than lower.

  Prognosis for partial lesions is reasonably good.

Immediate management

Cervical Stabilize the injury, e.g. support the head and neck with sandbags or cervical collar.

Do not permit flexion or rotation. Assign one person (the doctor) to this. Longitudinal traction with the head neutral has been recommended.

Dorsal/lumbar Transport prone or supine according to need. Designate helpers to control the pelvis, trunk, and shoulders. Roll the patient in one piece like a log if necessary.

Full neurological assessment and radiological investigation is performed in hospital. *Do not exacerbate spinal cord damage by careless movement of the patient.*

453

Reference

Piggot, J. and Gordon, D. S. (1979). Letter. *British Medical Journal*, **1**, ii, 193.

Acute pyogenic osteomyelitis

This is a disease of growing bones. It is common in infants and children but rare in adults unless they are immunosuppressed, or diabetic. The commonest infecting organism is *Staphylococcus aureus*. Others include *Streptococcus*, *Pneumococcus*, *Haemophilus influenzae*, *Brucella*, TB, Spirochaetes, and fungi. In children and adults with sickle cell disease the *Salmonella* group are the common infecting organisms, possibly due to haematogenous spread from devitalized gut to the sites of bone infarcts associated with the condition.

Clinical features

The patient is young, sometimes with a preceding history of trauma or infection (skin or respiratory). After a few days fever and malaise develop, followed by pain, swelling, and redness of the affected area. Often some treatment has been given and the child presents with a history of being 'a bit off-colour' and a limp (pseudo-paralysis).

In neonates there are two presentations: life-threatening septicaemia in which obvious inflammation of a long bone develops, or a more benign form in which symptoms are slow to develop but bone changes are extensive, often multiple bones being involved.

Pathology

The organisms settle near the metaphysis at the growing end of a long bone. Initially there is hyperaemia followed by suppuration within the medulla. Infective thrombosis with necrosis of bone develops (sequestrum formation). Repair begins after about 10 days. New subperiosteal bone is formed (the involucrum) which develops defects (cloacae) allowing spontaneous discharge of pus and fragmented sequestrum. Some dead bone is resorbed.

Radiology

- Plain films show no abnormality for 10 days.
- Technetium-99m bone scan with dynamic imaging of perfusion is effective in confirming the diagnosis early in the disease.
- Other means of scanning, e.g. gallium-67 and indium-111 are accurate but not generally available.
- Tomography and CT scanning can define the extent of sequestration and cavitation.
- Sinography can identify communication with underlying bone.

Radiographic features

- Soft tissue swelling is an early sign.
- Patchy lucencies develop in the metaphyses at around 10 days.
- Periosteal new bone formation may be seen.
- Involucrum formation is apparent at 3 weeks.

- The sequestrum appears dense compared to surrounding bone which is osteoporotic.
- Normal bone density occurs with healing.

Laboratory tests

The ESR and WBC are often high. Do blood cultures to attempt to identify the causal organism. Aspirate and culture exudates or septic joints.

Treatment

- Splintage provides pain relief.
- Antibiotics for 3-6 weeks. Give flucloxacillin elixir (150 mg/5 ml)—under 2 years 2.5 ml orally 6-hourly, 2-10 years 5 ml orally 6-hourly + fusidic acid 5 ml daily for 1-5 years; 10 ml 12-hourly over 5 years. Give antibiotics parenterally in severe cases and adults. Use ampicillin or co-trimoxazole for *H. influenzae* infections.
- Drainage: This is infrequently necessary, especially if anti-biotic treatment is begun early. Perform if there is continuing pyrexia or evidence of a subperiosteal abscess (overlying oedema).

455

Complications

Chronic osteomyelitis, metastatic infection, septicaemia. Suppurative arthritis, alteration of bone length due to epiphyseal damage.

Reference

O'Brien, T., McManus, F., MacAuley, P. H., and Ennis, J. T. (1982). Acute haematogenous osteomyelitis. *Journal of Bone and Joint Surgery*, **64**, B, 450-3.

Chronic osteomyelitis

This may follow acute pyogenic osteomyelitis, but is more commonly secondary to surgery or a compound fracture. It may also be chronic from the start.

Secondary to acute pyogenic osteomyelitis

This should be prevented by the early and adequate treatment of acute pyogenic osteomyelitis.

Features Sinus formation due to sequestra or resistant bacteria.

Treatment Conservative. Use dry dressings only. Sequestrectomy combined with Fucidin or complete excision of diseased bone with antibiotic cover may be indicated.

Recurrent acute attacks with spontaneous recovery may develop for years afterwards. Most patients do not need treatment unless an abscess forms which should be drained.

Recurrent attacks with sinus formation may eventually lead to amputation due to persistent foul discharge. Closed suction irrigation following excision of dead bone and affected soft tissue can be effective in keeping the area clean. The irrigation fluid may contain antibiotics. The disadvantage is early blockage of the system (often after a few days). Bone cement beads impregnated with gentamicin also play a role in keeping the area clean by releasing high concentrations of antibiotic from the beads as they are gradually withdrawn from the wound.

Secondary to trauma

Prevent this complication by immediate surgery in patients with compound fractures. Excise all dead tissue, scrub the bone ends, and give the patient benzylpenicillin 300–600 mg IV every 6 hours for 5–10 days. Add cefuroxime 750 mg every 6–8 hours IV.

If chronic infection does develop the treatment is as above, except that fracture stabilization must be achieved, usually with an external fixation device.

Secondary to joint replacement surgery

Prevent by strict aseptic technique and systemic prophylactic antibiotics (p. 112). Removal of the prosthetic joint may be necessary with replacement or an alternative form of arthroplasty.

Chronic osteomyelitis from the outset

Brodie's abscess This is usually near the metaphysis of a long bone. Some of the organisms become walled off, leading to cavities with well-defined sclerotic margins.

Treatment Operative drainage with excision of the abscess wall. Use prophylactic antibiotics.

Tuberculosis of bone is usually insidious, but may present with tenderness, soft tissue swelling, or joint effusion. Muscle atrophy develops. Spontaneous discharge of a cold abscess may lead to sinus formation and destruction of bone to spinal collapse and paraplegia (Pott's paraplegia—due to pressure from an abscess or bony collapse). This is the most serious complication, but is becoming less common.

Treatment Is with rest, traction to control deformity, and excision of dead bone. Arthrodesis of weight-bearing joints may be necessary. Antituberculous drugs should be given.

Syphilitic osteomyelitis is rare. Tertiary disease may lead to diffuse periostitis (with sabre tibia) or localized gummata with sequestra, sinus formation, and pathological fractures. In infants with congenital disease, epiphysitis and metaphysitis are typical. X-rays in children show areas of sclerosis near the growth plate separated by areas of rarefaction. In adults there is periosteal thickening with punched-out areas in sclerotic bone.

Treatment
- Confirm the diagnosis by serology.
- Immobilize to provide pain relief and prevent fracture.
- Give specific antibiotic chemotherapy.

Mycotic infection leads to bone granulomas, necrosis, and suppuration. Sites include the extremities of the long bone and the vertebrae.

457

Source Primary lung infection from coccidiomycosis, cryptococcosis, blastomycosis and, histoplasmosis.

Treatment
- Amphotericin B
- Surgical excision

References

Hedstrom, S., Lidgren, L., Torholm, C., and Onnerfalt, R. (1980). Antibiotic containing bone cement beads in the treatment of deep muscle and skeletal infections. *Acta Orthopaedica Scandinavica*, **51**, 863–9.

Raft, M. J. and Melo, J. C. (1978). Anaerobic osteomyelitis. *Medicine*, **57**, 83–103.

Peripheral nerve injuries

Around 9000 people each year in the UK are admitted to hospital with an injury to a peripheral nerve. Almost all require some form of surgical treatment.

Pathophysiology

Neuropraxia The nerve is stretched but remains anatomically intact. Recovery is complete, occurring in days or weeks. There is no axonal degeneration.

Axonotmesis The axon is divided but the connective tissue components remain intact. Wallerian degeneration occurs distally, to be followed by regeneration after a delay of about 10 days down the endoneural tube at the rate of 1 mm/day. This injury is associated with closed fractures, dislocations, and pressure.

Neurotmesis The nerve is completely divided or irreparably damaged. Common causes are open wounds, traction, ischaemia, and injection injury. Surgical repair is indicated if possible. This has a poor prognosis unless the nerve can be brought together so that axon regeneration to the correct end organ can occur.

Diagnosis

- Ask about injury. Is there an open wound, fracture, recent surgery or prolonged immobility? (e.g. 'Saturday night palsy'), see below
- Carry out a thoughtful examination. Know the normal anatomy and functions of each nerve. Use a pin. Compare the abnormal with the normal side.
- Anaesthetic skin looks shiny and does not sweat. Denervated skin does not wrinkle when immersed in water.
- There are specific clinical signs following injury to the nerves of the upper limbs.

Treatment

Closed injuries

- Injuries in continuity can be expected to recover spontaneously so exploration is not indicated.
- Compression injuries should be treated by removal of the compressive force, whether external (such as a plaster of Paris cast) or internal (e.g. carpal tunnel syndrome).

Open injuries

- *Primary repair* (suture) within 24 hours is ideal, but an uncontaminated operative field, adequate skin cover, and proper surgical equipment is required.
- *Secondary repair* is carried out after 7 days. The neuromas at the nerve ends are resected and the gap between them bridged by a nerve graft (usually the sural nerve).

Individual nerve lesions

Median nerve

Causes Injuries at the elbow or wrist (high and low)

Features
- *High injury* (elbow): Paralysis of forearm pronators and flexors of wrist except flexor carpi ulnaris and ulnar aspect of flexor digitorum profundus. There is wasting of the front of the forearm and a pointing index finger.
- *Low injury* (wrist): Paralysis of opponens pollicis. Wasting of radial half of hand. Sensory loss over radial 3.5 fingers.

Ulnar nerve

Causes Injuries at the elbow or wrist.

High Paralysis of flexor carpi ulnaris and ulnar aspect of flexor digitorum profundus. Hyperextension of little and ring fingers at the metacarpophalangeal joints. Paralysis of small muscles of the hand *except* the lateral two lumbricals and thenar muscles.

Low Paralysis of abductor digiti minimi and flexor digiti minimi, claw hand, wasting of intrinsic muscles. Loss of sensation over ulnar 1.5 fingers. Froment's sign (pinching paper between thumb and index leads to flexion of the distal interphalangeal joint).

Radial nerve

Causes Fractures and dislocations of the humerus or around the elbow. Crutch palsy (pressure in the axilla), Saturday night palsy (pressure on the radial groove from prolonged immobility with the arm hanging over the back of chair while asleep).

Features Wrist drop. Paralysis of triceps and brachio-radialis if the lesion is high. Sensory loss over dorsum of forearm and posterior aspect of first interdigital cleft.

Combined ulnar and median nerve lesions at the wrist

Causes Incised wounds, compression.

Features Claw hand, flattening of the radial aspect of the hand.

459

Brachial plexus injuries

Essence

These may occur at birth or during adult life. There is almost always a history of trauma.

Birth injuries result from difficult deliveries and are caused by traction on the plexus (e.g. breech delivery).

Types

- Erb–Duchenne (upper arm type) involves C_5 and C_6. The arm hangs at the side with the forearm pronated and the wrist flexed (the waiter's-tip position).
- Klumpke (lower arm type) involves C_8 and T_1. The hand is clawed due to intrinsic muscle paralysis, and if the sympathetic trunk is involved there is Horner's syndrome.

Treatment

Physiotherapy. Tendon transplants of triceps to biceps and wrist flexors to extensors may improve function if recovery is poor.

Prognosis

Spontaneous recovery is the rule for most lesions.

Adult injuries are all traumatic, and may be open (gunshot injuries) or closed (falls, motor cycle traction injuries) in which there is forced adbuction of the arm. The injury may occur to root, trunk division or cord.

N.B. Look for associated injuries, e.g. head, neck, abdominal. Local vascular injury may occur at the level of the lesion or more proximally in the case of traction injuries.

Treatment

- Pain relief. Treat associated injuries. Support the injured arm.
- Explore if there is evidence of vascular damage or if the injury is open.
- Carry out a primary repair if possible. If not, nerve grafting can be performed microsurgically at a later date.
- Localize the lesion if the injury is closed. Use plain X-rays to identify cervical or upper limb damage. Cervical myelography can then localize sites of avulsion or roots from the spinal cord (traumatic meningocoele). Other investigations include nerve conduction studies and the use of axon reflexes (If histamine is injected into the skin, a weal and flare indicate an intact spinal arc. A positive histamine test indicates a preganglionic lesion. If the axon reflex in interrupted the lesion is distal to the dorsal root ganglia (where the bodies of sensory fibres are located) and is therefore post-ganglionic.

Prognosis

- Recovery is slow and in many patients unsatisfactory. Poor prognostic features of complete lesions are pain, Horner's syndrome and the presence of cervical meningocoeles indicating complete root avulsion.

- Tendon transfers and shoulder arthrodesis may improve subsequent function.
- Amputation and prosthetic replacement is occasionally necessary.

461

Ganglion

A ganglion is a degenerative cystic swelling from an adjacent joint or synovial sheath containing clear visceral fluid.

Common sites

- The dorsum of the wrist originating from the scapholunate or mid carpal joint
- The radial aspect of the volar surface of the wrist in communication with the radio-carpal or radio-lunate joints
- The base of a finger from the digital flexor tendon sheath
- The dorsum of a dip joint of a finger
- The foot, and occasionally around the knee

Symptoms and signs

Pain, weakness of the affected joint in varying degrees. Patient complains of non-cosmetic appearance.

Transillumination is positive and is a useful diagnostic test. Needle aspiration, which often needs to be repeated, and injection with cortisone, is diagnostic and cures more than 60 per cent of patients.

Natural history and treatment

50 per cent will disappear spontaneously. Traumatic rupture (traditionally with the family bible) is associated with a high recurrence rate. Excision is the most successful treatment. Warn the patient that recurrence is possible even with excision.

Excision

- Carry out the procedure under general or regional anaesthesia.
- Use a tourniquet.
- Make sure your incision is large enough to enable you to identify the source of the ganglion from joint or synovial sheath.
- If the origin can be identified, transfix it with a catgut suture and excise the ganglion. If this is not possible try to dissect and excise the part of the capsule from which it arises.

Reference

Holm, P. C. A. and Pandey, S. D. (1973). The treatment of ganglia of the hand and wrist with aspiration and injection of hydrocortisone. *The Hand*, **5**, 63–8.

Treatment involves amputation through the bone proximal to the tumour and chemotherapy. If there are pulmonary metastases, treatment is by chemotherapy alone. Solitary metastases may be resected.

Prognosis 15 per cent of patients survive 5 years.

Chondrosarcoma occurs in older patients, commonly in a flat bone, e.g. rib, ileum. It usually arises *de novo*, but can arise in pre-existing chondromas. Spread is by the blood stream to the lungs, but local invasion is more usual and the tumour is slow growing.

Treatment is by wide excision or amputation. Tissue defects can be closed by prosthesis and/or myocutaneous flaps.

Osteoclastoma (giant cell tumour). This is rare below the age of 20 years. It occurs in the ends of long bones, causes osteolytic bone lesions, and may present as a pathological fracture.

Treatment is by local excision but recurrence is common. About 15 per cent metastasize to the lungs.

Ewing's tumour arises in the medullary cavity. It affects children and young adults, presenting as a hot swelling with associated pyrexia so that osteomyelitis may be suspected. It produces no bone and spreads rapidly via the blood to the lungs, liver, and skeleton.

465

Treatment radio- and chemotherapy.

Prognosis 35 per cent of patients survive 5 years

Other primary malignant tumours include fibrosarcoma and synovial cell sarcoma.

References

Pagani, J. J. and Libshitz, H. I. (1982). Imaging bone metastases. *Radiological Clinics of North America*, **20**, 545–9.

Simon, M. A. (1982). Current concepts review: Biopsy of musculoskeletal tumours. *Journal of Bone and Joint Surgery*, **64A**, 1253.

Low back pain

Most patients with low back pain have self-limiting disorders of a traumatic or degenerative nature. The investigation involves a clear history and physical examination.

Common causes of low back pain

- Lumbar disc prolapse
- Spondylosis (defect of neural arch due usually to stress fracture)
- Spondylolisthesis (spondylosis + associated disc degeneration leads to instability and displacement of the vertebral body forwards upon the one below)
- Coccydinia: Pain in the coccyx may be due to lumbosacral disc disease
- Metabolic bone disease: Osteomalacia, osteoporosis
- Inflammation: Ankylosing spondylitis
- Infective: Tuberculosis, staphylococcus
- Neoplasm: Osteoid osteoma, metastases, multiple myeloma

Pathology

The symptoms and signs increase in severity in the order articular (intervertebral joint and related structures) < dural (dural tension) < neurological (root pressure).

Articular The history should establish whether pain is intermittent (mechanical) or relentless (= sinister, e.g. malignancy, infection, etc.) and whether it is related to posture.
- Look for local muscle changes. In a muscle lesion there is pain which is worse on isometric or resisted contraction.
- Ligament tenderness is common but may be significant if it is the only sign.
- Joint lesions will cause pain at the extremes of range. Changes of osteoarthrosis and ankylosing spondylitis lead to symmetrical limitation of lateral flexion early. When there is internal derangement of an intervertebral joint the restriction of movement is usually asymmetrical.

Dural The pain is diffuse but does not radiate. Pain may be felt in a different segment of the trunk from the lesion.

Tests
- *Lasègue test*: spasm of the hamstrings of the affected side on straight leg raising—(L_5, S_1)
- *Femoral stretch test*: (L_3, L_4). (flex the knee with the patient prone).
  Both can be used to elicit the pain which the patient experiences.

Neurological The most common cause is lumbar disc prolapse. The pain is well delineated and corresponds to the anatomy of the affected root or its dermatome.
- S_1 and L_5 are most commonly affected. Pain is felt at the back of the thigh, the leg and side of the foot.
- L_4 affects the lateral aspect of the thigh and the front of the leg.
- L_3 affects the knee and anterior aspect of the lower thigh.

- Other features are pain on sneezing or bending. Numbness or paraesthesia is occasionally present.
- The earliest and most persistent neurological deficit is loss of a tendon reflex, e.g. S_1—ankle, L_{3-4}—knee, L_5—great toe.

Causes

Lumbar disc prolapse, spinal stenosis (reduction in the diameter of the lumbar spinal canal), cauda equina syndrome (compression can lead to irreversible loss of bladder control with weakness and wasting in S_1 (calves and hamstrings) with S_2 loss (buttocks) and loss of perineal sensation.

Investigations

History, clinical examination, spinal X-rays, myelography, CT scanning.

Treatment

Non-operative
- Rest, analgesics, traction, manipulation, weight loss, lumbosacral support. Exercise in the form of swimming or walking. Physiotherapy and psychological or psychiatric counselling may be required.

Operative
- Surgical removal of a prolapsed disc (by fenestration or microdiscectomy)
- Chemonucleolysis: dissolution of the prolapsed disc with chymopapain
- Nerve root decompression (neurogenic claudication)
- Spinal decompression (removal of lamina: spinal stenosis, spinal metastases)
- Lateral mass fusion (spondylolithesis)

Reference

Ransford, A. O. (1987). Backache. In *Current surgical practice*, Vol. 4 (eds J. Hadfield and M. Hobsley), pp. 31–46. Edward Arnold, London.

Paget's disease of bone (osteitis deformans)

This condition of unknown aetiology becomes increasingly common with advancing age. Although any bone may be affected, those in the spine, pelvis, thigh, shin, and skull are most often involved. The bones become softened but greatly thickened and are liable to pathological fractures.

Clinical features

Increased thickness of bones may be the only sign. The patient may complain that his hat does not fit, and the skull may be obviously 'bossed' or 'domed'. Subcutaneous bones such as the clavicle or tibia will also show changes if affected, the clavicles becoming enlarged and deformed and the shins 'sabre' and bowed in shape.

Others may complain of pain, dull and well localized, especially in weight-bearing bones.

Investigations

- Serum calcium and phosphorus are normal.
- Alkaline phosphatase is high due to increased osteoblastic activity.
- Urinary excretion of hydroxyproline is high. This reflects increased bone turnover.
- Isotope bone scan confirms increased uptake in affected bones.
- Radiology shows both sclerosis and osteoporosis. The cortex is thickened and the bones deformed. Pathological fracture is a feature. Normal bone architecture is lost.

Complications

- Pathological fractures
- Osteosarcoma develops in 5 per cent
- High-output cardiac failure may develop due to the increased vascularity of the Pagetic bone which acts functionally as an arteriovenous fistula.
- Deafness due to involvement of the bony foramina of the auditory nerve.
- Leontiasis ossea—thickening of the bones of the face (rare)
- Osteoarthrosis may develop in joints near deformed bone, but its incidence does not appear to be higher in patients with Paget's disease.
- Paraplegia due to vertebral involvement is rare.

Treatment

Calcitonin is effective in relieving pain and may also relieve neurological complications such as deafness due to involvement of the middle ear. Antibodies may develop with the use of porcine calcitonin so salmon calcitonin is more useful for long-term therapy because it is less immunogenic.

Dose 80–160 units 3 times per week subcutaneously or IM. Monitoring the urinary hydroxyproline gives an indication of the effect of treatment.

Side-effects Nausea, vomiting, taste aberrations, reactions at injection site.

The great toe

Hallux valgus

The obvious feature is prominence of the head of the first metatarsal with lateral deviation of the great toe due to the pull of the extensors. As time passes a protective bursa develops over the metatarsal head (bunion) and the great toe begins to crowd or even overlap its lateral neighbours.

Causes

- Congenital, often familial, related to metatarsus primus varus (the first metatarsal subtends medially and is rotated)
- Acquired in association with being overweight so that the foot becomes flattened

Symptoms

Even in cases of severe hallux valgus, there may be no symptoms and the patient may simply complain of the deformity and widening of the forefoot. When pain develops it is usually in the bunion itself due to inflammation or the development of a hammer toe, or in the metatarsal heads of the splayed forefoot.

Treatment

Conservative
- Correct footwear. Padding to protect the bunion.
- Intrinsic foot muscle exercises or foot support if there is metatarsalgia

Surgical
- Keller's operation. This is an excision arthroplasty. The proximal one-third of the proximal phalanx is excised and the head of the metatarsal trimmed.
- Simple bunionectomy and excision of the medial aspect of the prominent metatarsal head
- Various osteotomies. These are valuable in younger patients (under 50 years).

Hallux rigidus

The great toe is straight (not slightly valgus as is normal). There may be a callosity over the dorsum of the first metatarsophalangeal joint and the joint has limited dorsiflexion and plantar flexion.

Causes

- Congenital due to a short first metatarsal
- Acquired due to repeated trauma to the first metatarsophalangeal joint

Symptoms

Pain especially on walking. The condition is usually bilateral.

X-rays

Osteoarthrosis of the first metatarsophalangeal joint.

Treatment

Adolescents and young people
- Rocker sole to relieve pain. Persistent pain may be relieved by osteotomy.

Older age groups
- Rocker sole until an adapted shoe is fashioned
- Arthroplasty (Keller's)
- Arthrodesis

Arthroplasty

This is surgical refashioning of a joint to relieve pain, increase mobility, and provide stability.

Types

- Excision
- Interposition
- Partial or total replacement of one or both articular surfaces. This has been made possible by advances in bioengineering and the discovery of inert metal and plastic prosthetic components which can be fashioned to fit each other (usually the plastic component replaces the concave surface of the joint and the metal the convex). They can also be polished and sterilized.

Example: Total hip replacement

Indications

- Osteo- and rheumatoid arthritis when pain and stiffness affect sleep, work, and normal daily activities
- Multiple joint involvement
- Avascular necrosis of the head of the femur with deformity (elderly patients)

Prevention of infection The use of ultraclean air systems and prophylactic antibiotics during and after surgery have reduced the overall infection rate to less than 2 per cent.

Ultraclean air systems and exhaust body suits The air in a conventional operating theatre is filtered out so that there are about 20 changes per hour. By using a unidirectional laminar flow system, 300 or more air changes can be achieved per hour. This and the use of exhaust body suits which create a physical barrier between surgeon and patient (and therefore bacteria) reduce the incidence of infection in joint replacement surgery.

Prophylactic antibiotics These can be given systemically or locally, either by incorporation into the bone cement or infiltration into the wound.

Systemic broad-spectrum antibiotics, e.g. Cephaloridine (Keflin) 3 g IV are given on induction and at 6 and 12 hours postoperatively. Flucloxacillin or ampicillin are alternatives given with the premedication and continued orally for 14 days postoperatively. Local antibiotics like penicillin or chloramphenicol, instilled into the wound, also contribute to the reduction of wound infection but incorporation of antibiotics in bone cement remains unproven as a method of prophylaxis.

Procedure The surgical approach exposes the upper femur and acetabulum. The femoral head is removed. A cavity is reamed in the femur and the acetabulum is deepened to fit the acetabular cup of the prosthesis. Both the acetabular and femoral components are fitted and fixed with cement. The joint is then reduced and its stability tested. The wound is then closed in layers.

- Flail chest and pulmonary contusion. Many patients develop hypoxia and will need mechanical ventilation. Others can be treated with blood volume replacement with whole blood or fresh frozen plasma, intravenous diuretics (20 mg frusemide every 12 hours for 72 hours), methylprednisolone (30 mg/kg for 72 hours), frequent aspiration of pulmonary secretions and early ambulation
- Use aortography to diagnose vascular injury in stable patients. Thoracotomy is indicated if the patient is unstable

N.B. All patients with chest injuries need adequate analgesia. Use intermittent morphine administration or intercostal blocks.

Reference

Trinkle, J. K. and Richardson, J. D. (1981). In *Thoracic injuries in trauma* (eds D. Carter and H. C. Polk), pp. 66–98. Butterworths, London.

481

Genito-urinary trauma

This is a feature in only 3 per cent of patients with multiple injuries. Blunt or penetrating forces are responsible. Remember that iatrogenic injury is a risk in pelvic or abdominal surgery.

Renal injury

Suspect from the history or signs of fullness or guarding in the loin.

Diagnosis
- Intravenous urography + renal angiography if the kidneys are non-functioning. 90 per cent accurate
- Ultrasound
- CT scan (rarely available in A&E departments)

Management
- Conservative if the patient is stable with a functioning kidney. Bed rest, analgesia, and blood replacement. More than 80 per cent recover.
- Exploration in renovascular damage, unstable patients, and high-velocity missile injuries. If, at laparotomy a perinephric or retroperitoneal haematoma is found, leave it alone unless it is expanding.

Ureteric injury

The lower one third is often injured during pelvic surgical procedures. Blunt trauma may cause avulsion at the pelvi-ureteric or vesico-ureteric junction. High-velocity missiles may cause direct or cavitational injuries.

Management Immediate repair by end-to-end anastomosis over a stent gives good results in iatrogenic injury. Avulsion may be managed by repair or reimplantation into the bladder or calyces. High-velocity missile injuries should be treated by nephrectomy, cutaneous ureterostomy or primary repair depending on the extent of damage. Ureteric ischaemia may follow cavitation due to near-miss high-velocity missiles with subsequent sloughing several days later. Treat by percutaneous nephrostomy and late reconstruction after a few weeks.

Bladder injury

Suspect in penetrating trauma and pelvic fracture. Haematuria or blood at the external meatus are reliable signs. Confirm by cystography or retrograde urethrography.

Management
- Small extraperitoneal leaks: Catheter drainage
- Major ruptures: Formal repair with drainage

Urethral injury

- Causes: Road traffic accidents, iatrogenic trauma during catheterization or instrumentation
- Assess: Rectal examination may reveal an 'absent' or 'high' prostate membranous urethral injury with associated pelvic fracture. Water-soluble cysto-urethrography may also be valuable.

Hydatid disease

Causal organisms

Larval forms of the parasites *Echinococcus granulosus* and *Echinococcus multinodularis*.

Life cycle

The disease occurs in sheep- and cattle-raising areas of the world. Dogs are usually the primary host. They eat infected offal and the echinococcus parasite (about 1 cm long) develops in the dog's intestine. It consists of a head and three segments, the last of which contains hundreds of ova. These are passed onto grass, etc. by defecation. Sheep, cattle, or humans then ingest the ova which penetrate the duodenum to enter the portal circulation. About 80 per cent of the ova thrive in the liver with the development of hydatid cysts. They may also enter the general circulation forming cysts elsewhere (often the kidneys or lungs).

Clinical features

The infection is usually contracted in childhood and produces symptoms and signs in adult life. There is occasionally a palpable mass in the liver. As it enlarges, compression of the biliary ducts may produce jaundice. Rupture of the cyst into the peritoneal cavity causes peritonitis and shock. Cyst fluid also causes a severe allergic reaction with urticaria and oesinophilia.

Diagnosis

- This is relatively easy when the patient lives in an area where the disease is common. Ultrasound and CT scanning may be used to localize cysts. ERCP may demonstrate connections with the bile ducts.
- Casoni's test is positive in 80 per cent but gives many false positives.
- Indirect haemagglutination tests are most accurate.

Treatment and prevention

Medical Mebendazole 400 mg three times daily for 30 days.

Surgical Excision or aspiration of the cyst(s). Extreme caution must be taken to prevent peritoneal contamination. Black packs soaked in hypochlorite are placed around the liver to show up any daughter cysts or scolices.

The cyst is partially aspirated and partially refilled with hypertonic saline which is scolicidal. It is then carefully separated from the liver and the cavity closed or drained. Give mebendazole postoperatively.

Rupture

- Treat shock.
- Carry out peritoneal toilet.
- Give hydrocortisone before, during, and after surgery.
- The prognosis is poor.

Prevention Prevent access of dogs to infected offal and carcases.

Complications

Operative
- Sciatic nerve injury due to poor technique
- Improper positioning of the cup leading to dislocation or fracture of the femoral shaft
- Absorption of monomer cement leading to collapse or cardiac arrest

Postoperative
- Deep venous thrombosis
- Infection
  —early (within 3 months) (see prophylaxis)
  —late, due to blood-borne infection from elsewhere. Patients with prosthetic joints should be given prophylactic antibiotics before 'risk' surgery (dental, genito-urinary, biliary, gastro-oesophageal, and colonic procedures).

Late postoperative Loosening, periarticular calcification.

Long-term effects ?absorption of traces of metal.

References

British National Formulary (1988). Summary of antibacterial prophylaxis. Table 216, 193–4. British Medical Association and Royal Pharmaceutical Society of Great Britain, London.

Charnley, J. and Efferkhar, N. (1969). Post-operative infection in total prosthetic replacement arthroplasty of the hip joint. *British Journal of Surgery*, **56**, 641–9.

Pollard, J. P., Hughes, S. P. F., Scott, J. E., Evans, M. J. and Benson, M. K. D. (1979). Antibiotics prophylaxis in total hip replacement. *British Medical Journal*, **i**, 707–9.

12 Trauma

Vascular injuries

90 per cent result from penetrating trauma. Vascular injury from blunt trauma may be in relation to road traffic accidents or fractures. Seatbelt injuries may produce circumferential tears in the intima of the abdominal aorta. Iatrogenic causes include tight plaster of Paris splints, and inadvertent intra-arterial injection of sodium thiopentone or STD during injection of varicose veins. Prompt diagnosis and treatment are required to save life or limb.

Types of injury

Punctures, contusions, lacerations or transection. Punctures and lacerations usually continue to bleed. Contusions may lead to occlusion of a vessel due to intimal dissection flaps. Transection injuries may stop bleeding due to vessel retraction.

Venous injuries may follow penetrating trauma, or the vessel may be compressed by surrounding haematoma or soft tissue swelling. Oedema and distal venous congestion will cause gangrene. If there is arterial and venous injury an arteriovenous fistula may develop.

Assessment of the patient

Examination of traumatized patients *must* be meticulous. The state of the peripheral circulation must be established and frequently reassessed—peripheral pulses, temperature, colour, capillary and venous return. Note absent distal pulsation or pulsating haematomas. If there is swelling, make use of the Doppler probe. This can produce false positives. In shocked patients, carry out repeated assessments after resuscitation. If there are signs of impaired circulation, carry out *arteriography*.

Principles of management

- Control bleeding by direct pressure.
- Relieve shock by resuscitation (see shock p. 132).
- Evaluate the peripheral circulation. ?ischaemia—angiography unless obvious fracture is present.
- Formally explore puncture wounds. This requires expertise with wide exposure of the vessels proximally and distally to secure control.
- Reduce fractures and dislocations.

Complications of vascular injury

Early
- Haemorrhage
- Thrombosis
- False aneurysm

Late
- Secondary haemorrhage
- Arteriovenous fistula
- False aneurysm
- Vascular insufficiency
- Ischaemic muscle contracture

Indications for surgery

Signs of vascular insufficiency; continuing bleeding; expanding haematoma, vascular bruit.

opened, you may encounter massive haemorrhage. Be prepared by having blood ready. Deliver the small intestine. Evacuate clot. If bleeding is rapid, pack with large gauze packs. Remove them slowly and control each bleeding point in turn.

Management of specific injuries

Spleen Remove completely if it is shattered. If not extensively damaged try to preserve it by repair, partial splenectomy, or enclosing the organ in a mesh sac.

Liver Drain small lacerations with closed suction drains or a soft corrugated drain brought out through a single opening in the abdominal wall below the level of the laceration so that dependent drainage occurs. Explore deep lacerations and suture or ligate bleeding points, then drain. If unsuccessful, suture or apply an omental pack into the laceration and suture the edges with catgut. If none of these is possible hepatic lobectomy or segmentectomy should be undertaken. Temporary control of bleeding can be achieved by clamping the free edge of the lesser omentum (Pringle's manoeuvre). If haemorrhage is due to hepatic vein fracture, proximal and distal control of the vena cava may permit haemostasis by identification of the veins from within the liver substance.

479

Kidney If pulped, carry out a nephrectomy. Did the intravenous urogram confirm function on the other side? If possible, repair or perform only a partial nephrectomy.

Bowel Close small-bowel injuries with a single layer of interrupted sutures. If the mesentery is injured repair the defect and resect compromised bowel with end-to-end anastomosis. Mobilize the duodenum if injury is suspected. Perforations are easily missed in the extraperitoneal part. Perform a feeding jejunostomy if it has to be repaired. Convert caecal injuries into a caecostomy. In other sites the injured bowel should be exteriorized and a double-barreled colostomy fashioned. Resect massive injuries. Bring out a proximal colostomy if there is doubt about the anastomosis.

Stomach Close wounds with vicryl or PDS. Pass nasogastric tube.

Pancreas Resect a damaged tail and drain the area. Emergency pancreatico-duodenectomy may be required for injuries to the head.

References

Bewes, P. C. (1983). Open and closed abdominal injuries. *British Journal of Hospital Medicine*, **29**, 402–10.

Blumgart, L. H. (1980). Hepatic resection. In *Recent advances in surgery*, **10**. (ed. S. Taylor), pp. 1–26. Churchill Livingstone, Edinburgh.

Odling-Smee, W. (1984). Abdominal injuries. *Sugery*, **1**, 348–53.

Thoracic injuries

Thoracic injury accounts for 25 per cent of deaths from trauma. 50 per cent of patients who die from multiple injuries also have a significant thoracic injury. Open injuries are caused by penetrating trauma from knieves or gunshot. Closed injuries occur after blasts, blunt trauma, and deceleration. Road traffic accidents are the commonest cause.

Open injuries

Complications are pneumothorax, haemothorax, intrathoracic visceral and arterial damage, and infection.

Management guidelines Aim to establish an airway, stop bleeding, and restore the circulation.
- Take a brief history.
- Establish pulse, blood pressure, and two intravenous lines for rapid fluid replacement.
- Do haemoglobin, HCT, blood gases, group and cross-match.
- Examine the chest back and front. Are there associated injuries?
- Cover or close sucking chest wounds. If breath sounds are absent on the side of the injury and the patient is unstable, drain the chest.
- Record the drainage. It is often considerable initially but slows down as the lung expands.
- When the patient is stable, get an erect PA chest X-ray.
- Persistent shock or bleeding are indications for thoracotomy.
- Aortography, gastrografin swallow, peritoneal lavage may be indicated.

Closed injuries

Complications are rib fractures, flail segment, injury of the aorta and its branches, myocardial contusion or rupture, ruptured diaphragm or oesophagus. Blast injuries are associated with intra-alveolar haemorrhage, pulmonary haematoma, and hypoxia.

Management guidelines Establish the airway, assess ventilation, correct blood loss.
- If the patient remains persistently shocked despite attempts at resuscitation, thoracotomy or laparotomy are indicated.
- Diaphragmatic rupture can occur immediately or at any time up to weeks after injury. Look for bowel (stomach or colon) in the chest. Examine the diaphragm at all laparotomies for trauma.
- Suspect oesophageal rupture if there is deep subcutaneous emphysema from the mandible to the clavicle or mediastinal emphysema on the chest X-ray.
- Blunt cardiac injury can lead to tamponade or ECG changes from contusion which can mimic infarction. If tamponade is suspected (blood pressure down, jugular venous pressure up, QRS changes on ECG) carry out pericardial aspiration to confirm the diagnosis and buy time until thoracotomy. Most cases are due to a tear in the right atrium. Consider bypass surgery.

Management
- Suprapubic catheterization. Carry out contrast studies at 3–4 weeks. If there is a stricture it may be managed by urethrotomy or urethroplasty.
- If the patient requires a laparotomy the bladder should be opened and a catheter 'railroaded' through the rupture to splint the urethra and drain the bladder.

Penile injuries

May occur in industry, civil or military hostilities (e.g. explosions) and bizarre sexual practices (e.g. inserting the penis into a vacuum cleaner).

Management Repair fractures with interrupted nylon sutures. Evacuate haematomas. Carry out full debridement if there is foreign material. Spatulate the urethra to prevent stricture and provide skin cover with full-thickness grafts.

Testicular injury

Common in sports or street fights.

Management Exclude rupture with scrotal ultrasound. Evacuate expanding haematomas. Excise an obviously infarcted testis.

483

Reference
Cetti, N. E. (1988). Trauma to the genito-urinary tract. *Surgery*, **63**, 1492–8.

Head injuries

More than one million people each year present at hospitals in the UK with a head injury. Twenty per cent will be admitted to a primary care ward (general surgery, A&E, orthopaedics) of whom 60 per cent are discharged within 48 hours. Prevention of secondary brain damage could reduce the morbidity and mortality. A change in the patient's conscious level is the most important indication of secondary brain damage.

Types of brain damage

Primary occurs at the time of impact. It results in brain contusions, lacerations, and diffuse white-matter damage. Concussion implies transient reversible mild diffuse brain injury. Although originally it was believed to be unassociated with pathological changes at neuronal level, it is now believed that the difference between mild and severe brain injury is in degree only and that even in patients with concussion there has been neuronal loss (Oppenheimer 1968).

Post-traumatic amnesia is the time period from injury to the return of normal memory. It is a useful guide to the severity of head injury implying an optimistic outlook, if less than 24 hours.

Secondary brain damage includes haemorrhage, brain swelling, damage secondary to raised intracranial pressure, hypoxia, and infection. Therefore careful assessment of head-injured patients is vital to prevent or treat complications.

Assessment

The most important guide is a change in the patient's conscious level. This can be recorded by the Glasgow Coma Scale which has a low interobserver variability rate and is used by most neurosurgical units and primary hospitals in the UK (Gentleman and Teasdale 1981).

The Glasgow Coma Scale

Function	Response	Score
Eye opening	Spontaneous	4
	To speech	3
	To pain	2
	None	1
Best verbal response	Orientated	5
	Confused conversation	4
	Inappropriate words	3
	Incomprehensible sounds	2
	None	1
Best motor response	Obeys commands	6
	Localizes	5
	Flexes—normal	4
	—abnormal	3
	Extends	2
	None	1

Pupil size may be affected by hypoxia, alcohol, and hypotension. Defer its assessment until resuscitation is complete or sobriety has returned.

Investigations

CT scanning is most useful. Other measures include monitoring of intracranial pressure and evoked electrical potentials.

Management

Mild injuries in A&E Admit any patient for at least 24 hours observation if there is a skull fracture on X-ray or anything more than well-documented concussion (includes rhinorrhea, otorrhoea, and depressed skull fractures). All patients, or better their relatives or attendants, should be given a 'Head Injury' card upon discharge which carries instructions on what to do if there is vomiting, drowsiness, abnormal movement of limbs, etc.

Patients with an intracranial haematoma If there is deterioration in conscious level which cannot be attributed to extracranial causes suspect an intracranial haematoma and arrange transfer to a neurosurgical unit. Patients at risk should ideally be identified early and a CT scan arranged so that elective evacuation of clot can be performed. There are few situations which justify emergency burr holes.

485

Patients in coma without haematoma require intensive care. Increasing age and the extent of diffuse brain damage discriminate against a favourable outcome.

References

Gentleman, D. and Teasdale, G. (1981). Adoption of Glasgow Coma Scale in the British Isles. *British Medical Journal*, **283**, 408.

Oppenheimer, D. R. (1968). Microscopic lesions in the brain following head injury. *Journal of Neurology, Neurosurgery and Psychiatry*, **31**, 299–306.

Teasdale, G. and Jennett, B. (1981). In *Craniocervical Injury in Trauma* (eds D. Carter and H. C. Polk), pp. 40–54. Butterworths, London.

13 Gynaecological problems in general surgery

487

Gynaecological causes of an acute abdomen

Consider acute salpingitis, tubo-ovarian abscess, complications of functional ovarian cysts, adnexal torsion, abortion and ectopic pregnancy (p. 490).

Acute salpingitis

Common organisms are *E. coli*, anaerobic streptococci, *Bacteroides fragilis. Neisseria gonorrheae* and *chlamydia* are sometimes implicated. Usually both tubes are affected.

Risk factors Surgical procedures on the cervix, pelvic sepsis, after childbirth, abortion, intrauterine device (IUD). Venereal disease.

Clinical features Malaise, lower abdominal pain, temperature over 39 °C. Rebound tenderness in both iliac fossae. Purulent vaginal discharge. Vagina feels hot ('hot toobs'—American medical jargon).

Management FBC. MSSU. Pregnancy test. High vaginal swab. Treat with intravenous cefuroxime or gentamicin and metronidazole. Carry out laparotomy if there is doubt.

Complications Tubo-ovarian abscess. Treat with antibiotics as above. Surgery is indicated for rupture.

Complications of ovarian cysts

Haemorrhage Most are follicular or corpus luteum cysts. Manage the patient expectantly. Resolution occurs in 4–8 weeks. Review the patient regularly.

If extensive haemorrhage is found during surgery for a presumed appendicitis, simple cystectomy with oversewing of the excised area and suction drainage should be undertaken.

Diagnosis
- Pregnancy test negative suggests corpus luteum rupture.
- Pregnancy test positive ?ectopic pregnancy. Abdominal ultrasound may exclude ectopic pregnancy. Laparotomy is mandatory if there is sufficient doubt.

Torsion occurs at any age but it is uncommon in postmenopausal women, usually in association with ovarian cysts or neoplasm.

Clinical features Colicky or constant lower abdominal pain with tenderness or a mass on the affected side usually accompanied by vomiting.

Diagnosis Pelvic ultrasound. Laparoscopy.

Treatment Laparotomy. Determine ovarian viability. Carry out ovarian cystectomy or oophorectomy according to the findings. Examine the other ovary. Exclude neoplasm. If this is suspected in a patient over 40 years old carry out bilateral salpingo-oophorectomy ± hysterectomy. Hormone replacement therapy

Surgical procedures

- Punctures and clean lacerations: primary closure
- Ragged lacerations and intimal flaps: excision and reconstruction with venous patch or excision with anastomosis
- Intramural haematoma, vascular wall disruption, intimal disruption: excision and interposition, graft

N.B. Irrigate the distal vascular tree with heparinized saline. Interposition grafts are necessary if the repair causes tension, stenosis or there is extensive local sepsis.

Abdominal trauma

Causes

- *Blunt*: road traffic accidents, industrial injuries, sporting accidents
- *Penetrating*: stab wounds, low- and high-velocity missile injuries

N.B. Resuscitation may have to be carried out concurrent with examination.

History

Trauma of sufficient force. Ask about shoulder-tip pain (splenic injury). Ask about the position of the victim at the time of injury if penetrating.

Examination

Is the patient shocked? Note the presence of patterned bruising suggestive of severe pressure. Measure the abdominal girth and mark the site for future measurements. Note whether there is distension, localized tenderness, guarding, or rigidity. 'Spring' the pelvis. Do a rectal examination. Is there blood on the finger? Ask the patient to pass urine if conscious. Failure may imply urethral injury. Are there associated injuries, e.g. head, rib, cervical spine, etc?

Investigations

- Chest X-ray (erect if possible): Look for diaphragmatic injury, rib fracture near the spleen, subphrenic air suggestive of ruptured viscus.
- Erect and supine abdominal X-ray: Look for splenic or renal shadows, the position of the gastric bubble and whether it is displaced. The psoas shadow, is it obscured by haematoma? Look at the pelvis. Is it fractured?
- Peritoneal lavage is an accurate method of establishing the presence of bleeding.
- Four quadrant tap with a no. 1 needle has a high false positive rate and is not so valuable.
- Intravenous urography, contrast radiography, arteriography, and CT scans are occasionally indicated.

Indications for laparotomy

- Eviscerations
- Gunshot wounds (90 per cent have intra-abdominal damage)
- Copious blood on peritoneal lavage
- Patterned abrasions or shoulder-tip pain
- Continuing shock despite resuscitation
- Subphrenic gas
- Signs of spreading peritonitis

Laparotomy

Use a vertical paramedian or midline incision. Make it separate from any stab wound. Examine every organ in order (see laparotomy, p. 710). Note your findings. Remember that often more than one organ is damaged. When the peritoneum is by subcutaneous implantation of oestradiol 100 mg and testo-sterone 100 mg should be given every 6–9 months to prevent symptoms of premature menopause.

Abortion

Criminal abortion is now infrequently encountered but can lead to septic complications like pelvic peritonitis and endotoxic shock which may call for a general surgical opinion.

Ectopic pregnancy

The acute abdomen is more difficult to diagnose in women than in men. Keep the possibility of ectopic gestation in mind. It is an important cause of premature death in young women.

Pathophysiology

Implantation of the fertilised ovum anywhere other than the uterine mucosa is an ectopic pregnancy. 95 per cent are in the Fallopian tube. The ampulla is the most common site, followed by the isthmus. Some occur in the ovary, peritoneum, or intestinal portion of the tube. The chorionic villi erode the tubal wall, leading to bleeding and separation of the ovum with abortion into the peritoneal cavity or tubal rupture.

Risk factors

Acute or recurrent salpingitis, tubal ligation (ectopic pregnancy can occur after this procedure), tubal repair, previous ectopic pregnancy.

Clinical features

A history of menstrual abnormality is frequently present. Symptoms and signs are variable. Fever is not a feature. Pain is the most common symptom but may be constant, intermittent, localized, or diffuse. Shoulder tip pain may represent intra-peritoneal blood and diaphragmatic irritation. An adnexal mass may be present in 30 per cent of patients.

Acute tubal rupture causes severe abdominal pain with hypovolaemic shock and shoulder tip pain.

Diagnosis

- Ultrasound may identify a gestational sac or free fluid in the peritoneal cavity.
- Pregnancy testing may be positive in more than 25 per cent.
- Culdocentesis (aspiration of blood through the posterior vaginal fornix). A positive result justifies laparotomy.
- Laparoscopy is reserved for non-shocked patients in whom there is no clear-cut indication for laparotomy.

Differential diagnosis

- Abortion (there is usually amenorrhoea more than 8 weeks)
- Salpingitis (temperature is often greater than 39 °C)
- Appendicitis, cystitis, complications of ovarian cysts

Treatment

Laparotomy ± simultaneous blood transfusion. Identify the site of the pregnancy and perform salpingectomy (partial or total depending on the findings, the patient's condition, and her wishes for future pregnancies). Evacuation of the pregnancy with tubal repair is possible in selected patients but requires an experienced microsurgeon.

Prognosis

Impaired fertility almost always follows. Further ectopic pregnancy is a risk.

14 Tropical diseases

Surgery of tropical diseases

Although the principles of surgical care are the same in the tropics as in developed countries there are some important differences.

- In tropical and subtropical countries resources are often extremely limited.
- There are major transport problems.
- Treatment is often complicated by underlying disease such as anaemia, malnutrition, and sickle cell disease.
- There are differences in cultural attitudes to disease, and ignorance of surgical possibilities and trust in modern techniques amongst patients.
- The presentation of disease may be modified by treatments prescribed by indigenous 'medical' practitioners.

Therefore, surgery in tropical countries requires flexibility both in clinical management and in operative technique.

Doctors and surgeons irrespective of where they work are expected to be knowledgeable on common tropical diseases both in diagnosis and treatment.

The surgical complications of typhoid and paratyphoid

Chloromycetin is effective in treating typhoid and paratyphoid, and complications are consequently uncommon. When they do occur it is important to review the adequacy of specific chemotherapy as well as dealing with the complication.

Chloromycetin destroys the organisms responsible for the production of vitamin B complex, which must be replaced.

Intestinal perforation

This can occur during the first, second, or third weeks. In the first week there may be multiple tiny perforations along the anti-mesenteric border of the ileum at the site of Peyer's patches. There is usually a fibrinous exudate.

In the third week perforations tend to be solitary and are occasionally the first intimation of the disease. The perforation is usually in a typhoid ulcer parallel to the long axis of the gut, in the lower ileum. In paratyphoid B, large-bowel perforation may occur.

Surgical closure of the perforations and chloromycetin therapy is indicated in most cases.

Other complications of typhoid and paratyphoid

- Paralytic ileus is common
- Intestinal haemorrhage must be distinguished from purpura with intestinal symptoms and intussusception. Urgent blood transfusion and Widal reaction should be carried out.
- Cholecystitis with perforation can occur. Typhoid bacilli in the bile of patients with chronic cholecystitis can produce a carrier state which may persist for the rest of the patient's life. The bacilli are excreted in the faeces or urine and lead to isolated outbreaks of the disease wherever the carrier goes. One such affected person was the cook 'Typhoid Mary' who infected food wherever she worked.
- Phlebitis especially of the left common iliac vein
- Genito-urinary typhoid
- Bone and joint infection
- Myositis
- Parotitis
- Laryngitis with airway obstruction

Amoebic liver abscess

This is a complication of amoebic hepatitis which is secondary to amoebic dysentery.

Pathology

From focal lesions in the colonic wall *Entamoeba histolytica* gains access to the upper and posterior portion of the right lobe via the portal vein. The organism then produces local necrosis proportional to the size of the colony, the resistance of the host, and extent of secondary infection. The pus is characteristically chocolate-coloured due to liver cell liquefaction. It usually contains staphylococci, streptococci, *E. coli*, and *E. histolytica*.

The liver enlarges, usually in a cephalad direction, and the abscess may rupture into the pleural, pericardial, or peritoneal cavities or perforate into a hollow viscus.

Clinical features

Most occur in young men following an attack of amoebic dysentery. The patient loses weight, becomes anaemic with rigors, pyrexia, and night sweats.

There is rigidity and tenderness in the right upper quadrant and lower chest, right shoulder-tip pain, and right-sided basal changes including pleural rub.

Chest X-ray

Elevation of the right cupola of the diaphragm is characteristic.

Investigations

Liver scan. Aspiration of pus—mobile amoebae may be seen.

Treatment

Medical
- Flagyl 800 mg three times daily for 7–10 days
- Emetine 40 mg intramuscularly daily for 7 days

Surgical Surgical drainage is indicated when the abscess is not controlled medically or by repeated aspirations and when the abscess ruptures or extends to form a left lobe abscess.

At laparotomy the wound and abdominal packs should be soaked in Emetine to prevent cutaneous amoebiasis. The cavity is opened and aspirated. Its walls are cleaned with swabs but no drain is inserted to prevent cutaneous amoebiasis.

Filariasis

The causal organism *Wuchereria bancrofti* is transmitted to humans by the bite of many genera of mosquitoes. The disease is widespread in tropical and subtropical areas (India, Africa, China, the West Indies, and Australia).

Effects

The filarial worms lead to acute and chronic lymphadenitis and lymphangitis, especially of the lower limbs and genitalia. Chronic lymphatic obstruction results with the development of gross lymphoedema and elephantiasis of the legs. (Massively oedematous lower limbs usually below the knee.)

Treatment

Acute lymphadenitis
- Rest of the affected part
- Antibiotics (ampicillin 500 mg by intramuscular injection twice daily for 10 days) are used to treat secondary infection caused by beta-haemolytic streptococcus and *Staphylococcus aureus*.
- Specific antifilarial drugs, diethylcarbamazine citrate (Hetrazan, Banocide) 2 mg/kg orally three times daily for 3 weeks
- Surgical drainage of abscesses

Chronic lymphatic oedema (elephantiasis) There is no satisfactory operation. In one procedure the abnormal subcutaneous tissue is excised and the affected area is covered with a split skin graft. A variation is to excise the skin of the leg in long strips. The subcutaneous tissues are then excised and the skin reapplied to the denuded tissue. A plaster of Paris dressing is applied. The results are satisfactory but certainly not cosmetic.

Sickling disorders

The substitution of valine for glutamic acid in the sixth position peptide of the B chain leads to **sickle cell anaemia** (homozygotes) or **sickle cell trait** (heterozygotes). Sickle cell S-haemoglobin forms crescent-shaped rods when in the reduced state, hence the name.

The conditions are found in countries where malaria is endemic, and the trait confers a degree of protection against falciparum malaria.

Clinical importance

Patients with sickle cell anaemia suffer from vaso-occlusive episodes due to low oxygen tension in the tissues. The abnormally shaped red blood cells cannot pass through arterioles and capillaries. Thus infarction, pulmonary emboli, arthralgia, and intercurrent infections (TB or *Salmonella osteomyelitis*) are frequent complications. Few survive more than 40 years.

Patients with sickle cell trait have no clinical disabilities but may suffer from sickling episodes when at altitude or flying in unpressurized aircraft. The sickling tendency may lead to splenic infarction.

Equally, vaso-occlusive episodes may occur if hypoxia results from a general anaesthetic.

Precautions

Risk patients must be identified before elective, if not all, surgical procedures.

- Carry out a 'sickling test' in suspected patients. Add a reducing agent (e.g. CO_2) to an unstained drop of blood. Homozygote blood sickles in a few hours. Heterozygote blood takes up to 24 hours.
- If the sickling test is positive, ensure adequate oxygenation during general anaesthesia and prevent dehydration.

503

Schistosomiasis (bilharzia)

Causal organisms

Schistosoma haematobium (affinity for vesical plexus), *S. mansoni* (inferior mesenteric vein), *S. japonicum* (superior mesenteric vein).

Infestation occurs when bathing or standing in infected water. The intermediate host is a snail (*Bullinus contortus*) which inhabits slow-running water throughout Africa, Syria, Israel, Iraq, and Arab countries. Multiplication of larvae occurs in the snail. They become free-swimming and enter the skin of the legs of humans bathing, washing, or working in the water. They then shed their tails and are swept by the bloodstream to all parts of the body. Only those which reach the liver survive and develop into male and female worms living on erythrocytes. When they are mature they leave the liver and swim against the bloodstream to reach the venous plexuses and veins for which they have an affinity. The worms mate and the ova pass from the veins into the lumen of the intestine or bladder to be expelled in faeces or urine. The eggs hatch and the ciliated larva (miracidia) enter the snail within 26 hours.

S. mansoni and S. japonicum produce intestinal ulcers, polyps, inflammation, and peritonitis. Fibrosis and cirrhosis of the liver may occur with portal hypertension. The brain, lungs, and spinal cord may also be involved. The patient experiences symptoms ranging from malaise to fever and abdominal discomfort, distension (due to ascites), and pain. Bloody diarrhoea is a feature.

S. haematobium produces urinary symptoms such as increased frequency. The bladder becomes ulcerated, with sloughing of areas of mucosa and subsequent contraction. There is an association with squamous cancer.

Diagnosis

- Centrifuge and examine microscopically an early morning urine. Demonstration of living eggs confirms infestation. Dead eggs imply past infection.
- Examination of the stools
- Examination of biopsy of bladder or rectal mucosa

Treatment

Medical Antimony preparations are effective in the early stages of infestation before fibrosis has set in. Antimony sodium tartrate is used IV on alternate days in doses of 3–120 mg/day. Metriphonate and praziquantel are also effective. The dead worms take many months to be expelled.

Surgical The need for surgery arises from the development of complications, e.g. portal hypertension, urethral stricture, granulomas of the gut. Each surgical procedure should deal specifically with the complication.

Malaria

This disease is endemic in Africa, South-east Asia, and South America. Attempts to eradicate it have been unsuccessful because of economic factors and the development of resistance to DDT of mosquitoes and resistance of the malarial parasites to drugs.

Causal organisms

- *Plasmodium falciparum*: malignant tertian malaria
- *Plasmodium vivax*: benign tertian malaria
- *Plasmodium ovale*: benign tertian malaria
- *Plasmodium malariae*: quartan malaria

Clinical features

The incubation period is 10–14 days. Intermittent fever with sweating, chills, or rigors is characteristic. This is followed by an afebrile period and the reappearance of the fever in a definite pattern according to the species. The parasites are intra-erythrocytic and may lead to splenomegaly and occasional jaundice.

Diagnosis

Based on a history of exposure (e.g. tourists) and the clinical features. Microscopy of a stained blood smear increases the chance of finding parasites.

Treatment

Uncomplicated malaria A 3-day course of antimalarial drugs is sufficient to relieve symptoms and effect the disappearance of parasitaemia as follows: initial dose chloroquine (adults) 600 mg orally followed by 300 mg 6 hours later then 300 mg daily for 2 days.

Prophylaxis and suppression A dose appropriate to age should be taken once a week on the same day each week (see BNF, infections section).

Complications

Relate mainly to falciparum-induced malaria.
- *Blackwater fever*: Hyperpyrexia, coma, acute renal failure, haemoglobinuria
- *Malarial crises*: Patients may develop hyperpyrexia and haemolytic anaemia with severe headache, delirium, and death (cerebral malaria)

Treatment

Rehydrate the patient. Transfuse blood and give steroids for cerebral malaria. Drug treatment is with quinine hydrochloride 500–600 mg in 15 ml saline IV over 10 mins/8 hours (1500–1800 mg daily for 7–10 days).

Prognosis

Untreated vivax malaria subsides in 10–30 days but may recur intermittently. Intercurrent infection worsens the prognosis. Untreated falciparum malaria is frequently fatal.

Malaria in surgical practice

There are varying degrees of immunity in the indigenous population. Malaria can present as an acute abdominal emergency with pain, pyrexia, backache, and vomiting or as unexplained pyrexia or deterioration postoperatively. In visitors to the UK from tropical countries fulminant forms are common.

Leprosy (Hansen's disease)

Leprosy is endemic in much of Africa, Southern Asia, the Far East, and southern America. It affects about 15 million people and is caused by an acid-fast bacillus, *Mycobacterium leprae*. It is usually contracted in late childhood or adolescence, the likely source of infection being nasal discharge from infected patients and not the skin lesions.

Pathology

The bacilli infiltrate the dermis, upper respiratory tract, and peripheral nerves predominantly, although they also accumulate in the liver, bone marrow, kidneys, and spleen. The dermis becomes replaced to varying degrees by bacilliferous tissue which destroys pigment-forming cells, hair follicles, sweat, and sebaceous glands. A characteristic of leprosy is its predilection for peripheral nerves.

Classification

The type of lesion which develops depends on the immune response of the host:

Tuberculoid leprosy There is good cell-mediated immunity and bacilli are scant in numbers.

Lepromatous leprosy Cell-mediated immunity is depressed. Bacilli are present in large numbers.

Indeterminate leprosy Patients with this type of disease are unstable and may drift towards lepromatous leprosy if untreated. With treatment they will shift towards the tuberculoid disease.

Clinical features

Consider the diagnosis of leprosy in any patient who presents with a combination of neural and dermatological disorders.

Tuberculoid leprosy causes sharply localized lesions, usually only affecting one part of the body. It may present as a small area of hypopigmentation or as a dry, hairless, and anaesthetic lesion. Thickening of peripheral nerves is common, especially the ulnar and peroneal nerves. Deformity occurs early but is localized.

Lepromatous leprosy is symmetrical and extensive, presenting as a widespread hypopigmented or erythematous rash of the face, limbs, and trunk. Associated iritis is common. There is general malaise, fever, and joint pain. The skin and peripheral nerves become thickened, initially painful then areas of anaesthesia develop with neuropathic ulceration.

Diagnosis

Based on
- A history of contact
- The clinical findings
- Histological confirmation of *M. leprae*

In practical terms, infected tissue is usually obtained by taking a smear with a scalpel blade inserted into the pinched skin of the eyebrow or earlobe. The tissue fluid obtained is stained with a modified Ziehl–Neelsen stain.

Treatment

- Dapsone (diaminodiphenylsulphone) 50–100 mg daily
- Combination chemotherapy is often used to reduce the incidence of dapsone resistance, e.g.
  - —Rifampicin 600 mg once per month
  - —Dapsone 100 mg daily
  - —Clofazimine 50 mg daily + 300 mg once per month

Treatment continues for at least 2 years and often for life. Contacts receive BCG vaccination or prophylactic dapsone.

Surgery is indicated to correct:
- *Primary deformities* caused directly by the disease, e.g. thickening of the skin, paralysis of the eyelids, paralysis of the hands and feet.
- *Secondary deformities* relating to anaesthetic skin. This involves educating the patient to look after himself and prevent injury. Anaesthetic limbs may become severely damaged and amputation may be necessary.

Ascariasis

The roundworm, *Ascaris lumbricoides*, commonly infests the intestines of children and adults in eastern and south-eastern Africa, Sri Lanka, southern and south-eastern India, and Bangladesh. The incidence of the condition is related to the state of development of the sewage systems.

Clinical features

The patient may present with symptoms of indigestion and weight loss. Pallor and abdominal colic are common. Large worm loads lead to anaemia, hypoproteinaemia, and protein-losing enteropathy.

Diagnosis

The ova are demonstrated by examining the stools.

Treatment

- Piperazine: 1 dual dose sachet repeated after 14 days (adults) 1/3 sachet (3–12 months) 2/3 sachet (1–6 years) 1 sachet (over 6 years)
- Mebendazole: 100 mg twice daily for 3 days (adults and children over 2 years)

Piperazine causes flaccid paralysis of roundworms and threadworms and thus permits their expulsion by peristalsis.

Mebendazole immobilizes the worms by disrupting their transport systems.

Complications

Roundworm intestinal obstruction can occur due to impaction in the terminal ileum or right colon. Depending on the worm load, this may lead to intussusception or mechanical obstruction. Often there is an abdominal mass which changes position from day to day. A plain abdominal X-ray or barium meal may demonstrate radiolucent lines within a dense shadow which represents individual worms in a large worm load. Subacute obstructions may be treated with antihelminthics. Laparotomy is indicated in complete obstruction. If the intestines are not inflamed the worm load is squeezed into the caecum and colon where it will be removed by peristalsis. Alternatively it may be removed via an enterostomy. Any non-viable gut should be resected and continuity restored by end-to-end anastomosis.

Peritonitis due to perforation of the bowel should be differentiated from peritonitis associated with typhoid ulceration and incidental roundworm infestation. Both conditions will require to be treated.

Roundworm appendicitis

Biliary and pancreatic ascariasis Roundworms can enter the biliary tract through the ampulla of Vater causing bile duct strictures, liver abscess, cholangitis, and empyema of the gall-bladder. Treatment involves the use of antispasmodics to allow

the sphincter to relax. The worm load is then killed by anti-helminthics. Worms may also be removed by ERCP or at laparotomy.

Roundworm granuloma

Guinea worm infestation

This freshwater-borne disease is common in the Middle East, south-east Asia, certain parts of Africa, and India. The infestation enters the stomach and duodenum via drinking water contaminated with the freshwater arthopod cyclops which consumes the larva of the Guinea worm (*Dracunculus mediensis*).

Pathology

From the stomach the larvae eventually find their way to the connective tissue of the abdominal wall where they mature and mate. The females then migrate to areas of the body likely to be submerged in water such as the lower leg where the eggs are laid. The worm produces a proteolytic toxin which leads to blister formation. When the blister bursts the worm has to be physically extruded, usually by winding it round a matchstick a few turns a day.

Complications

Cellulitis, abscess and sinus formation, allergic conjunctivitis, osteomyelitis, arthritis.

Soft tissue radiology

When the worms die they may be visible as calcified areas in the subcutaneous tissue of the soles of the feet, legs, groin, scrotum, and back.

515

Threadworms (*Enterobius vermicularis*)

This nematode infests the large bowel, particularly the sigmoid colon, rectum, and anal canal.

Clinical features

There is chronic and sometimes severe anal irritation with associated excoriation of the perianal skin. Vaginitis and urethritis are associated complications.

Diagnosis

Proctoscopy and anal or rectal mucosal smear. (This is done using a wooden spatula to the end of which adhesive tape has been applied, adhesive surface outwards. The skin or mucosa is touched with the tape and smeared on to a slide, thus enabling the diagnosis to be made.)

Treatment

Treat the whole family.
- Piperazine: 1 dual dose sachet repeated after 14 days orally 1/3 sachet (3 months–1 year), 2/3 sachet (1–6 years) 1 sachet (over 6 years)
- Mebendazole: 100 mg orally repeated after 2–3 weeks (over 2 years)

15 Organ transplantation

Guidelines on obtaining permission for organ transplantation

- Responsibility for the identification of potential donors lies with the consultant under whose care the patient is.
- Confirm the diagnosis of brain death (p. 36).
- Seek your consultant's permission before approaching the relatives about organ donation.

General criteria for potential donors

- The patient is brain dead with an intact circulation.
- Death is due to head injury, cerebral/subarachnoid haemorrhage, primary brain tumour, cardiac or respiratory arrest.
- The following should be absent: malignant disease (except primary brain tumour), systemic sepsis, prolonged shock, primary renal disease, severe chronic hypertension.

Specific organ criteria

- Kidney: Age 2–70 years with acceptable renal function
- Heart: Age 1 month–50 years with no known cardiac disease
- Heart–lung: As above. No pulmonary disease or trauma. PO_2, PCO_2 levels acceptable on less than 50 per cent inspired oxygen.
- Liver: Age 1 month–55 years. No known liver disease, drug addiction, or hepatitis B.
- Cornea: Age—all adults up to 12 hours after death provided there is no history of corneal disease or infection

Contact with the transplant team

This is the consultant's responsibility, which may be delegated to a junior doctor. Contact may be made before brain death is diagnosed. The co-ordinator will be able to give advice on criteria for organ donation, how to obtain consent from the next of kin or cohabitees, the coroner's involvement if necessary, screening for HIV and hepatitis B, and blood-grouping the donor.

Consent and counselling of relatives

'Staff need to decide in the light of the individual circumstances who is best qualified to approach the relatives. ... It would normally be a hospital doctor of some seniority and experience in carrying out personal interviews. ... There may be occasions though when the person best qualified to be given this responsibility is a senior nurse, a chaplain or the family doctor.' *Cadaveric organs for transplantation* HMSO (1983).

The relatives should not be approached until death has occurred, and confidentiality must be maintained. Basic counselling and follow-up support should be given to all donor families. The transplant co-ordinator can help with these arrangements.

Reference

Cadaveric organs for transplantation—a code of practice including the diagnosis of brain death (1983). HMSO, London.

Useful addresses

The British Organ Donor Society, Balsham, Cambridge CB1 6DL. Telephone 0223 893636. Members comprise the families of organ donors.

CRUSE (local branches in most areas) 126 Sheen Road, Richmond, Surrey TW9 1CR. Telephone 01 940 4818/9047.

Organ transplantation

This has revolutionized the treatment of organ failure by substituting healthy organs from donors (alive or dead) for end-stage organs in recipients. Avascular organs such as cornea or cartilage provoke little or no host reaction and are widely used. The successful transplantation of organs like the kidney, liver, heart, and lung depend not only on competent surgical techniques but also on a favourable immunological environment.

Organ donors

- *Living*: Kidney, pancreas, bone marrow
- *Cadaveric*: Kidney, pancreas, cornea, heart, lungs, liver

Although the establishment of definitive criteria for brain death have increased the availability of potential donors, the demand for organs still outstrips the supply and the harvesting of multiple organs from a single donor has become common. The subsequent cold storage of certain organs (e.g. kidney, pancreas) permits their transport to recipients and implantation 12–48 hours later. The heart and liver must be transplanted within a few hours. When a heart–lung transplant is to be carried out, both donor and recipient are operated upon in the same theatre.

Tissue matching

The homograft reaction can be reduced by a combination of accurate tissue matching between donor and recipient and the use of immunosuppressive therapy. ABO compatibility is essential. The main histocompatibility antigens which are on the surface of host cells are of the human leucocyte associated type (HLA). These are genetically determined, and currently four subgroups have been defined—A, B, C, D. Absolute compatibility gives the best chance of a successful transplant, but in practice at least two matching HLA antigens between donor and recipient are acceptable for living donors. This is less important for cadaver organs, but matching the common HLA-A, HLA-B, and HLA-C antigens as closely as possible is accepted practice.

Immunosuppression

Some form of immunological control is necessary unless there is absolute histocompatibility as in donation from an identical twin. Agents used include azothiaprine (Imuran), corticosteroids, cyclosporin A, antilymphocytic serum, and monoclonal antibodies.

Complications

Pre-existing disease Hypercalcaemia is frequently present in patients with renal disease. It is due to secondary hyperparathyroidism which may become tertiary even after correction of the renal problem. Treatment is by subtotal parathyroidectomy before transplantation.

Technical problems Most transplantations are not technically difficult. However, leaks and stenosis of the vascular and ureteric anastomoses can occur.

Immunosuppression
- Infection of some kind occurs in 40 per cent of patients.
- Metabolic disease, e.g. diabetes mellitus, osteoporosis may be steroid-related.
- *De-novo* malignant disease, especially lymphoma, is more frequent.
- Poor wound healing, Cushing's syndrome, and psychosis do occur and are usually steroid-related.
- Peptic ulceration and pancreatitis may be drug-related.

Rejection From time to time patients may experience a 'rejection crisis' heralded by flu-like symptoms. Counselling helps patients to recognize this. Successful control may be achieved by increasing the dose of steroids.

Renal transplantation

This is the cheapest and most effective treatment for end-stage renal failure. Minimal ischaemia renal damage is essential for success.

Definitions

Warm ischaemic time is defined as the period from the cessation of circulation until perfusion of the organ with perfusion fluid (see below) at 5–8 °C. Cold ischaemic time is the period from perfusion with perfusion fluid to the establishment of the new circulation within the transplant. Total ischaemic time is the period throughout which the organ is deprived of its circulation and thus comprises both warm and cold ischaemic times.

Donor

Living donor A nephrectomy is performed removing as much artery and vein as possible with a good length of ureter. The patient receives 5000 units of subcutaneous heparin twice daily for 3 days to prevent pulmonary embolus. The recipient is prepared in an adjacent theatre. An incision is made in the lower quadrant parallel to the inguinal ligament, the external iliac vessels are exposed and the renal artery anastomosed end-to-side to the external iliac or end-to-end to the internal iliac artery. The renal vein is anastomosed to the external iliac vein and the ureter implanted into the bladder.

Cadaver donor Brain death is confirmed. Ventilation and heart-beat are maintained. The patient is taken to theatre. When multiple organs are to be removed the kidneys and liver are prepared by cannulation of the aorta, vena cava, and portal circulation, and are perfused with cold perfusion fluid while the heart is being removed. They are then removed and transported packed in a plastic container containing saline or dextrose which is kept at 5 °C by placing it in crushed ice. Alternatively the artery may be connected to a pulsatile hypothermic perfusion machine.

Postoperative care

Immunosuppression Cyclosporin A is given intravenously for 24 hours in a dose of 4 mg/kg in 500 ml saline, then orally (12 mg/kg/day). It may be used alone or in combination with azathiaprine and prednisolone.

Urinary output is measured hourly. Fluid replacement is via a CVP line with dextrose saline solution. Measured and insensible losses (up to 500 ml) must be accurately replaced ml for ml.

Complications

Acute tubular necrosis, rejection, urinary tract obstruction, renal infarction.

Management of oliguria

This ranges from clearing a blocked urinary catheter to the assessment of renal vascular perfusion with isotope renography or digital vascular imaging. Ultrasound is particularly valuable in the detection of perirenal collections or urinary tract obstruction. Percutaneous needle biopsy will confirm the presence of rejection.

Results

- Cadaver kidney 1 year survival rate is 80 per cent.
- Live donor kidney 1 year survival rate is 95 per cent.

Perfusion fluid

Usually consists of albumin with added benzyl penicillin, hydrocortisone, glucose, and electrolytes.

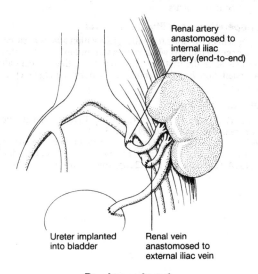

Renal artery anastomosed to internal iliac artery (end-to-end)

Ureter implanted into bladder

Renal vein anastomosed to external iliac vein

Renal transplantation

Liver transplantation

The liver is the least susceptible to graft rejection of all transplanted organs.

Indications

- Parenchymatous liver disease (e.g.biliary atresia, primary biliary cirrhosis, Budd–Chiari syndrome)—the commonest indication
- Primary malignant disease of the liver provided the presence of metastases is excluded by CT scan and venocavography

Procedure

The liver is placed in the bed of the removed organ. Usually the donor organ is harvested at the same time as the kidneys and heart. It must be transplanted within 6 hours of removal and is stored cold.

Anastomoses

- Donor vena cava to recipient vena cava end-to-end above and below the liver
- End-to-end anastomosis of portal vessels
- Donor coeliac artery to recipient common hepatic artery
- Biliary tract anastomosis with T-tube drainage
- An alternative biliary anastomosis makes use of the donor gallbladder as a conduit between the two biliary tracts.

Complications

Sepsis is the commonest cause of death. Rejection occurs in 20 per cent of patients.

Perioperative mortality and survival

18 per cent of transplanted patients die within a week of surgery. 30 per cent of patients will survive a year or more. The mortality rate falls after the first year. Survival more than 8 years after transplantation is documented, as is normal pregnancy and delivery.

References

Calne, R. Y. and Williams, R. (1979). Liver transplantation. *Current Problems in Surgery*, **16**, 3–44.

Guillou, P. J. (1987). Organ transplantation; the present position. In *Current surgical practice*, Vol. 4, (eds J. Hadfield and M. Hobsley), pp. 193–216. Edward Arnold, London.

Pancreatic transplantation

The procedure is justified in juvenile onset diabetic patients with
renal failure who have had a successful renal transplant. The aim
is to provide endogenous insulin and prevent microangiopathic
complications.

Technical difficulties

Whole-gland transplants have been abandoned for segmental
grafts, usually of the tail placed extraperitoneally in the opposite
iliac fossa to that of the kidney. The main difficulty is how to deal
with the pancreatic duct. It may be occluded with latex or
neoprene, but drainage into a Roux loop of jejunum or the
stomach is more successful. As a result of such difficulties,
isolation of islet cells with subsequent injection into the liver via
the portal vein and subcapsular renal injection have aroused
considerable interest.

Complications

- Sepsis
- Non-healing of anastomoses due to immunosuppression
- Fistula formation
- Rejection

One-year transplant Survival: 30 per cent.

References

Calne, R. Y. (1984). Paratopic segmental pancreas grafting: a
    technique with portal venous drainage. *Lancet*, **i**, 595–7.
Groth, C. G. *et al.* (1982). Successful outcome of segmental
    human pancreatic transplantation with enteric exocrine
    diversion after modifications in technique. *Lancet*, **ii**, 522–4.

Heart and heart–lung transplantation

Heart transplantation

Indications End-stage heart disease not responsive to medical treatment in patients under 50 years old with no infections and minimally elevated pulmonary vascular resistance.

The heart ± lungs are removed while still beating at the time of multiple organ harvesting. They are then stored at 4 °C and implanted into the recipient as soon as possible. Although ABO compatability is essential, HLA mismatching does not adversely affect transplant survival.

The donor heart is appropriately fashioned so that the recipient pulmonary veins can be anastomosed to the left atrium and the recipient right atrium to the donor right atrium. The aortas and pulmonary arteries are then anastomosed to the corresponding vessels of the donor heart.

The procedure is carried out under cardiopulmonary bypass.

Survival rates
- 75 per cent of patients are alive after 1 year.
- 50 per cent of patients are alive after 5 years.

Heart–lung transplantation has been successfully performed in patients with pulmonary vascular hypertension due to either primary causes or secondary to cardiac disease.

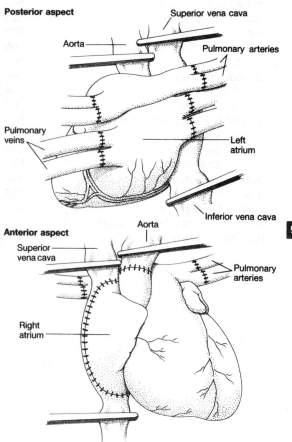

Posterior aspect

Aorta

Superior vena cava

Pulmonary arteries

Pulmonary veins

Left atrium

Inferior vena cava

Anterior aspect

Aorta

Superior vena cava

Pulmonary arteries

Right atrium

531

Heart and heart–lung transplantation

16 Surgical oncology

Natural history of malignant disease

Epidemiology

Incidence rate is the proportion of a defined population developing cancer within a stated period.

Prevalence rate is the proportion of a defined population developing cancer at any one point in time.

The population 'at risk' must be clearly defined and will depend on factors which may be geographical (e.g. cancer of the stomach is common in Japan, whereas cancer of the breast and colon are much commoner in the USA and UK); occupation (e.g. chimney-sweeps' cancer); dietary; environmental; related to smoking or alcohol (cancer of the lung and oesophagus); related to sex and sexual activity; related to radiation (skin cancers, ?leukaemias).

The age structure of the population should be stated, e.g. 'cancer of the rectum is commonest in those aged 50–70 years, with a male:female ratio of 5:3.'

Natural history of tumour growth

The growth of a tumour is dependent on:
- Cell cycle time
- Growth function
- Cell loss fraction

The combination of these gives the *tumour doubling time*. In tumours like leukaemias the doubling time remains remarkably constant: the cell mass increases proportionally with time. This is *exponential growth*. This phenomenon is not so marked in solid tumours, where the doubling time slows as size increases. This is referred to as *Gompertzian* growth. Illustrations of these two theories of tumour growth are graphically presented.

'Cure' of cancer

534

In clinical practice 'survival' times are quoted, most often as 5-year or 10-year rates. These vary according to the stage of the cancer, its type, aggressivity, and response to treatment.

Treatment

Knowledge of the natural history permits rationalization of treatment which, though urgent, is rarely carried out as an emergency. When a patient presents with suspected malignancy, establish the diagnosis and plan investigations and treatment. Ideally this should be within 2–4 weeks after first presentation. Keep the patient well informed throughout.

Most cancers are treated by surgery, radiotherapy, chemotherapy, or endocrine ablation. Often combinations of all of these are necessary.

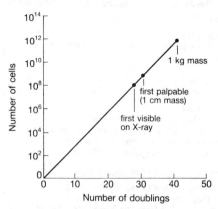

Exponential growth

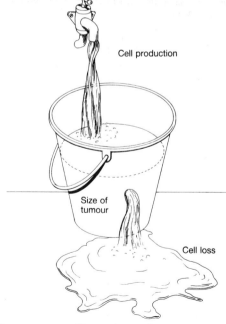

Gompertzian growth

Natural history of malignant disease (*cont.*)

Side-effects of chemotherapy and radiotherapy

(not experienced by all patients)

Chemotherapy
- Metallic taste for 2–3 days after treatment
- Mucusitis (mouth)
- Nausea, vomiting, diarrhoea
- Rashes and ulceration of the skin
- Hair loss (not all drugs cause this)

Radiotherapy
- Skin rashes, moist desquamation
- Nausea, mouth infections especially candidiasis
- Local burning pain to the irradiated area
- Cystitis, diarrhoea (pelvic irradiation)

Useful information for patients

Radiotherapy—your questions answered
Chemotherapy—your questions answered
Both are available from the Royal Marsden Hospital, Fulham Road, London SW3 6JJ.
The British Association of Cancer United Patients and their Families and Friends (BACUP) 121/121, Charterhouse Street, London EC1M 6AA.

Malignant tumours: grading and staging

The extent and degree of malignancy of a tumour are established clinically by the clinician to provide an indication of the prognosis and to act as a guide for the type of treatment necessary. They are also important for comparability between centres and in clinical trials.

Methods

Three methods are used, clinical, pathological, and histological. Pathological staging is used almost exclusively in clinical trials. The examples given are for carcinoma of the breast.

Clinical staging

Stage I: Lump less than 5 cm, not adherent to muscle or chest wall.
Stage II: As in I, but there are mobile ipsilateral axillary nodes.
Stage III: Lump greater than 5 cm, fixed to skin or muscle, ipsilateral fixed axillary nodes, supraclavicular nodes, oedema of the arm. There may also be peau d'orange.
Stage IV: Lump greater than 5 cm, fixed to skin or chest wall with node involvement and distant metastases.

Pathological staging The TNM classification is recommended by the International Union against Cancer. The clinician assesses three facts: What is the extent of the tumour? (T). What is the node status? (N). Are there metastases? (M).

Primary tumour (T)	*Nodes* (N)	*Metastases* (M)
T1S—Preinvasive cancer	N0—No nodes	M0—No metastases
Paget's disease	N1—Mobile ipsilateral	M1—Metastases
No tumour	axillary nodes	including skin
T1—Tumour less than 2 cm	N1a—Contain tumour	involvement
No skin fixation	N1b—Do not contain	beyond the breast
T2—Tumour > 2 cm,	tumour	and involvement of
< 5 cm. Skin tethered.	N2—Fixed ipsilateral	contralateral nodes
No muscle fixation	axillary nodes	
T3—Tumour > 5 cm. Skin	N3—Ipsilateral supra or	
infiltrated or ulcerated.	infra clavicular nodes	
Pectoral muscle fixation	Oedema of the arm	
T4—Any size of tumour.		
Direct extension to chest		
wall or skin		
T4a—Tumour fixed to chest		
wall		
T4b—Oedema, infiltration,		
skin ulceration, nodes		
T4c, T4a + T4b		

Variations of both systems are used to classify other cancers.

Histological grading May also give a guide to the behaviour of a cancer. This makes use of the degree of differentiation of the tumour.
Grade 1: represents the least malignant tumours
Grade 2: 25–50 per cent of the lesion is undifferentiated
Grade 3: 50–75 per cent of the lesion is undifferentiated
Grade 4: more than 75 per cent of the cells are undifferentiated

In all methods there is a degree of observer variation which reduces the accuracy of staging. When there are nodes, sampling and histological proof of involvement should be undertaken to exclude reactive changes. Dukes Classification (p. 571).

Tumour markers

Tumour markers are useful to assess the response of various tumour to treatment and to identify possible recurrence.

Types

CEA	carcino embryonic antigen
AFP	alpha-fetoprotein
HCG	human chorionic gonadotrophin
PAP	prostatic acid phosphatase
CA 125-EIA	cancer antigen enzymeimmunoassay
CA 125-RIA	cancer antigen radioimmunoassay
CA 19.9-RIA	cancer antigen radioimmunoassay
Beta-2-M-RIA	beta-2 microglobulin RIA
ER	oestrogen receptor
TdT	terminal dioxyribonuclear transferase
SCC	squamous cell carcinoma antigen

Primary site of cancer	Most frequent metastases
Breast	lung, liver, bone, brain
Lung	liver, bone, brain
Colo-rectum	liver, lung, bowel wall
Pancreas	liver, lung, peritoneum
Liver	lung
Stomach	liver, lung, bone
Prostate	bone
Testis	liver, bone, brain
Uterus	liver, lung, bone
Ovary	liver, lung
Thyroid	lung, bone

Normal values

CEA	≤3 ng/ml in non-smokers; 10 ng/ml in smokers
AFP	≤ 20 ng/ml
VHCG	≤ 3 mIU/ml
PAP	≤ 2 ng/ml
CA 125	≤ 35 U/ml
CA 19.9	≤ 37 U/ml
ER NEGATIVE**	≤ 20 fmol/mg protein
ER POSITIVE**	> 20 fmol/mg protein
SCC	≤ 2 ng/ml

** values established with ER-EIA

Reproduced by kind permission of Abbott Diagnostics Division

Clinical use of tumour markers

Clinical use of tumour markers

Carcinoma	Histologic type	Tumour marker	Diagnosis	Staging	Prognosis	Monitoring
Breast	Adenocarcinoma	CEA	0	00	00	000
	All types	ER(2)			000	
Lung	Adenocarcinoma + small-cell carcinoma	CEA	0	0	00	000
	Squamous cell carcinoma	SCC	000	00	00	000
		(TA-4)				(1)(2)
Colorectum	Adenocarcinoma	CA 19,9 + CEA	(1)	(1)	(1)	000(3)
		CEA	00	00	000	000
Pancreas	Adenocarcinoma	CA 19,9	000			000
		CEA	0		0	00
Liver	Hepatocellular carcinoma	AFP	000		0	000
	Metastasis from primary tumours	CEA	00			000
Stomach	Adenocarcinoma	CA 19,9 CEA	00			000(1)
						00

		Marker				
Prostate	Adenocarcinoma	PAP	0		00	000
Testis	Nonseminoma	AFP and/or βHCG	000	00	000	000
	Seminoma	βHCG	000		000	000
Uterus	Adenocarcinoma	CEA	000			00
	Choriocarcinoma	βHCG	000	000	000	00
Ovary	Mucinous tumours	CEA		00		000
	Epithelial tumours (except mucinous)	CA 125			00	000
	Germ cell tumours	AFP and/or βHCG	000			000
Thyroid	Medullary carcinoma	CEA + Calcitonin	000	00	00	000
Leukaemia	Acute lymphocytic	TdT	000			000(1)
	Acute undifferentiated					
Cervix, head, and neck	Squamous cell carcinoma	SCC (TA-4)	00	00	00	000

0   Useful
00   Important
000   Very important

(1) Preliminary clinical data
(2) Also used for therapy selection
(3) Improved sensitivity and lead time prior to clinical diagnosis

Clinical use of tumour markers (*cont.*)

Recommended schedule for monitoring (1, 2)

Carcinoma	Tumour marker	Preoperative(3)	1st postoperative year	2nd year	3rd–5th year
Breast	CEA	Once	Every course of therapy then quarterly	Quarterly	Twice/year
Lung	CEA + SCC	Once	Every course of therapy then quarterly	Quarterly	Once/year after the 5th year Twice/year
Colorectum	CA 19,9 + CEA	Once	Postoperation, then monthly for 6 months	Quarterly	Twice the 3rd year then once/year
Pancreas	CA 19,9 + CEA	Once	Monthly for 6 months then quarterly	Quarterly	Twice/year
Liver	AFP	Once	Monthly for 6 months then quarterly	Quarterly	Twice/year
Stomach	CA 19,9 and/or CEA	Once	Every course of therapy then quarterly	Quarterly	Twice the 3rd year then once/year

Prostate	PAP	Once	Quarterly	Once/year
Testis	AFP + βHCG	Once	Daily for 1st week then monthly	Twice/year
Uterus	CEA	Once	Post-treatment then quarterly	Twice the 3rd year then once/year
Ovary	CEA CA 125 AFP and/or βHCG	Once quarterly	Every course of therapy then	Twice/year
Thyroid	CEA	Once	Postoperation then quarterly	Twice the 3rd year then once/year
Cervix, head, and neck	SCC	Once	Postoperation every course of treatment then quarterly	Twice/year

(1) Advanced carcinomas or carcinomas with poor prognosis may require more frequent monitoring
(2) Markers should be reassayed within one month whenever they begin rising
(3) Control value for the patient

Malignant tumours of the skin

The skin is the commonest site for the development of carcinomas and melanomas. 90 per cent occur on exposed surfaces.

Aetiology

Exposure to sunlight. Racial factors, e.g. caucasians develop skin cancer more commonly than negroes. Highest incidences are in caucasians living close to the equator. Chronic infection. TB of the skin, leprosy, chronic sinuses (osteomyelitis), gravitation ulcers (Marjolin's), and unhealed burns may undergo malignant change. Industrial, e.g. chimney-sweeps' carcinogens, skin cancer in early radiologists.

Types

Basal cell carcinoma (BCC) (rodent ulcer). Arises in the 'mask area' of the face between eyebrows and mouth. Can occur elsewhere. 50–70 years. Firm disc-like nodule or ulcer with a pearly rolled edge. Fails to heal. Low rate of metastases but causes extensive local damage. Treatment is by adequate local excision which may need skin grafting or transposition flap.

Squamous cell carcinoma (SCC). Common in those over 50 years. May be a nodule, wart, or ulcer with hard, everted, raised edges. May occur elsewhere but more common in solar keratoses or the lips of smokers. Metastasises via lymphatics. Those of lip, ear, genitals metastasize early.

Treatment:
- Excision—clear the tumour by 1 cm of normal tissue on all sides
- Block dissection of neck for lymphatic metastases
- Radiotherapy when wide excision is not possible or as a preliminary measure for block dissection of the neck.

Prognosis 90 per cent of patients remain disease-free at 5 years.

Malignant melanoma Arises from melanocytes. Less common than BCC or SCC. Unpredictable course. Most arise in previously normal-looking skin. 30 per cent may arise from pre-existing moles. Suspect if a pigmented lesion shows rapid enlargement, bleeding, ulceration, increased or decreased pigmentation, itching, satellite lesions (small lesions around the original growth suggest skin metastases).

Types
- Lentigo maligna melanoma (LMM) (Hutchison's freckle): Occurs on the face in older patients. Least aggressive.
- Superficial spreading melanoma (SSM): Younger people. Discoloration of skin with itching, bleeding and 'spilling' of pigment from the lesion.
- Nodular melanoma (NM): Most aggressive. Can occur anywhere, especially genitals or anus.

Spread Enlargement occurs by surface spread and protuberance. Dermal lymphatic invasion leads to 'satellite' lesions and lymph node involvement. Blood-borne spread leads to hepatic and pulmonary metastases.

Treatment
- Excise down to deep fascia with a wide margin of normal tissue depending on the size of the lesion (up to 5 cm may be necessary in the line of lymphatic spread).
- Close the defect with a skin graft or flap.

Follow-up Examine regional lymph nodes every 3 months. Excise *en bloc* if there is clinical or lymphangiographic involvement.

Other treatment
- Cytotoxic chemotherapy parenterally or by isolated limb perfusion (phenylalanine mustard group drugs)
- Radiotherapy for lesions in sites unsuitable for surgery. This is never curative.

 Spontaneous regression may occur in melanoma but is rare.

Prognosis for malignant melanoma

Depends on dermal invasion. The depth of invasion (Breslow's thickness) is seen from the basal layer of the epidermis. Metastases are never seen if this is less than 0.75 mm but risk increases with increased thickness greater than 2 mm.

Salivary gland tumours

Salivary gland tumours are rare. They account for 0.4 per cent of all malignant tumours. 80 per cent arise in the parotid gland, 80 per cent are benign and 80 per cent of benign parotid tumours are pleomorphic adenomas (mixed tumours).

Clinical presentation

Most patients present with a slow-growing lump in the affected gland. Pain, anaesthesia, or trismus and facial palsy imply malignancy.

Pleomorphic adenoma

This occurs in men and women equally, aged 30–50 years. It is composed of epithelial cells which form a mucous matrix at one time supposed to be cartilage. This was the reason for the term 'mixed'. The tumour grows slowly and has no true capsule, so that strands of tumour epithelial cells protrude into normal surrounding tissue. Local extension may be widespread with recurrence if excision is incomplete. Malignant change may develop after 10–30 years.

Warthin's tumour (adenolymphoma)

Usually affects men over 50 years. It is benign. It presents as a slow-growing soft swelling. Recurrence is unusual after adequate excision.

Mucoepidermoid tumour

A low-grade malignancy composed of sheets of epidermoid cells and cystic spaces. Most grow slowly to invade locally and metastasize to neck lymph nodes, lungs, and skin. Clinically they feel hard.

Cylindroma (adenoid cystic carcinoma)

Comprised of myoepithelial cells which take on a glandular appearance. It is rare and grows slowly, but eventually nerve involvement leads to pain and facial anaesthesia. Perineural extension over long distances is common. The tumour is usually both hard and fixed.

Acinic cell carcinoma

Women are more often affected than men. The cells arrange themselves in serous acini. The tumour is slow-growing but may metastasize unexpectedly.

High-grade malignant tumours like squamous cell carcinomas, adenocarcinomas, and undifferentiated carcinomas rapidly invade the complex anatomy of the parotid region leading to pain and trismus. Clinically there may be skin fixation or ulceration with facial nerve palsy and invasion of the external auditory canal.

Treatment

Benign parotid tumours Excision of the parotid gland superficial to the facial nerve (superficial parotidectomy).

Benign tumours in other salivary glands Excision of the entire gland.

Malignant tumours Radical excision with sacrifice of the facial nerve in the case of the parotid gland. The procedure may be accompanied by block dissection of the neck.

Complications

Warn your patients about facial nerve damage in order to obtain their informed consent.
- Facial nerve paralysis: This may be treated by nerve grafting.
- Frey's syndrome: Facial flushing and sweating of the skin innervated by the auriculo-temporal nerve when the patient salivates. Most people can tolerate their symptoms.

Prognosis

Recurrence of benign tumours may develop 20 years after surgery, especially in the patient where enucleation rather than superficial parotidectomy is sometimes performed.

The 5-year survival rate for all these malignancies is around 60 per cent.

Oral cancer

The commonest encountered in practice is squamous carcinoma of the lips, mouth, and tongue. The tumour spreads to the lymphatics of the neck. It is a disease of the elderly, more common in men. However, the incidence is declining in both sexes.

Causes

The cause is unknown but the following are known to be predisposing factors:

Leukoplakia (white patches)
Smoking
Mechanical irritation (e.g. poorly fitting dentures)
Excessive alcohol consumption
Syphilis
Chronic hyperplastic candidiasis
Betel nut chewing
Excessive sunlight

Pathology

Most arise on the lower lip. Other sites are the undersurface or lateral border of the tongue, mucosa of the cheek, mandible or maxillary alveolus, retromolar areas, or palate. Local infiltration and ulceration is common and metastatic spread is to the submandibular or preauricular lymph nodes then the other lymph nodes of the neck. Ulceration and severe sepsis is common.

Clinical features

Cancer of the lip usually presents as an ulcer which does not heal. Cancers of the tongue may cause pain, slurring of the speech and severe fetor oris. Sometimes a hard lump is all that is felt. Cheek cancers may appear as ulcers or speckled leukoplakia (white patches interspersed with reddened areas due to *Candida* infection). Always arrange to biopsy such areas. Other presentations include loosening of the teeth, chronic infection, the erroneous diagnosis of a dental abscess, and oral bleeding.

Diagnosis

Examination (under anaesthetic if necessary) and biopsy.

Investigations

- X-rays of bones near the tumour. Photographs of the tumour.
- Chest X-ray. ECG
- FBC. Clotting studies, U&E. Tests for syphilis (VDRL, TPIT)
- CT scan

Treatment

Stages I and II—radiotherapy or local surgery. Block dissection of the neck ± mandibular resection. Reconstruction of the mandible can be achieved by a radical forearm free flap or pectoralis major myocutaneous flap plus underlying rib.

Prognosis is related to the stage of the disease when the patient first presents. Good prognostic features include:
- Site of disease: Cancers of the lip and cheek have a better prognosis than palate or maxillary alveolus.
- Sex: Women do better than men.
- Age: Patients between 40 and 50 years survive longer than older patients. However, patients under 40 years do badly.
- Histology: Patients with well-differentiated cancers survive longer
- Absence of bone invasion implies a better prognosis.

Thyroid cancer

Most cancers of the thyroid are primary carcinomas. They account for 0.5 per cent of all malignancies. Metastases are rare but can be blood-borne from breast, colon, kidney, and melanomas.

Aetiology

- Endemic goitre is associated with follicular carcinomas but the risk is small.
- Irradiation of the neck in childhood increases risk.
- Medullary cancers are occasionally familial and may be multiple endocrine adenomas syndrome-related.

Types

Four main types are described:

Papillary carcinoma Younger patients, < 30 years. Small primary tumour which metastasizes to cervical lymph nodes ('lateral aberrant thyroid' is an erroneous diagnosis which implies secondaries in the neck from a papillary cancer). Relatively good prognosis with more than 10-year survival for 50 per cent of patients.

Follicular carcinoma Resembles normal thyroid. Spreads by blood to lung and bone. Metastases may be susceptible to ^{131}I therapy. 40 per cent 10-year survival.

Medullary carcinoma Affects a wide age range. Arises from parafollicular (C) cells. May be familial and associated with adrenal phaeochromocytoma or mucocutaneous polyps. Variable prognosis.

Anaplastic carcinoma Rapidly growing. Early metastases via lymphatics and bloodstream. Most do not concentrate iodine. Poor prognosis.

Clinical features

Thyroid cancers can present as a diffuse or nodular swelling, a solitary nodule of the thyroid, or lymph node metastases in the neck. Cragginess, fixity, and hoarseness imply extracapsular spread. Papillary cancers tend to occur in patients < 30 years, follicular those > 30, anaplastic in the elderly. Early cases of medullary cancer may be detected by high calcitonin levels.

Diagnosis

- Mainly clinical
- Isotope scanning ('cold' nodule may imply malignancy)
- US scan may reveal areas of suspicion in the gland or metastases
- Fine needle aspiration biopsy (FNAB) if available

Treatment

Papillary Extracapsular thyroidectomy and postoperative oral thyroxine to suppress TSH (0.1–0.4 mg daily). Use ^{131}I for metastases and inoperable cases.

Follicular Extracapsular thyroidectomy. Use ^{131}I in therapeutic doses for metastases.

Anaplastic Total thyroidectomy and irradiation of the neck. Inoperable tumours may be treated by radiotherapy.

Medullary Total thyroidectomy and irradiation of neck. Irradiate inoperable tumours.

Guidelines for house surgeons

- Obtain informed consent. Warn the patient about possible nerve palsies, especially recurrent laryngeal.
- Request thoracic inlet X-rays to exclude retrosternal extension and tracheal involvement.
- Do pre- and postoperative serum calcium levels.
- Have the vocal cords checked pre- and postoperatively.

Tumours of the thymus

The thymus originates from the third pharyngeal pouch. It comprises two lobes and lies in the anterior mediastinum. Although it is epithelial in origin it consists of lymphoid tissue.

Function

The thymus is large at birth, increases in size during childhood, and then degenerates to become fibrous tissue and fat. It is the site where T-lymphocytes acquire their immunological competence to be disseminated later in other lymphoid tissue.

Tumours are benign or malignant and include epithelial, lymphoid, and teratomatous neoplasms. About 25 per cent of thymic tumours are associated with myasthenia gravis or paraneoplastic syndromes.

Clinical features

Most produce no symptoms and are discovered by routine chest X-ray as a mass occupying the anterior compartment of the superior mediastinum. Some may lead to superior mediastinal compression.

Treatment

All thymic tumours should be excised because of the risk of invasive characteristics. Sternal splitting gives good access. Radiotherapy is used to treat invasive lesions, from which only a biopsy should be taken.

Prognosis

60–80 per cent of patients with thymomas will survive 5 years. Those under 35 years with myasthenia gravis have a poorer prognosis.

Follow-up

Patients with invasive disease need regular review.

Cancer of the breast

Breast cancer is the commonest malignant disease of women in the western world. Each year in the UK 20 000 women develop the disease and 10 000 die from its effects. The risk for any individual woman is 1 in 14. This is doubled if a first-degree relative is affected. Despite intensive research the cause remains unknown.

Risk factors

- *Increased risk*: Early menarche (< 12 years); increased animal fat diet around puberty; affected first degree relative; social class 4 or 5
- *Decreased risk*: Multiple pregnancies at early age (< 25 years); breast feeding

Natural history

Variable and relatively unpredictable. Commonest in women 50–60 years. Incidence is reduced during menopause. Recurrence or the reappearance of residual disease can develop as long as 15–20 years after treatment. Large tumours and more than four involved lymph nodes at presentation have a worse prognosis. Histologically well-differentiated tumours and those with oestrogen receptor activity have a good prognosis.

Types

Most are infiltrating ductal cancers. 10 per cent are invasive lobular cancers which have an increased risk of metachronous and bilateral second primaries.

Spread

- Local to skin, muscle, and chest wall
- Lymphatic by embolism or penetration to internal mammary, axillary nodes, supraclavicular nodes and occasionally to other breast
- Blood spread to pelvis, vertebrae and long bones. Ovary, adrenal, brain and peritoneum may also be involved in secondary spread.
- Liver and lung metastases occur late.

Presentation

There is usually a painless lump in the upper outer quadrant where most breast tissue is. Look for nipple retraction, skin fixity, asymmetry of the breasts. Palpate the lump. Is it craggy, fixed to skin or muscle? Is there nipple discharge? Often the patient will be able to demonstrate this by pressing one particular quadrant. Are there axillary nodes? If so, are they fixed or mobile?

Diagnostic tests

- Fine needle aspiration. If negative, proceed to excision biopsy.
- Mammography has 95 per cent accuracy. Of value in screening the opposite breast and synchronous ipsilateral lesions.

- Trucut biopsy is usually unnecessary with good cytology.
- Excision and frozen section biopsy (becoming increasingly less used)

Investigations

ESR, FBC, LFTs, oestrogen receptors (not available everywhere), bone scan, CT scan (super staging by these methods not yet of proven value).

Treatment

- Small lumps < 2 cm (unless there is evidence of aggressive behaviour, e.g. *in situ* changes, synchronous primary tumour)—wide local excision (lumpectomy) and radical external beam radiotherapy to breast and regional lymph nodes. The women should be permitted to choose either lumpectomy or simple mastectomy. Screening detected lumps may be cured by mastectomy.
- Lumps 2–5 cm without fixation (I and II)—simple mastectomy + node sampling. If the nodes are involved, give postoperative radiotherapy.
- Fixed lumps > 5 cm may be treated by local radiotherapy or mastectomy (which is preferable since it avoids the risk of fungation).

Drugs

Tamoxifen (anti-oestrogen) should be given as an adjunct to all groups and in recurrence; L-phenylalanine and 5-fluorouracil are of value in those with high oestrogen levels, over 50 years, or with metastases. Combinations of cyclophosphamide, methotrexate, and 5-fluorouracil (CMF) are suggested only as an adjunct in premenopausal women with more than four positive nodes. Pre- and perimenopausal women who do not respond may benefit from oophorectomy. If oestrogen receptor positive 60 per cent respond, if negative 30 per cent. Endocrine ablation can also be achieved by aminoglutethimide. Quadruple chemotherapy may be of value in advanced symptomatic disease, producing high frequency but short-lived remission rates. Radiotherapy is useful for symptomatic bony metastases.

557

References

Chetty, U. and Forrest, A. P. M. (1986). Breast conservation. *British Journal of Surgery*, **73**, 599–600.

Gazet, J.-C., Rainsbury, R. M., Ford, H. T., Powles, T. J., and Coombes, R. C. (1985). Survey of treatment of primary breast cancer in Great Britain. *British Medical Journal*, **290**, 1793–5.

Scottish Breast Cancer Trials Committee (1987). Adjuvant tamoxifen in the management of operable breast cancer: the Scottish trial. *Lancet*, **ii**, 171–5.

Bronchial carcinoma

This is the commonest cancer of men and the second commonest in women in the UK, where it accounts for 100 000 deaths each year. It is poised to overtake breast cancer as the commonest malignancy in women.

Prognosis

Only surgical resection offers hope of long-term survival, but 3 out of 4 patients have inoperable disease. Of those who have putatively curative resection, 30 per cent survive for 5 years and 17 per cent for 10 years.

Aetiology

Cigarette smoking, atmospheric carcinogens, e.g. pollution, carcinogens at work, asbestos, radioactivity, nickel.

Pathology

60 per cent are central near the hilum and may lead to obstructive symptoms. 40 per cent are peripheral and may cause local pressure on sympathetic chain and brachial plexus (Pancoast's syndrome).

Lung cancers may also elborate hormones which produce misleading symptoms.

Spread occurs locally to involve hilar and tracheobronchial nodes, leading to pressure on the great vessels or bronchi. Spread throughout the pleural cavity leads to pleural effusions. Distant metastasis occurs commonly to liver, bone, brain, and kidney. Frequently the metastasis is the presenting feature.

Histology

Four types: squamous cell (30 per cent), undifferentiated (25 per cent), and oat-cell cancer (5 per cent) usually arise centrally from metaplastic changes in the main cell bronchi. Adenocarcinomas (40 per cent) usually arise peripherally and are rapidly becoming the most common form.

Clinical features

- Weight loss, anaemia, anorexia
- Persistent cough, episodic haemoptysis, dyspnoea, recurrent or unresolved pneumonia, lung abscess, hoarseness of the voice
- Peripheral neuritis due to a peripheral tumour or enlarged node
- Mood changes (cerebral metastases) pathological fracture (bone metastases)
- Endocrine disorders like Cushing's syndrome, hypercalcaemia simulating hyperparathyroidism and increased ADH secretion with hyponatraemia
- Thrombophlebitis migrans
- Finger clubbing and hypertrophic pulmonary osteodystropathy (new bone laid down on the phalangeal shafts leads to a severe type of finger clubbing)

Diagnosis

- Chest X-ray
- Sputum cytology, aspirate of pleural effusion
- Needle biopsy
- Bronchoscopic appearance and cytology brushings

Treatment

- *Surgical resection*: Lobectomy or pneumonectomy (only 1 in 4 patients). Mortality is 4 per cent for lobectomy, 8 per cent for pneumonectomy.
- *Radiotherapy*: Little value as a cure but useful in relieving symptoms of cough, pain from bone metastases, engorgement of superior vena cava obstruction.
- *Cytoxic drugs*: Alone or in combination have little to offer except in oat-cell tumours.

Assess operability

CT scan, endoscopy, isotope scanning, and US scan (detects distant metastases). Avoid unnecessary thoracotomies by pre-operative staging.

Contra-indications to surgery

- Superior vena cava obstruction
- Blood-stained pleural effusion
- Recurrent nerve palsy. Phrenic nerve paralysis.
- Widespread metastases

Prophylaxis

The only effective way of reducing the death rate is to persuade people not to smoke.

References

Belcher, J. R. (1981). Carcinoma of the bronchus. In *Current surgical practice*, (eds J. Hadfield and M. Hobsley), pp. 276–91. Edward Arnold, London.

Berlin, N. I., Buncher, C. R., and Fontana, R. S. (1984). Early lung cancer detection: Summary and conclusions. *American Review of Respiratory Diseases*, **130**, 565–70.

Faber, L. P., Jensik, R. J., and Kittle, C. F. (1984). Results of sleeve lobectomy for bronchogenic carcinoma in 101 patients. *Annals of Thoracic Surgery*, **37**, 279–85.

559

Carcinoma of the oesophagus

The commonest symptom is difficulty in swallowing. Sometimes the patient can accurately indicate the level at which the food tends to stick, but more commonly the site indicated is well above the tumour.

Clinical features

Regurgitation of food may occur as well as saliva. Persistent coughing due to laryngeal irritation and excess oesophageal content may disturb sleep, leading to aspiration of infected material.

Symptoms

Rapid emaciation and anaemia occur. Symptoms of hoarseness from laryngeal palsy or foul breath due to bronchial involvement may represent extension of the growth.

Clinical examination

The patient is usually an elderly male who is anaemic and poorly nourished. The highest incidence is in those over 70 years old.

Radiology

The characteristic features are of a long, tortuous stricture irregular in its upper and lower margin with variable degrees of dilatation above.

Diagnosis

Confirmation is by biopsy or brushing by flexible oesophago-gastro duodenoscopy. 90 per cent are squamous carcinoma, 10 per cent are adenocarcinoma which tend to affect the lower third only.

Spread

- *Direct*: Submucosal and transmucosal
- *Lymphatic*: By intramural lymphatic permeation and embolization
- *Blood-borne*: To liver, bone, and skin

Treatment

Surgery Tumours below the diaphragm can be radically excised through a left thoracoabdominal approach with oesophageal-jejunal anastomosis-en-Roux or oesophagogastrostomy.

Carcinoma above the diaphragm can be excised by the Lewis Tanner procedure. A third stage may be added to gain clearance and the oesophagogastric anastomosis performed in the neck.

Upper-third tumours may be treated by transhiatal resection or by the three-stage procedure using stomach, colon, or small bowel interposition either intrathoracically or retrosternally.

Radiotherapy Generally, this is unsatisfactory but it may be used when it may relieve symptoms of dysphagia in 50 per cent of patients. The remaining 50 per cent will be unable to complete their treatment due to progressive disease. Pre-operative radiotherapy also remains of unproven value. Adeno-carcinoma is radio-resistant.

Unresectable carcinoma—options
- Intubation, e.g. with Celestin tube
- Retrosternal bypass by colon, jejunum, or stomach
- Endoscopic relief of dysphagia and control of bleeding may be achieved by use of Nd: YAG laser

The average survival is 4 months after operation. In resectable tumours survival is rarely more than 18 months.

References

Ellis, F. H., Gibb, S. P., and Watkins, E. (1983). Oesophagogo-gastrectomy for palliation of carcinoma of the oesophagus. *Annals of Surgery*, **198**, 531–40.

Groisser, V. A. (1984). YAG laser therapy of gastrointestinal tumours. *Gastrointestinal Endoscopy*, **5**, 311–12.

Groves, L. K. and Rodriguez, A. A. (1973). Treatment of carcinoma of the oesophagus and gastric cardia with concen-trated preoperative irradiation followed by early operation. *Annals of Thoracic Surgery*, **15**, 333–8.

Cancer of the stomach

This is the commonest GI cancer after those of the colon and rectum, affecting 60/100 000 per annum, 3 men:1 woman in the 40–70 year age group. It accounts for 9000–11000 deaths per year.

Epidemiology

There is worldwide variation in incidence, which is especially high in Japan and other coastal countries where dietary nitrate content is high. Hypothesized causes are eating smoked fish or pickled and heavily salted foods, use of food preservatives, occupation (especially mining, asbestos, and rubber workers).

Associations

Pernicious anaemia. Blood group A. Positive family history. Gastric polyps, atrophic gastritis. Partial gastrectomy, cholecystectomy, pyloroplasty, chronic biliary gastritis, acanthosis nigricans.

Pathology

50 per cent involve the pylorus, 25 per cent lesser curve, 10 per cent cardia, 15 per cent are multifocal. They may be ulcerating, polypoidal, superficial spreading, diffuse, or localized thickening (leather bottle, linitis plastica).

Histology

90 per cent are adenocarcinomas. Others include lymphoma and squamous.

Spread

To lymph nodes along both curvatures to coeliac axis, para-aortic nodes, and portal fissure. Metastases in liver and lung are due to portal venous spread. Transperitoneal spread leads to ascites and ovarian tumours (Krukenberg). Colon, pancreas, and spleen may become involved from local spread, Thoracic duct permeation leads to enlarged left supraclavicular nodes (Troisier's sign, Virchow's nodes).

Clinical features

Diverse. Most patients have disease for 6 months before presentation. Weight loss, anorexia, dysphagia, postprandial bloating, diarrhoea, nausea, and vomiting are common symptoms. Epigastric mass, jaundice, hepatomegaly, ascites, obstruction, enlarged ovaries may be noted on examination.

Diagnosis

Barium studies and gastroscopy + biopsy provide the diagnosis in 95 per cent.

Population screening

Allows early detection and improves prognosis with reported 90 per cent 5-year survivals in Japan, but this may reflect a different disease process.

Treatment

Curative or palliative. 80 per cent of patients present too late for curative surgery. Check the liver for metastases by liver scan, CT or US scan. If there is involvement gastrectomy is useless unless to stem bleeding or correct perforation or obstruction.

Curative surgical criteria

No metastases in liver or peritoneum. No cancer in the cut edge of the stomach. The extent of the gastric resection is total or partial but entails removal of the greater and lesser omenta with reconstitution of the GI tract by gastroduodenal, gastrojejunal or gastro/oesophageal–jejunal pouch anastomosis.

Relative curative criteria

As above, except that nodal involvement is present. These cancers may be treated by radical R2/R3 resections which involve removal of stomach, nodes, and spleen.

Palliative surgery

May involve partial resection, bypass (e.g. with Celestin tube) or reconstruction. The aims are to reduce pain, vomiting, dysphagia, and bleeding.

Prognosis

15 per cent survive 5 years, 60 per cent survive 5 years if disease is curative. Screening may be an answer (cf. Japan).

References

Cushieri, A. (1986). Cancer of the stomach. In *Recent advances in surgery*, (ed. R. C. G. Russell), **12**, pp. 125–42. Churchill Livingstone, Edinburgh.
Silverberg, E. (1985). Cancer statistics. *Cancer*, **35**, 19–35.

Small-bowel tumours

These are rare.

Benign

Adenomas and leiomyomas can cause obstruction, intussusception, or bleeding.

Lipomas may cause intussusception.

Peutz–Jegher's syndrome is familial hamartomatous polyposis of the jejunum associated with melanotic spots of the mouth and lips. The polyps can lead to bleeding or intussusception.

Premalignant conditions

Crohn's disease is associated with small-bowel adenocarcinoma which tends to develop proximally.

Coeliac disease with lymphoma.

Adenomas with adenocarcinoma (as in familial adenomatous polyposis).

Leiomyomas with leiomyosarcoma.

Malignant tumours

Account for less than 2 per cent of gastrointestinal malignancies and usually present at an advanced stage.

Adenocarcinoma is the commonest, found more often in the jejunum. It presents as intestinal obstruction or melaena by which time the tumour is advanced with metastases present.

Sarcomas are often multiple. Patients give a long history of anaemia and weight loss. Perforation of advanced disease is a common presentation.

Lymphoma may present as a palpable mass or small-bowel perforation.

Treatment

Of small-bowel tumours is resection. Palliation may be achieved by bypass or chemotherapy.

Carcinoid syndrome

Is caused by metastasis of carcinoid tumours of the ileum or appendix to the liver. 5-Hydroxytryptamine is released by the tumour along with prostaglandins and kinins. These lead to attacks of severe facial flushing associated with diarrhoea and asthmatic attacks.

Treatment is radical resection of primary tumours and partial hepatectomy if metastases are confined to only one lobe of the liver.

Diagnosis of small-bowel tumours

- Double-contrast barium small-bowel enema or meal
- Fibreoptic endoscopy
- Incidental finding at laparotomy
- On-table fibreoscopy may detect multiple tumours.
- If bleeding carry out angiography or isotope scanning.

Prognosis

Less than 30 per cent of patients survive 5 years.

Reference

Taggart, D. P., McLatchie, G. R., and Imrie, C. W. (1986). Survival of surgical patients with carcinoma, lymphoma and carcinoid tumours of the small bowel. *British Journal of Surgery*, **73**, 826–8.

Liver tumours

The commonest tumours of the liver are metastases transported by the hepatic artery. The usual organs of origin are pancreas, colon, stomach, oesophagus, and breast. More than 30 per cent of patients who die of malignant disease have hepatic metastases.

Primary benign tumours

Haemangioma (usually cavernous) may be treated by radiotherapy or hemihepatectomy if causing symptoms but there is always a risk of bleeding.

Hepato- and cholangioadenomas are usually harmless. The former may be multiple and related to the contraceptive pill (focal nodular hyperplasia).

Primary malignant tumours

Hepatoma arises from liver cells. It accounts for 80–90 per cent of primary liver tumours. Rare in Europe but common in Africa, Japan, China, Mozambique (100/100 000). The tumour varies greatly in aggressivity but may be multicentric especially when associated with cirrhosis (15 per cent of cirrhotic patients develop hepatoma). Other associations include hepatitis B, the contraceptive pill, abuse of anabolic steroids, and aflatoxins present in maize and wheat contaminated by the fungus *Aspergillus flavus*.

Cholangiocarcinoma is rare and rarely resectable. It is associated with chronic ulcerative colitis, primary sclerosing cholangitis, and choledochal cysts. Histologically it is an adenocarcinoma.

Clinical features

Anorexia, weight loss, ascites, pain, and jaundice. The patient is 20–40 years and may present with an abdominal mass. Growth is frequently rapid. Alpha-fetoprotein is elevated in about 90 per cent of patients and is a useful marker.

Diagnosis

US scan, needle biopsy, ERCP, PTC, CT scan, radioisotope scan, coeliac axis angiography.

Treatment

- Solitary hepatic metastases can be treated by hemihepatectomy, hepatic lobectomy, or palliative bypassing and stenting.
- Cholangiocarcinomas may respond to radiotherapy or Whipple's procedure when they affect the lower end of the common bile duct.
- Primary hepatomas affecting one lobe may be treated by hepatic lobectomy. Hepatic transplantation is occasionally carried out if there is no extrahepatic spread. Chemotherapy and radiotherapy are of no value.

Prognosis

- Hepatic metastases: Survival is 6 months (mean) after diagnosis
- Primary hepatic tumours: Survival is 4 months (mean) after diagnosis

References

Cady, B. (1983). Natural history of primary and secondary tumours of the liver. *Seminars in Oncology*, **10**, 127–34.

Lee, N. W., Wong, J., and Ong, G. B. (1982). The surgical management of primary carcinoma of the liver. *World Journal of Surgery*, **6**, 66–75.

Lutwick, L. I. (1979). Relation between aflatoxins and hepatitis B virus and hepatocellular carcinoma. *Lancet*, **i**, 755.

Cancer of the pancreas

Carcinoma of the exocrine pancreas has an incidence of 12/100 000 of the population in the UK. This is rising more in the USA than the UK, where it is now the fourth-commonest cancer causing death. It afflicts men and women equally and is rare below 40 years, most cases occurring in patients over 60 years. It is difficult to diagnose early, so that few patients at presentation can have corrective treatment.

Epidemiology

Risk factors include cigarette smoking, diabetes mellitus, industrial pollutants, chronic alcoholism (often associated with heavy cigarette smoking), and coffee ingestion.

Pathology

Most (75 per cent) are adenocarcinomas resembling the pancreatic ductal cell. Other cell types each make up less than 5 per cent of the total, e.g. cyst adenocarcinoma. 60 per cent involve the head, 20 per cent the tail and 20 per cent the ampulla, but combinations are not uncommon. Spread is local to involve spleen, adrenal glands, and transverse colon. Lymphatic spread has occurred to regional lymph nodes, liver, lungs, and peritoneum in 80 per cent of patients by the time of surgical exploration.

Clinical features

Pain and jaundice are common. 60 per cent have backache worse at night. Jaundice may imply advanced disease with bile duct and possible liver involvement, or early disease in localized ampullary lesions. Early symptoms include weight loss, fatigue, malaise, dyspepsia, and pruritus. On examination the gall bladder may be palpable but painless. 'If in a case of painless jaundice the gall bladder is palpable the jaundice is unlikely to be due to stone disease' (Courvoisier's Law). Stone disease rarely leads to distension because there is usually associated gall bladder fibrosis. There may be hepatomegaly or a palpable epigastric mass. Thrombophlebitis migrans occurs in 10 per cent presenting as emboli. Splenic vein thrombosis may lead to splenomegaly in 10 per cent of patients.

Diagnosis

US and CT scan with needle biopsy are successful in 90 per cent. ERCP is very accurate: pancreatic duct obstruction, duct stenosis, and necrotic cavity formation are typical of pancreatic cancer. Cytology at ERCP, operative cytology/frozen section and on-table US scan are also helpful.

Treatment

Pancreatico-duodenectomy (Whipple's procedure) for operable lesions. This involves a partial gastrectomy, duodenectomy, partial pancreatectomy, distal choledochectomy ± cholecystectomy. GI continuity is restored by gastroenterostomy, Roux-en-Y jejunocholedochotomy, and pancreaticojejunostomy. There is also a pylorus-preserving variation. Operative mortality is 10–30 per cent.

Total pancreatectomy may be indicated for multifocal disease.

Regional pancreatectomy may be indicated for multifocal disease of its surrounding structures both venous and arterial which are reconstructed at the end of the procedure. The resectability rate is 40 per cent but the procedure is still too new to evaluate survival. Enucleation of ampullary lesions is sometimes possible.

Palliative decompression of the biliary tract by triple bypass or stenting at endoscopy, the correction of duodenal obstruction, and pain management are all that can be offered in most cases.

Prognosis

In untreated patients the average survival is 9 weeks from diagnosis. In patients with operable lesions 10 per cent will survive 5 years.

References

Allen-Mersh, T. J. and Earlam, R. J. (1986). Pancreatic cancer in England and Wales: surgeons look at epidemiology. *Annals of the Royal College of Surgeons of England*, **68**, 154–8.

Further, J. G. (1973). Regional resection of cancer of the pancreas: A new surgical approach. *Surgery*, **73**, 307–20.

Cancer of the colon and rectum

These are the commonest GI cancers, accounting for 20 000 deaths per annum in the UK. They present after middle life. Colonic cancer is commoner in women, rectal in men.

Epidemiology

Animal fats, food preservatives, dyes, and bacterial breakdown of faecal steroids are implicated. High-fibre diets reduce the risk. Selenium and calcium may be protective. Familial adenomatous polyposis coli predisposes to cancer of the colon.

Associations

All are adenocarcinomas. Grossly they are cauliflower-shaped, annular, tubular, or ulcers. Spread is local around or through the bowel wall and to other viscera when fistulas may form. Lymphatic spread is to epi-, paracolic nodes then to nodes around main vessels. Venous spread occurs late to liver.

Rectal cancers are ulcerated, well to poorly differentiated adenocarcinomas. Spread is local to perirectal fascia, prostate, seminal vesicles, bladder, vagina, and uterus. Lymphatic spread is usually upward to pararectal nodes. Late venous spread leads to metastases in liver and lungs predominantly, but adrenals and brain may also be involved.

Investigations

FBC, FOB, flexible and rigid sigmoidoscopy and biopsy, double-contrast barium enema, colonoscopy. 75 per cent of rectal cancers can be felt with the finger. 'If you don't put your finger in it, you'll put your foot in it.' Never fail to examine the rectum if cancer is a possibility.

Clinical features

Depend on site.
- *Caecum and ascending colon*: Iron-deficiency anaemia, palpable mass in 50 per cent (increasing in incidence)
- *Descending colon*: Change of bowel habit. Rectal bleeding
- *Sigmoid and rectum*: (80 per cent of patients) Change of bowel habit, incomplete defecation, tenesmus, spurious diarrhoea

25 per cent of large-bowel neoplasms present as emergencies with obstruction, (caecum and recto-sigmoid junction), haemorrhage (which is rarely massive), perforation (due to closed-loop obstruction or penetrating tumour), or peritonitis.

Treatment

Elective
- *Caecum and ascending colon*: Right hemicolectomy
- *Transverse colon*: Segmental resection with anastomosis
- *Descending colon and sigmoid*: Segmental resection + anastomosis
- *Rectum*: Anterior resection (resection + anastomosis); abdomino-perineal resection if lesion is too low

Stapling devices may permit a more distal anastomosis, but not at the expense of limited resection.

Palliative procedures involve transrectal resection with the resectoscope or cautery. Radiotherapy may be of value to reduce tumour mass and bleeding.

Emergency Improve fluid and electrolyte balance. Do primary resection and anastomosis if possible with on-table colonic lavage. If you cannot resect at the first operation, find someone who can. Right transverse loop colostomies should be a thing of the past. Carry out Hartmann's procedure in peritonitis, ischaemia, and gross obstruction.

Prognosis relates to Dukes modified classification.

Stage	Extent of growth	5-year survival
A	Confined to bowel	80 per cent
B	Penetration of wall. No nodes	60 per cent
C	Regional nodal involvement	30 per cent
C1	Local pararectal	
C2	Nodes on blood vessels up to site of division	
D	Distant metastases	5 per cent

90 per cent are operable even for palliation. Operative mortality is 2–5 per cent.

References

Bolin, S., Nilsonn, E., and Sjodhal, R. (1983). Carcinoma of the colon and rectum—growth rate. *Annals of Surgery*, **198**, 151–8.

Killingback, M. J. (1985). Indications for local excision of rectal cancer. *British Journal of Surgery*, **72** (supplement), 554–6.

Radcliffe, A. G. and Dudley, H. A. F. (1983). Intraoperative antegrade irrigation of the large intestine. *Surgery, Gynaecology and Obstetrics*, **156**, 721–3.

Blood transfusion and cancer recurrence

Intraoperative blood transfusion has been associated with a higher recurrence rate of solid tumours including cancers of the colon, rectum, lung, breast, kidney, vulva, cervix, stomach, and soft tissue sarcomas. Although the evidence remains controversial it has been hypothesized that the stored plasma fraction of whole blood causes earlier tumour recurrence in some instances.

General effects of blood transfusion

- Correction of anaemia and hypovolaemia
- Preoperative transfusion improves the survival time of renal transplants.
- Transfusion of a husband's blood to his wife may prevent spontaneous abortion.
- Blood transfusion may be associated with increased postoperative infection.
- Blood transfusion may increase the risk of transmission of HIV and hepatitis B virus.
- Blood transfusion has been associated with tumour remissions, especially melanoma.
- Blood transfusion also stimulates an immune response (humoral and cellular) in the recipient which may explain its beneficial effect in renal transplant but its possible adverse effect in cancer recurrence. Humoral factors include antibodies which inactivate T-cell clones, blocking antibodies which inactivate immunoglobins and lymphocyte toxins which kill lymphocytes. Cellular factors induced are non-specific and specific suppressor cell activity, impairment of macrophages, and a reduction in natural killer cell activity.

Patients with colorectal cancers who receive only 1 unit of whole blood have a much greater risk of tumour recurrence than those receiving either ≤ 3 units of red cells or no transfusion. This is independent of all other prognostic tumour and patient factors. It is probable that the plasma components of whole blood initiate an adverse immunological reaction.

Guidelines for blood transfusion of cancer patients

- Transfusions of whole blood should be avoided or delayed.
- Red blood cell concentrates should be used (≤ 3 units given on < 4 separate occasions).
- Autologous blood transfusion may reduce the immunological response elicited by blood donated from an unrelated person.
- Red cells washed free of plasma may reduce the effects.
- If the patient suffers haemorrhage and is at immediate risk appropriate blood transfusion should be given as the risks of transfused blood are still hypothetical.

References

Blumberg, N., Heal, J., Chuang, C., Murphy, P., and Agrawal, M. (1988). Further evidence supporting a cause and effect relationship between blood transfusion and earlier cancer recurrence. *Annals of Surgery*, **207**, 410–15.

Hamblin, T. J. (1986). Blood transfusions and cancer: anomalies explained? *British Medical Journal*, **293**, 517–18.

Cancer of the prostate

This is the fifth-commonest malignancy of males in the UK affecting men over the age of 65 years. Many patients will die of other causes like stroke or myocardial infarct, particularly if oestrogen therapy is instituted, rather than the cancer itself. It is therefore not always clinically important.

Aetiology

The only clear factor is the presence of testosterone. Men with testicular failure have a low rate of prostatic cancer.

Pathology

It is an adenocarcinoma which usually arises in the posterior lobe of the prostate below the level of the ejaculatory ducts. Spread is local through the capsule to involve adjacent structures like the rectum. Blood spread by the pelvic veins to the pelvic occurs early. Lymphatic spread by pararectal and lymphatics passing over the seminal vesicles leads to retroperitoneal, mediastinal and occasionally supra clavicular nodal involvement.

Clinical features

The patient is usually elderly. Some are asymptomatic, a hard nodule being discovered on rectal examination. There may also be:

- A history of urinary incontinence, dysuria, outflow obstruction, bladder irritation
- A history of arthritis or rheumatism due to metastatic disease
- Back pain, sciatica, bone pain, or pathological fracture due to metastases
- The incidental finding of prostatic cancer by the pathologist after prostatectomy. Rectal examination and transrectal biopsy.

Diagnosis

Histology. Bone scan for metastases. Acid phosphatase which is elevated in 90 per cent of patients with metastases and 45 per cent with disease confined to the prostate. Prostatic specific antigen may also be detected in the blood, and is a tumour marker.

Radiology may reveal typical sclerotic secondaries.

Management

- Small intracapsular tumours. Many patients have a normal life expectancy without treatment.
- Diffuse or poorly differentiated. Radical prostatectomy with removal of prostate and capsule and seminal vesicles and vas deferens is advocated by some centres. Radiotherapy is advocated by others.

- Metastatic disease or locally advanced is the commonest presentation. Patients with no metastases receive radiotherapy. Those with metastases receive endocrine therapy by subcapsular orchiectomy or stilboestrol (but side-effects like gynaecomastia and myocardial infarct make this drug less acceptable). Long-acting luteinizing hormone (LH) analogues suppress testosterone production, are more acceptable to the patient, and may be as effective as orchiectomy. 85 per cent have a good initial response.

Prognosis

Cure is not possible. Remission occurs in 85 per cent with hormonal manipulation. Less than 20 per cent survive 5 years.

Reference

Rustin, G. J. S. (1987). Testicular tumours and urinary tract malignancies. *Medicine International*, **40**, 1685-8.

Testicular tumours

These are the commonest malignancies of young men (25–34 years) but are still rare. There are 600 new cases each year in the UK.

Pathology

90 per cent arise from the germ cells. 10 per cent are lymphomas (older men) and Leydig or Sertoli cell tumours (uncommon).

Germ cell tumours are classified as seminomas with a peak incidence at 35 years or teratomas with a peak incidence at 25 years.

Seminoma commences in the mediastinum testis. As it enlarges it compresses normal tissue so that the enlarged testis is smooth and firm. Spread is lymphatic to para-aortic nodes. Histologically the tumour is arranged in sheets of rounded cells. Lymphocyte infiltration is a good prognostic sign. Seminomas produce placental alkaline phosphatase which is detectable in the serum and can be used as a tumour marker.

Teratoma commences in the rete testis, originating from totipotential cells. The tumour varies greatly in size but maintains the shape of the testis and often feels slightly irregular. Histologically, elements of the three embryonic layers may be present including multicystic tissues at varying degrees of maturation, trophoblastic tissue producing human chorionic gonadotrophin (HCG), embryonal carcinoma, and extra-embryonic mesoderm, including yolk sac remnants, which produces alpha-fetoprotein.

Presentation

Most patients present with a painless swollen testicle. 25 per cent present with pain and relate the lump to recent trauma. 10 per cent present with symptoms due to metastases.

Associations

Undescended testis is a feature in 10 per cent.

Management

- Take blood for tumour markers—HCG, alpha-fetoprotein. Do routine haematology. Store sperm.
- Explore any solid testicular mass through an inguinal incision. Clamp the cord with a soft clamp before delivering the testis and splitting it open. If tissue looks grossly abnormal do an orchidectomy.
- Repeat tumour markers postorchidectomy. Do a CT scan of chest and abdomen. These are important staging investigations. If there are para-aortic nodes, lymphatic dissection may be indicated.

Seminomas are treated with radiotherapy to the paraortic and ipsilateral pelvic nodes and cytotoxic drugs as for teratomas.

Teratomas are treated with varying combinations of the chemotherapeutic drugs, viz. etoposide, vinblastine and methotrexate. Surgery may be used to deal with residual tumour or nodes after chemotherapy.

Most patients who enter complete remission appear to be cured.

Prognosis

- *Seminoma*: 90–95 per cent survive 5 years
- *Teratoma*: survival depends on the histological type and varies between 60–90 per cent 5-year survival

Reference

Rustin, J. S. (1987). Testicular tumours and urinary tract malignancies. *Medicine International*, **40**, 1685–7.

Cancer of the penis and scrotum

These are uncommon cancers which are usually squamous cell carcinomas.

Aetiology

Cancer of the scrotum was one of the first industrial cancers to be described. Fortunately, the chimney-sweep's scrotum and mule-skinner's scrotum are now of historical interest only.

Cancer of the penis usually affects uncircumcised males. It is related to poor hygiene with accumulation of smegma which is carcinogenic. It occurs most commonly in the elderly.

Pathology

Both cancers spread to the regional inguinal lymph nodes which may also be enlarged by concomitant infection.

Clinical features

Carcinoma of the scrotum presents as a nodular or ulcerating squamous carcinoma. Cancer of the penis may be a papillary or ulcerating squamous carcinoma. The patient may report a bloody penile discharge or a lump on the shaft.

Diagnosis

Is based on the clinical appearance and biopsy.

Treatment

Cancer of the scrotum is treated by excision. Bleomycin is also effective.

Cancer of the penis. Early cases respond to bleomycin. Superficial radiotherapy is often very successful so that amputation is unnecessary.

Late cases are treated by amputation.

Prognosis

10–20 per cent of patients survive 5 years.

579

Cancer of the kidney

Benign tumours are rare. Consider all clinically recognized neoplasms of the kidney as malignant. The commonest malignant tumour is renal cell carcinoma (hypernephroma, Grawitz tumour). It affects men more often than women. Most patients are > 50 years.

Pathology

When incised the carcinoma appears encapsulated and semi-opaque or dull white. It usually involves the upper pole, is often vascular, and may extend into the renal vein or inferior vena cava. Blood-borne tumour emboli lead to pulmonary 'cannon-ball' secondaries. Lymphatic spread is less common but occurs after invasion of the capsule to hilar and para-aortic nodes. Histologically the carcinoma comprises sheets of cells with small nuclei and attempts at differentiation to tubules.

Tumour grading

Grade I: Well-differentiated papillary carcinoma
Grade II: Moderately differentiated
Grade III: Anaplastic

Clinical features

Haematuria, loin pain and abdominal mass are common. Other presentations include PUO, anaemia, weight loss, malaise, hypertension, polycythaemia and 'clot' ureteric colic.

Investigations

- *Intravenous urography ± tomography* is the commonest diagnostic test.
- *US scan, CT scan* may be diagnostic and indicate vena caval involvement.
- *Arteriography ± therapeutic embolization* indicates the degree of vascularity. Therapeutic embolization reduces the vascularity of large tumours.

Treatment

- *Nephrectomy* via a loin incision unless the tumour is large when the transabdominal or thoraco-abdominal approach may be used. This permits ligation of the renal blood vessels before the tumour is handled so reducing the spread of tumour cells into the circulation.
- *Radiotherapy* is palliative for pain due to metastases.
- *Progesterone therapy* gives a 15 per cent response.

Prognosis

In patients whose disease is localized to the kidney, nephrectomy offers a 65 per cent survival rate.

Transitional cell tumours of kidney and ureter

These are identical to bladder tumours. Three stages of pathological malignancy (G1, G2, G3) are described. The cell types are squamous (associated with stasis), adenocarcinoma (associated with infection—rare), and transitional cell. Spread is initially local and then later by lymphatics. Clinical presentation is with haematuria or pain. Intravenous or retrograde ureto-graphy shows a filling defect.

Treatment

- Many are multifocal. Treatment is by local resection (if appropriate) or nephroureterectomy with removal of the ureter including its site of entry into the bladder.
- Chemotherapy as for bladder cancer may be given systemi-cally or by injection into the ureter.

Cancer of the bladder

This is the commonest form of urological cancer. It is 50 times more common in the transitional cell lining of the bladder than in the mucosa of the renal pelvis or ureter. 80 per cent are transitional cell carcinomas but chronic irritation from stones or infection may cause squamous changes.

Aetiological associations

Cigarette smoking in 30 per cent. Industrial exposure to aromatic amines in rubber and dye industries. Urinary tract infection with *Bilharzia* in Africa and Middle East.

Pathology

Most are papillomata or carcinomata showing varying degrees of differentiation. All should be regarded as potentially malignant. Histologically they are papillary 70 per cent (well-differentiated) or solid (usually anaplastic).

Clinical features

The patient usually presents with painless haematuria. This always requires urgent investigation. Early detection may permit the removal of potentially curable bladder tumours.

Diagnosis and investigation

- FBC. U&E
- Urine—bacteriology, cytology (permits accurate follow-up in unfit patients)
- Intravenous urography may demonstrate a filling defect in the bladder or hydronephrosis associated with ureteric involvement at the orifice.
- Cysto urethroscopy + biopsy
- Bimanual examination under general anaesthetic before and after therapeutic endoscopy may allow the tumour to be staged on the TNM system (see p. 538) and assess the effectiveness of treatment.
- Chest X-ray, CT scan, bone scan for evidence of secondaries

Treatment

T1 and T2 tumours can be removed by transurethral resection or cystodiathermy. Regular check-ups are required for the rest of the patient's life. 50 per cent recur after treatment and if multiple may respond to intravesical therapy with ethoglucid, mitomycin, thiotepa, or doxorubicin hydrocholoride.

Tumours infiltrating muscle superficially may respond to total or partial cystectomy with urinary diversion (ileal conduit) ± radiotherapy. Tumours with deep invasion (T3) rarely respond to these measures, but cisplatin and methotrexate have produced remissions.

Prognosis

Early tumours are curable if the muscle is deeply invaded (T3) less than 40 per cent survive 5 years.

Screening for large-bowel cancer

Populations at risk

- Asymptomatic relatives of patients with colorectal cancer
- Patients over 50 years with large-bowel polyps. Most cancers arise in polyps and most polyps occur in the > 50 year age group.

Methods

History, physical examination, and rigid sigmoidoscopy Rigid sigmoidoscopy has a yield 10× greater than physical examination (digital examination). In a retrospective study of 100 patients with colorectal cancer no difference in bowel habit was noted compared to controls. Nevertheless public education is still probably important.

Flexible endoscopic screening Flexible sigmoidoscopy in apparently healthy subjects does not result in a large increase in the number of polyps or cancers diagnosed but both flexible sigmoidoscopy and colonoscopy are effective in detecting large-bowel polyps in asymptomatic people > 50 years (34 per cent may have polyps).

Faecal occult blood (FOB) screening—(still controversial) Patients who are positive should be rescreened after dietary restriction. If positive a second time full barium enema and colonoscopy should be undertaken. 80 per cent of this second time positive group may have neoplastic disease.

Tumour marker screening

pH Faecal pH is higher in patients with colorectal cancer than in controls.

Serum markers Carcinoembryonic antigen (CEA), carbohydrate antigen 19.9 (CA 19.9). CEA and CA 19.9 offer no advantage either singly or together in the detection of early disease. Both may give false positive results in pulmonary and liver disease. They are widely used outside the UK. (see p. 540).

Flow cytometric DNA content (controversial) When applied to colonoscopic biopsies this may be helpful in the diagnosis of malignant and premalignant conditions. Survival in colorectal cancer may be poorer in patients who have abnormal DNA tumour content.

Radioimmuno localization Monoclonal antibodies have been demonstrated to localize in xenografts of human colon cancer and in primary and secondary colorectal cancers.

Conclusion

Screening is suggested in at risk groups using flexible endoscopic techniques in conjunction with FOB testing. Monoclonal antibodies hold hope for the future and need further evaluation.

Reference

Armitage, N. C. and Hardcastle, J. D. (1986). Neoplasms: Screening and diagnosis (review article). *Current Opinion in Gastroenterology*, **2**, 16–21.

Screening for breast cancer

Hypothesis

Early detection of breast cancer improves prognosis. Reductions in the mortality of breast cancer of between 30–40 per cent have been achieved by screening but are not likely to be sustained at this level.

Methods

Clinical examination (CE) has the advantage that it can be carried out by properly trained health professionals, e.g. practice nurses, radiologists, community nurses, etc. as well as doctors. It should be reserved for symptomatic patients who present and patients in whom screening shows abnormalities.

Breast self-examination (BSE) if practised regularly, may detect breast cancer while it is still at an 'operable' stage of 1–4 cm in diameter. The technique should be demonstrated or described. All four quadrants of the breast must be examined, including the retroareolar area and the axilla. It must be practised weekly so that subtle changes are not missed and is useful as a supplement to mammography and public health education.

Mammography is now established as both the most sensitive and specific method. It is much superior to other methods detecting cancers in a ratio of 6:1 compared to CE. It is also capable of detecting cancers *in situ* of < 2 cm in diameter, and impalpable lesions. It is safe with radiation doses of between 0.05 and 3 rad per exposure (single oblique view for screening or two view for diagnosis).

Breast screening by mammography

Population screened Currently 50–64 age group. In these women the greatest reduction in mortality has been obtained from treatment. Women with dysplasia or a positive family history should also be screened even if < 50 years.

Frequency Every 2–3 years. More frequently in high-risk patients.

Cost Each cancer detected costs £4000 as at 1988.

Advantages Early detection permits more conservative surgery (controversial). Easy local control may reduce the need for radiotherapy. Death in younger women may be postponed for 10 years. Avoidance of unnecessary biopsies.

Disadvantages of screening programmes in general
- Cancer phobia
- Anxiety
- Poor adaptation to the diagnosis and treatment of disease in asymptomatic patients

A national breast cancer screening programme is currently being introduced in the UK to be fully implemented in the 1990s.

References

Baum, M. (1982). Will breast self-examination save lives? *British Medical Journal*, **284**, 142–3.

DHSS (1986). *Breast Cancer Screening*. HMSO, London.

Roebuck, E. J. (1986). Mammography and screening for breast cancer. *British Medical Journal*, **292**, 223–6.

17 The breast

Clinical examination of the breast

An accurate history of the complaint is essential. What are the symptoms? When do they occur, e.g. premenstrually? Is there a family history of breast disease? Establish the age at first pregnancy, and whether the patient breast-fed. Is there a nipple discharge or recent change in the appearance of the nipple, e.g. retraction? What is the current menstrual status? Is the patient taking oral contraceptives?

Examination

This should be done with the patient sitting facing you and also lying down. Often the patient can demonstrate the lump—occasionally by leaning forward.

Sitting Ask the patient to undress and sit facing you. Note the shape and contour of the breasts as she sits with her arms by her sides, on her hips, and elevated. Is there asymmetry, nipple retraction, tethering of the skin, obvious lumps?

As the patient to place her hands on your shoulders. Palpate the axillae for nodes.

Lying down Ask the patient to lie down and place each hand in turn behind her head. Put a pillow under the shoulder on the side being examined. Ask the patient to find the lump or abnormality. Palpate each quadrant of the breast in turn between the fingers and chest wall. Do this on each side. Repeat axillary palpation supporting each arm in turn.

There is a lump!

Note size, shape, site, consistency. Is it fixed to skin or underlying muscle? Ask the patient to place her hands firmly on her hips to contract the pectoral muscles. Does the lump move now? If not, it is fixed to muscle. Pinch the skin overlying the lump. Does it move freely or is it tethered?

Nipple discharge?

Ask the patient to reproduce this. Note the colour. Test it for blood. Which quadrant/area of areola is it extruded from? Do this by compressing each quadrant in turn (usually the patient will guide you) or compressing the nipple and retroareolar tissue at each quadrant. Send discharge for cytological examination. If the duct can be identified a ductogram can be performed.

Cystic or solid?

If there is a lump, attempt to aspirate it with a 20 ml syringe and needle. Steady the mass. If it contains fluid it is usually brown. If it is solid maintain suction whilst inserting and withdrawing the needle at various points in the lump. Smear the aspirate on to a slide. Air dry and send for cytology (fine needle aspiration biopsy—FNAB).

Closed biopsy

Can be achieved by FNAB (ideal), high-speed drill biopsy or Trucut needle biopsy performed under local anaesthetic.

Further information regarding the nature of breast lumps can be obtained by referring the patient for mammography.

If there is doubt about the nature of a breast lump, repeat FNAB, seek a second opinion, or admit the patient for excision biopsy.

Explain at all times to the patient what you intend to do and what your findings are.

Benign breast disease

Most consultations for patients with breast lumps are because of benign disease.

N.B. If there is doubt about a breast lesion, arrange for excision biopsy and paraffin section.

Presentation

Solitary lumps

Fibroadenoma occurs in women between 15 and 35 years and presents as a discrete smooth mobile mass often referred to as a 'breast mouse'. Although usually only 1–3 cm, much larger fibroadenomas have been reported. Excision leads to cure.

Cystosarcoma phylloides is a rare tumour which resembles an intracanalicular fibroadenoma. It presents as a large tumour (> 5 cm) which is mobile. It is locally aggressive, and 25 per cent metastasize. Treatment is by mastectomy and reconstruction.

Fat necrosis, sclerosing adenosis, confluent periductal mastitis may give rise to localized firm swellings. The history suggests benign disease. Excision biopsy with a margin of normal tissue is the correct action.

Acute and chronic abscess. Acute abscesses occur in the puerperium. Treatment is incision and drainage. Chronic abscesses result from inadequate drainage or the late administration of antibiotics. Without biopsy, this cannot be distinguished from carcinoma.

Breast cysts may be single or multiple. Pain may be experienced before periods. Most can be adequately treated by aspiration. If there is blood or recurrence after 6 weeks, the cystic area should be excised.

Discharge from the nipple

Eczema of the nipple is usually bilateral and may be due to simple adenoma, but Paget's disease in which an underlying intraduct cancer invades and ulcerates the nipple skin has to be excluded. All bleeding nipples should have scrapings sent for cytology or biopsied. Paget's disease has a good prognosis, as the cancer is sometimes *in situ*. In patients with galactorrhoea consider the possibility of a prolactinoma.

Duct ectasia presents as a nipple discharge from one or more duct openings. It is due to duct dilatation with surrounding periductal mastitis. The discharge is usually green or brown in women of 40–50 years. There may be associated nipple retraction and chronic inflammation. Treatment is excision of the affected ducts and surrounding breast tissue.

Mammillary fistula (uncommon) In this condition a fistulous tract in communication with a major duct opens in the para-areolar region. It affects women in their 30s and is treated by excision of the affected duct system and fistula (Adair's operation).

Duct papilloma leads to bleeding from the nipple. Usually solitary it can be excised with a portion of the duct wall—microdochectomy. If multiple, total ductal excision or even mastectomy may be necessary.

Lumpy painful breasts This is most often due to benign mammary dysplasia (BMD) or cyclical changes. In BMD there are multiple pathological changes including adenosis, cystic change, and ductal proliferation (syn = fibrocystic disease, chronic mastitis). Typically pain and nodularity is cyclical. Thickening is most common in the upper outer quadrant or axillary tail. It is not related to breast cancer when women from 20–50 years are affected. Treatment involves aspiration of cysts, excision of painful lumps, diuretic and hormonal drugs (e.g. danazol 100 mg twice daily for 6 months). In rare cases subcutaneous mastectomy is indicated.

N.B. If a lump is clinically and mammographically benign with negative FNAB cytology (or Trucut if FNAB not available), it may be left behind.

18 Practical hints for housemen

Endotracheal intubation

This procedure is indicated in certain acute respiratory disorders and patients with cardiac arrest, and is commonly performed in the anaesthetic room of surgical theatres before surgical operations. It is here that expertise can be acquired.

Equipment required

All medical and nursing staff should know where the resuscitation set is to be found in the ward. The essentials are a laryngoscope, endotracheal tube and inflation bag, and 10 ml syringe to inflate the cuff. They should be checked regularly to ensure that they are in good working order.

Procedure

In patients with cardiac arrest or who are unconscious for other reasons there is no need to use a muscle relaxant. Remove any dentures, and position the patient to lie supine. Insert the laryngoscope, pushing the tongue to the patient's left. Hold the prelubricated endotracheal tube (size 8 or 9 mm females, 9 to 11 mm for males)* in the right hand and pull the laryngoscope upwards to expose the epiglottis and vocal cords. When you see them, slip the tube between them into the trachea. Inflate the cuff immediately to secure the tube and prevent aspiration of vomitus or gastric contents, and connect the tube to the inflation bag. The chest should then move uniformly in response to manual ventilation.

When relaxation is required, an anaesthetist should be contacted. In most hospitals he is a member of the 'crash call team'. He will give a combination of muscle relaxant and an induction agent to anaesthetize the patient who will subsequently need to be ventilated. This is a skilled procedure, which should not be performed without expert help or supervision.

Most endotracheal tubes are removed in 24–48 hours. If longer-term ventilation is required, nasotracheal intubation may be used or tracheostomy considered.

Helpful hint Theory is no substitute for practice. It is within the capability of all medical students and housemen to become skilled at endotracheal intubation. Contact the anaesthetic department of your hospital and arrange to attend some surgical lists where endotracheal intubation can be learned under expert guidance in the anaesthetic room, particularly during lists of short procedures during which many patients are treated.

* For children under 4 years of age, the internal diameter of the tube = ($\frac{1}{4}$ age) × 3.5 mm.
For children over 4 years of age, the internal diameter of the tube = ($\frac{1}{4}$ age) × 4.5 mm.

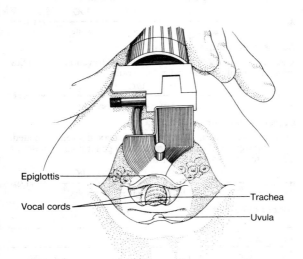

Epiglottis

Vocal cords

Trachea

Uvula

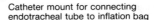

Catheter mount for connecting
endotracheal tube to inflation bag

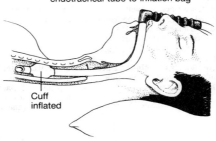

Cuff
inflated

Endotracheal intubation

Equipment for local anaesthesia and venepuncture

The needles and anaesthetic agents described are representative of those used in most hospitals. Many are in everyday use.

Needles

These are gauged in descending numerical order as the base of the needle increases. They are also colour-coded:

	25G	23G	21G	19G	
smallest					→ largest
	orange	blue	green	white	

Abbocaths, Venflons

These are regularly used for establishing intravenous lines. They have numerical gauging and colour coding:

	22G	20G	18G	17G	16G	14G	
smallest							→ largest
	blue	pink	green	yellow	grey	brown	

Butterfly needles

Their use is indicated when repeated access to a vein is needed, e.g. for injection of several drugs (as in general anaesthesia or intermittent antibiotic administration):

	25G	23G	21G	19G	
smallest					→ largest
	orange	blue	green	white	

Local anaesthetics

Anaesthetic	Indication
Xylocaine Safe dose = 200 mg plain/4 hour period 0.5% 1% 2% (the lower the percentage, the higher the volume which may be infused)	Most minor local anaesthetic procedures
Lignocaine Hydrochloride safe dose = 200 mg/4 hours, 2.13% w/v	Most minor local anaesthetic procedures
Xylocaine 1 per cent with adrenaline 1:200 000 (dangerous in digital anaesthesia and in patients with thyroid or cardiac disease) Safe dose = 7 mg/kg/ 4 hours	For local anaesthetic procedures in vascular areas (e.g. face)
Lignocaine hydrochloride 2 per cent with adrenaline 1:80 000	For local anaesthetic procedures in vascular areas
Ethyl chloride spray	Topical local anaesthesia e.g. drainage of an external plexus haematoma (anus)

Procaine hydrochloride 3%	Local anaesthetic procedures.
Octopressin (Citanest) (produces methaemoglobinaemia. Increasingly less used)	Can be used in large volumes (check literature)
Marcain 0.5% with glucose (Marcain heavy steripack)	Spinal anaesthesia, epidurals
Marcain 0.5%	Local anaesthetic procedures Dental anaesthesia

Local anaesthesia

Local anaesthesia is of particular value in carrying out most minor surgical procedures such as the excision of skin lesions and minor digital surgical procedures. Do not use adrenaline-containing anaesthetics in areas supplied by end arteries. It also offers an effective method of pain relief in patients with digital injuries or traumatic amputations. It can be extended to provide regional anaesthesia by blocking major nerve plexuses and cutaneous nerves.

Anaesthetic agents

Lignocaine, xylocaine, procaine in preparations of 0.5–2 per cent are very useful. When these preparations are used with adrenaline this reduces bleeding and prolongs anaesthesia (solutions of 1 in 200 000 are satisfactory) but should not be used in the digits, ears, penis, or in inflamed tissue because of the risk of gangrene. In this situation ethylchloride might be used for an incision.

Safe maximum dosages

20 ml of 1 per cent solution without adrenaline. 50 ml of 1 per cent solution with adrenaline. These are adult doses and should be reduced in old people.

Overdose may lead to neurological changes (fits) or cardiovascular changes (arrhythmias).

Patient preparation

A full, clear explanation of the procedure must be given, pointing out that after the skin prick the infusion of the anaesthetic agent can produce a stinging sensation after which numbness of the part occurs. The patient should also be warned that he may still be able to appreciate pressure, but that there should be no pain. If the patient does feel pain further injection of local anaesthetic may be necessary.

Technique

Clean the skin with an antiseptic (aqueous for mucous membranes, alcohol-based for skin). Always scrub your hands and wear a gown, mask, and gloves. Insert a fine needle intradermally and raise a bleb before infusing more deeply. When local anaesthesia is being used to excise a skin lesion, it is usual to inject the agent at several sites around the lesion. With each insertion withdraw the plunger of the syringe to ensure that a blood vessel has not been entered. When anaesthetizing an open wound the needle may be inserted at various sites along the edge of the wound withdrawing the plunger with each insertion before injecting the anaesthetic. When anaesthetizing digits the needle should be inserted into the web space on either side of the digit, directed anteromedially towards the bone so that the local anaesthetic may be infused around the digital nerves (ring block). Often a red rubber catheter is used as a tourniquet to control bleeding. This should be removed after 30–45 minutes.

If the procedure is going to take longer than this the tourniquet should be released for 5 minutes and then reapplied. This must be removed at the end of the procedure.

Excision of benign cutaneous lesions

The lesion should be anaesthetized with a local anaesthetic containing adrenaline (1 in 200 000) to provide as bloodless a field as possible (face and scalp only). A good rule of thumb is to send all specimens for histological examination.

Preoperative preparation

Most lesions can be excised via an elliptical incision. The skin should be stretched in various directions prior to this, to ensure that the incision will lie in the skin lines. The ellipse should be marked out using a marking pen. Wedge-shaped excisions may be indicated when the lesion is on the lip, the nostril, ear, or eyelid. Circular incisions may be used where the facial skin is closely applied to underlying cartilage. Examples include the nasal tip or the anterior surface of the auricle. Such lesions can then be closed with a full-thickness graft.

Instruments

Fine skin hooks, sharp scissors, and detachable scalpel blades sizes 11 and 15 are required. Use fine sutures with needles which have a cutting edge. Fine and small-tooth tissue forceps and a light needle holder with smooth jaws are also required.

Operation

Elliptical incision Use a size 15 blade to cut through the skin at right angles to its surface. This permits the corners to be cut out more exactly and prevents x-shaped overcuts at the corners of the ellipse. Undermine the wound edges to enable the wound to be closed without tension using skin hooks and a single sweep of a size 15 blade in the superficial subcutaneous tissue.

Closure Two layers. The deep layer is approximated with inverted absorbable sutures (vicryl, Dexon). The skin edges with fine interrupted monofilament non-absorbable (nylon, prolene 3/0).

Wedge excision Use a size 15 blade to cut the skin surface. Change the blade to a size 11 and use a sawing motion so that this can be passed completely through the tissues.

Closure The wound can be closed in three layers from the deepest part of the wedge outwards using interrupted nylon to skin and ensuring that in the case of the lip the vermilion cutaneous border is accurately re-aligned.

Circular excision Use a size 15 blade. Incise the skin vertically down through subcutaneous fat to superficial fascia. Ensure haemostasis by applying artery forceps and ligating any bleeding points with fine ties (vicryl, Dexon 4/0).

Closure The defect can be closed by a split skin graft or alternatively by sliding triangular subcutaneous pedicle flaps to close the circular defect and closing the two remaining triangular defects as a V–Y plasty.

18 Practical hints for housemen
Excision of benign cutaneous lesions

Venepuncture

This everyday procedure is an essential part of the houseman's repertoire which should be learnt as an undergraduate. It must be performed safely and as painlessly as possible.

Equipment

A syringe, needle, venous tourniquet, and sterile swabs should be at the ready. These, with the prelabelled tubes for the blood, may be carried in a small tray.

Preparation

The hands should be clean and the site for venepuncture on the patient should be cleaned with an alcohol swab (Steriswab, Mediswab, etc.).

Approach

Be reassuring. Explain the procedure clearly. Keep the needle out of sight (especially with children). Ask the patient to look away.

Sites for venepuncture

For routine sampling the ideal site is the cubital fossa in the forearm. The cephalic and basilic veins are suitable. For injection of anaesthetic agents or repeated sampling the veins of the dorsum of the hand may be used, a butterfly needle being inserted.

Applying the tourniquet

This should be gently applied proximal to the site of venepuncture. Do not keep the tourniquet applied when taking blood for serum calcium estimation. Release it when the needle has been inserted.

Technique

The needle may be straight or bent and should be applied to the skin obliquely in line with the vein. Loss of resistance is felt when the lumen is entered.

A 21 gauge (green) needle is best. Once inside the vein, withdraw the blood slowly to prevent haemolysis. With vacuum tubes the blood is withdrawn at a steady rate, but accurate venepuncture is required. The new Becton Dickinson syringes have a Luer lock which engages with a $\frac{1}{4}$ turn of the hub of the needle. This reduces the risk of the needle becoming detached from the syringe, and so prevents spillage.

When sufficient amount of blood has been removed, release the tourniquet, apply a swab to the venepuncture site, and withdraw the needle while exerting pressure on the site. Ask the patient to maintain pressure. Dispose of the needle in a Burn Bin. This is a solid cardboard bin which seals automatically and is incinerated when full.

Precautions

Do not spill the blood on your skin or stab yourself with the needle. Both AIDS and serum hepatitis can be contracted in this manner (see above for technique).

Take care too not to apply the tourniquet too tightly. This is unnecessary.

Difficulties

In old patients with tortuous veins immobilize the vein with your fingers. In some patients, e.g. hypotensives, a blood pressure cuff may be necessary as a tourniquet.

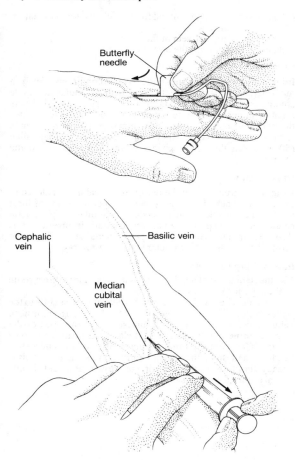

Venepuncture

Cannulation of a central vein

Indications

- Measurement of central venous pressure (CVP) in the management of shock or monitoring procedures
- Regulation of large volumes of transfused fluids or in the intensive care unit in severe sepsis or multiple organ failure
- Parenteral nutrition (must be inserted under strict asepsis for long-term access)

Veins used

- The basilic vein in the cubital fossa is suitable for cannulation with long catheters
- The internal jugular vein of the neck
- The subclavian vein (the best site)

Methods

Insert all central venous lines under aseptic conditions, ideally in the operating theatre. There are many cannulas available commercially. A 16 G needle is most frequently used. Those retained in position by a teflon or plastic outer cannula reduce the risk of cannula embolization which is present if the cannula is introduced through a metal needle. Have isotonic saline (500 ml) 'run through' a giving set ready to connect to the cannula.

Basilic vein

Read the instructions on the use of the IV cannula enclosed in the cannula pack. These are precise. Inject a few ml of local anaesthetic. In effect the procedure is similar to peripheral cannulation, except that once the cannula is inserted it is threaded up the arm to the superior vena cava. Check the position of the tip of the catheter with a chest X-ray.

Internal jugular vein

Use the right internal jugular vein, as it gives a more direct route to the superior vena cava.

The patient lies on the table with his head turned slightly to the opposite side. The table is then tilted 15–25° to congest the neck veins. Inject a few ml of local anaesthetic at the apex of the triangle formed by the sternal and clavicular heads of sterno-mastoid. Carry out a test venepuncture with a syringe and a 19 G needle to determine the angle of entry with the cannula. This is usually about 45° to the skin directed downwards and slightly lateral to the carotid pulsation.

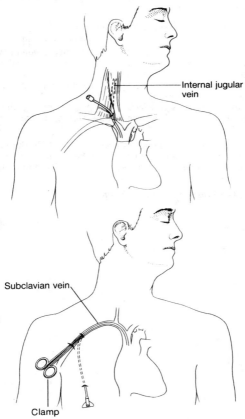

Internal jugular vein

Subclavian vein

Clamp

The cannula is pulled through the subcutaneous tunnel

Cannulation of a central vein

Cannulation of a central vein (*cont.*)

Subclavian vein

Approach this vein by the infraclavicular route. Access is made easier with slight Trendelenburg tilt. Insert the needle at the mid-clavicular point, directing it towards the sternoclavicular joint. In practice this means upwards under the clavicle at an angle of about 45°. For IV nutrition a cuffed silicone rubber catheter is used (Hickman). Make sure that nursing staff know how to maintain sterility. In many hospitals a TPN surveillance team will do this. The feeding line should be used for no other purpose, and the site of insertion should be dressed with a closed antiseptic dressing.

Checking for successful venepuncture

- Connect the saline infusion to the cannula and apply a clip to the tubing. Lower the bag below the level of the patient. Blood should now flow back down the infusion tubing, indicating successful venepuncture.
- Check the position of the tip of the cannula by chest X-ray and adjust as required.
- If the position is satisfactory apply antiseptic to the skin puncture site and a transparent sterile dressing (polyurethane, Opsite).

Tunnelling If the cannula is to be used for TPN the risk of infection may be reduced by bringing the cannula out at some distance from the site of venepuncture by fashioning a subcutaneous tunnel.

Complications see nutrition in surgery (p. 94).

Arterial blood sampling

This investigation provides information on the pH status of the blood and the partial pressures of oxygen (pO_2) and carbon dioxide (pCO_2). There are two methods of sampling:

- Direct arterial puncture
- Indwelling arterial catheter for multiple samplings in patients who require regular monitoring (usually carried out in intensive care units)

Direct puncture

The radial artery at the wrist is the safest site. Puncture of the femoral artery in the groin may lead to dislodgement of atheromatous plaque with embolization distally. Use of the brachial artery with subsequent intimal drainage may lead to distal ischaemia.

Requirements

- One 10 ml syringe free of all air, heparinized with 1 ml of sodium heparin solution (5000 units/ml) and needle to collect the blood
- Local anaesthetic solution (1 per cent), syringe, and needle
- Antiseptic solution

Procedure

Forewarn the laboratory of the sample's arrival. Explain the procedure to the patient. Wash your hands thoroughly and clean the skin over the wrist where the radial artery can be palpated with Hibitane or Mediswab. Inject about 1 ml of local anaesthetic. Palpate the radial artery with the forefinger and middle finger of the left hand. Separate your fingers and insert the needle vertically between them. Steady the syringe now and aspirate the blood gently. Often a small spurt of blood with indicate that the needle is in the lumen. Remove the needle and press a swab over the puncture site. An assistant should maintain the pressure for about 5 min. Immediately take the syringe to the laboratory, rotating it gently to mix the heparin with the blood. Special syringe caps are provided by some laboratories.

Insertion of an intercostal drain

This technique is indicated in pneumothorax or haemo-pneumothorax.

Equipment required

The correct side should first be identified by chest X-ray, but there may not be time for this in an emergency. Explain to the patient what you are going to do and why. Have an Argyle drain with trochar available, antiseptic solution, swabs, needles and syringe, local anaesthetic, gauge 1/0 silk, Prolene, or nylon, scalpel, and dressing pack. In addition an underwater seal system, drainage tubing, and clamp should be available. Use an aseptic technique. Wear gloves and gown.

Sites of insertion

- Outside the mid-clavicular line in the second intercostal space (not cosmetic for female patients)
- Posterior axillary line via the eighth or ninth intercostal space

Identification of site

The manubrium sterni is at the level of the second costal cartilage. Count the spaces from this level to identify the site.

Procedure

For pneumothorax a single drain inserted through the second interspace is usually all that is required. When there is an accompanying haemothorax, drains are often inserted at both sites—the lower level to drain the blood.

Local anaesthetic is infiltrated into the skin over the second interspace. Then deeper infiltration of the muscles and parietal pleura can be achieved by inserting the needle into the space over the rib (not under it) to prevent neurovascular damage. Use the scalpel to cut the skin transversely, enough to permit insertion of the drain. Then insert the trochar and cannula (Argyle drain). Considerable pressure may be required and it is wise to hold the trochar and cannula with both hands, the right pushing and the left acting as a stop when the trochar penetrates the pleura. The left hand should be about two inches (5 cm) from the tip. When the pleura has been penetrated there is considerable 'give'. Withdraw the trochar whilst threading the drain into the pleural cavity to the mark on the drain. Apply the clamp across the drain and suture the skin using a purse-string suture. Leave the ends untied so that the suture can be tied on removal of the drain. Now connect the underwater seal drainage system, ensuring that it is below the level of the patient (otherwise water may run into the pleural cavity). Secure the chest drain to the skin with sutures. Apply the dressing and release the clamp. Are there bubbles in the drainage system?

Get a check chest X-ray to determine the position of the drain.

Removal of the drain

In uncomplicated cases the lung usually re-expands in 24–48 hours, but may take longer. Take daily erect chest X-rays to check expansion. When complete clamp the drain and remove it 24 hours later, with a further chest X-ray. Seal the wound with the purse-string suture and a dry dressing.

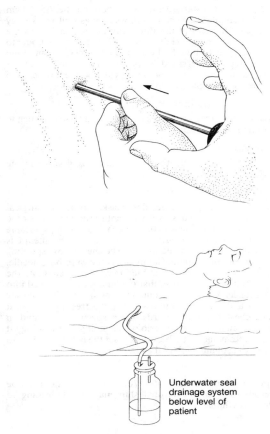

Underwater seal
drainage system
below level of
patient

Insertion of an intercostal drain

Aspiration of fluid from the pleural cavity

Pus and other fluids can be aspirated from the pleural cavity under local anaesthesia.

Preliminary investigations

A chest X-ray, posterior–anterior and lateral, will demonstrate the site of the fluid which is usually in the most dependent part of the chest posteriorly.

Percuss the chest to determine the upper level of dullness before starting.

Equipment required

Have at hand, on a sterile tray, a wide-bore needle 7.5–10 cm long with a 50 ml syringe and three-way tap as well as a kidney dish or jug. One end of the three-way tap is attached to the syringe, another to the needle and the third to sterile tubing to drain the aspirated fluid. Make sure that the mechanism of the three-way tap is understood (to prevent inadvertent pneumo-thorax). Have ready any antibiotic or cytotoxic agent which is to be instilled into the pleural cavity.

Position of the patient

The patient sits erect and places his arms on a table or bench in front of him (see diagram).

Site of aspiration

This is usually in the eighth or ninth intercostal space in the posterior axillary line.

Procedure

Use aseptic technique, scrub, don mask, gown, and surgical gloves, clean the patient's skin with antiseptic and use sterile drapes. Infiltrate the skin with 0.5 per cent–1 per cent, xylocaine (10–20 ml) down to pleura intercostally. As the pleura is infiltrated the accuracy of the siting can be checked by aspirating a small amount of the fluid. Gently insert the wide-bore needle, apply the three-way tap so that fluid may be aspirated into the syringe, then close the tap so that the fluid may be flushed into the receptacle. As the fluid reduces in amount, gently withdraw the needle and cover the wound with a sterile dressing. Carry out a check chest X-ray afterwards. Keep specimens of fluid in universal containers for bacteriology, histology, or cytology if necessary. Remember it is best to contact the laboratory first to ascertain the volume of fluid they require.

Warning

Aspiration of more than 600–700 ml at a time may be dangerous. Sudden mediastinal shift may result, leading to cardiac arrhythmias.

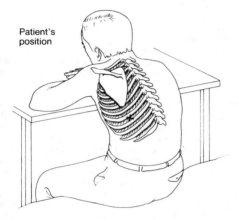

Patient's position

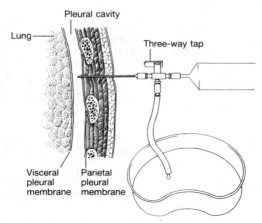

Lung

Pleural cavity

Three-way tap

Visceral pleural membrane

Parietal pleural membrane

Aspiration of fluid from pleural cavity
N.B. sites of aspiration—8th/9th intercostal space

Abdominal paracentesis

This is a useful diagnostic technique. It is also indicated as a therapeutic measure to relieve discomfort in patients with increasing ascites (often due to malignant disease) and administer cytotoxic agents intraperitoneally. The method is also of use in the diagnosis of abdominal trauma.

Equipment required

Local anaesthetic agents, syringe and needle, swabs, silk sutures (2/0 or 3/0), antiseptic solutions, scalpel blade, trochar, and cannula or peritoneal catheter. These, with adequate drainage tubing and collecting bag, may be made up in a sterile pack.

Sites for drainage

- Midway between symphysis pubis and umbilicus
- Midway between umbilicus and anterior superior iliac spine in either iliac fossa. (Remember the course of the inferior epigastric arteries.)

Procedure

Explain clearly to the patient what is to be done. Prepare the skin with antiseptic and infiltrate a few ml of local anaesthetic into the chosen site. Make a small nick in the skin with the scalpel and *gently* insert the trochar and cannula or peritoneal dialysis catheter (Tenckhoff silicone-rubber catheter which has multiple perforations along its tip). When the peritoneal cavity has been entered there is a distinct feeling of 'giving way'. Remove the trochar or introducer from the peritoneal catheter and connect the cannula or catheter to the collecting system. Insert a suture into the skin and tie it around the drainage catheter. Apply a dressing or tape swab around the catheter.

Specimens of fluid should then be sent to the laboratory for histology, bacteriology, biochemical analysis, or cytology. When the patient has a very distended abdomen drain the fluid at a rate of 2 l/day (rapid decompression can produce shock as in chest aspiration) by using a clip to control the catheter flow rate. Ask the patient to change position as the flow rate slows down by lying on his side to let the fluid gravitate. If the tube blocks it may be flushed with sterile saline or manipulated to reposition it within the peritoneal cavity.

Peritoneal lavage

Following surgery for widespread peritonitis, and in some patients with severe acute pancreatitis, peritoneal lavage may be undertaken to irrigate and debride the peritoneal cavity. Two catheters are used. One, a Tenckhoff peritoneal dialysis catheter, is inserted deep in the pelvis; the other, either a catheter or drain, is inserted in one or other paracolic gutter depending on the site of maximum sepsis. Normal saline or peritoneal dialysis fluid with added antibiotics (see peritonitis) is introduced through the pelvic catheter and drained out through that in the paracolic gutter.

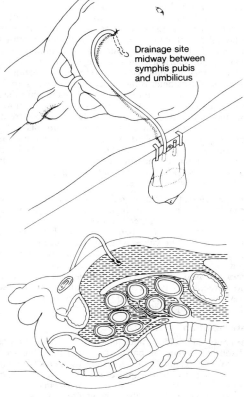

Drainage site
midway between
symphis pubis
and umbilicus

617

Abdominal paramentesis

Urethral catheterization

This procedure is performed to relieve retention of urine, to monitor renal output, as a preliminary to surgical procedures, or occasionally to obtain urine for bacteriological analysis.

Types of catheter

The most commonly used is the Foley catheter which is retained in the bladder by an inflatable balloon. (The smallest reasonable size is recommended, e.g. size 14 for adult males.) Others such as the Gibbon and whistle-tip catheters are useful when prostatic obstruction is present.

Technique

Catheterization of the male patient A full explanation must be given and strict asepsis observed. The patient lies back with his legs apart. After scrubbing and gloving up the doctor lays sterile towels (often a single paper towel with a hole in the centre to accommodate the penis) around the penis. A sterile swab is then used to hold the shaft of the penis while the prepuce is retracted and the glans cleaned with antiseptic. The urethra may then be anaesthetized and lubricated by inserting sterile lignocaine gel into it. This comes in a tube with nozzle for easy insertion. The bag containing the catheter is then opened to reveal its tip. This is then inserted into the urethra with the right hand while the left hand pulls the penis on the stretch in a horizontal position. At the same time the catheter is progressively introduced from its sterile bag by feeding it out of the open end. In this way the catheter itself is not touched. When the bladder has been entered urine will begin to flow and the balloon may be inflated with sterile water in a volume which conforms to the instructions on the side of the catheter. The collecting bag may then be attached and the towels removed.

Catheterization of the female patient It is sensible to have a chaperone for this procedure (if a male) or better still for a female nurse or doctor to catheterize the patient. The patient lies on her back with her knees bent, heels together, and thighs abducted to expose the vulva. Aseptic procedure is observed, gloves, cap, and mask being worn. The labia are parted with the left hand and cleaned with antiseptic swabs held in forceps. Each swab is used once and the labia cleaned in an antero-posterior direction to avoid contamination from the perineum. The urethral orifice is identified, the prelubricated catheter passed, and the balloon inflated.

Catheter fixation

In some confused or elderly male patients the catheter may have to be fixed to prevent the patient pulling it out. Replace the prepuce over the glans. Cover the tip of the penis with sterile gauze impregnated with antiseptic cream. Then apply strapping longitudinally along the dorsum and ventral aspects of the penis over the gauze and apply it to the catheter along its length so that the overlapping edges stick together.

Intermittent catheterization may be carried out by cooperative patients whose bladders leak when overfull. The procedure is safe provided the patient uses a 'clean' technique. It is one method of maintaining dryness in patients with overflow incontinence.

Indwelling catheter A narrow silicone-rubber catheter should be used. The main complications are infection and blockage. They should therefore be changed regularly. It often helps if the urine is acidified with vitamin C and mandelamine. A high fluid intake maintains a dilute urine. Indwelling catheters are of value in some patients with urinary incontinence.

Suprapubic cystostomy

This permits drainage of the bladder when urethral catheterization is impossible or contra-indicated.

Contra-indications

Patients with bladder tumours.

Essentials

- The bladder must be distended to midway between the pubis and umbilicus.
- Avoid the peritoneum.
- Use a balloon catheter of adequate size (16F Ingram trocar catheter).

Procedure

Read the instructions on the Ingram trocar catheter. Scrub up. The patient lies supine. The suprapubic area is prepared and draped. Local anaesthetic is infiltrated in the midline two finger breadths above the pubic symphysis, using a lumbar puncture needle (12–16 FG). Infiltrate deeply and finally aspirate a little urine from the bladder. Note the direction in which the needle is pointing.

Make a small stab wound incision over the anaesthetized area. Insert the trochar catheter at right angles over the right hand, using the left hand to control its depth of penetration once the bladder has been entered. Now advance the catheter into the bladder and withdraw the trochar. Inflate the catheter balloon with about 5 ml water and connect the catheter to a drainage bag.

Drain the bladder fully and adjust the flange until it makes contact with the skin. Suture it in position with 3/0 nylon or Prolene to prevent accidental removal of the catheter.

Sigmoidoscopy

The sigmoidoscope is a rigid steel or plastic hollow tube about 25–30 cm in length and up to 2 cm in diameter. It comprises an outer hollow tube, an obturator which is withdrawn after the instrument has been inserted, and a light source which is fibre-optic in modern instruments. There is an attachment on the lens piece for the insufflation of air to open up the lumen ahead of the instrument.

Preparation of patient

The patient should have the procedure explained fully and give consent. Preparation of the bowel is often unnecessary. A disposable enema may help in cases of difficulty, but may also have the reverse effect sometimes.

Position of patient

The patient lies in the left lateral position with the hips flexed and the buttocks raised on a sandbag or pillow.

Procedure

Carry out digital examination of the rectum. If it is loaded, defer sigmoidoscopy until the bowel has been prepared with a disposable enema.

Lubricate the sigmoidoscope with K-Y jelly. Disposable plastic sigmoidoscopes are lubricated under running warm tapwater. With the obturator in place introduce it gently through the sphincter to about 5 cm by pointing it towards the umbilicus. Remove the obturator and advance under direct vision after attaching the eyepiece, insufflator, and light source. Keep the lumen in view at all times as you pass the instrument upwards to its full length. Negotiation of the rectosigmoid junction requires considerable experience and skill. If the patient suffers discomfort do not persist.

Note the appearance of the mucosa, the presence of contact bleeding, ulceration, or neoplasm. Take biopsies of suspected abnormalities for histological examination using the specialized biopsy forceps provided. Polyps can be removed by snaring and diathermy.

It requires considerable experience to interpret what you see.

N.B. If a biopsy has been taken the patient should not have any type of enema for at least a week, for risk of perforation of the bowel or air embolus.

Carry out proctoscopy last. This is best for banding and injection procedures.

Flexible sigmoidoscopy This fibre-optic instrument is now in routine use. It is longer than the rigid sigmoidoscopy (up to 40 cm) and good bowel preparation is required. Like oesophago-gastro-duodenoscopy, considerable training is required to become skilful.

Colonoscope This instrument is even longer than the flexible sigmoidoscope and can be passed through the anus, a terminal or loop colostomy, or an ileosotomy. It is especially useful in assessing the sigmoid and descending colon which may present difficulties radiologically due to the two segments overlying each other.

Indications

- Assessment of the colon in patients with suspected malignant disease (can be passed to the caecum)
- Follow-up assessment of patients with resected malignancy
- Assessment and biopsy of inflammatory bowel disease
- Screening of patients with adenomas (and their removal)
- Assessment and biopsy of diverticular disease (which may mimic carcinoma on occasions)
- Removal of polyps
- Decompression of sigmoid volvulus
- Investigation and decompression of pseudo-obstruction

623

Upper GI endoscopy—technique

Checklist

- Ensure that the patient is fully prepared, has received an adequate explanation, and has given consent.
- Ensure that the endoscope is working properly.
- Ensure that the endoscope has been cleaned.

Technique of insertion

Insert a butterfly needle into a vein on the dorsum of the hand. Spray the fauces with anaesthetic. Ask the patient to lie in the left lateral position. Insert a mouth guard if he has teeth. Give the patient sedation. This varies according to preference, but often 10 mg diazemuls IV slowly is adequate (large-dose midazolam is an alternative but after sedation patients need care). Buscopan may also be given (20 mg IV) to provide smooth muscle relaxation. Wear gloves during the procedure.

Hold the instrument controls in the left hand. Insert the index finger of the right hand into the mouth and pass the tip of the endoscope under the finger. Now gently pull the tongue and the tip of the endoscope forwards. Gently advance the endoscope, asking the patient to swallow. *Do not use force to pass the instrument.*

An alternative method is to insert the instrument without directing it with the right index. The instrument is inserted into the mouth and aligned with the pharynx under direct vision. It is passed over the tongue with the tip flexed. As it passes down the tip is returned to neutral and the patient asked to swallow.

Inspection of upper GI tract

Oesophagus Look for webs or carcinoma in the upper one third. Gently inflate after clearing mucus by suction. The gullet usually distends so that a clear view can be obtained. Look for landmarks like the cardiac impulse. In the lower oesophagus check the mucosal folds for inflammation due to oesophagitis and note the distance from the incisor teeth. Check also for hiatus hernia. Deep inspiration by the patient helps to identify the position of the diaphragm. If there is an obvious lesion pass the biopsy forceps down the biopsy channel and take a specimen for histology (preferably brushings for cytology which is safer).

Stomach Follow the lumen at all times. Inflate as required. On entry note obvious abnormalities like abnormal folds, tumours, blood, bile, retained fluids or foods. Aspirate excessive fluid to prevent inhalation. Now inspect in turn the whole lumen, the greater and lesser curves, the roof of the pyloric antrum (by flexing the tip), and the fundus by increasing flexion of the tip and rotating the instrument over more than 90°. Carry out biopsy or brushings if indicated. You will be able to see the proximal part of the 'scope entering the cardia. This is a useful point for orientation. Now advance the 'scope to inspect the pylorus. Note its mobility and the presence of deformity or inflammation.

Duodenum Enter the duodenum by keeping the pylorus in the centre of your view. The up-and-down control with the left hand assists advancement. Once inside the duodenal cap carry out a full inspection, withdrawing the tip and depressing it to achieve this.

To examine the remainder of the duodenum the 'scope usually has to be rotated 90° to the right using the up-and-down control to see the lumen.

Withdrawal Reinspect the lumina of the duodenum, stomach, and oesophagus on withdrawal by rotating the tip. Aspirate excessive air from each area once you are satisfied with the examination.

19 Anatomy for surgeons

The thyroid and parathyroid glands

The thyroid gland

This comprises right and left pear-shaped lobes which lie at the sides of the larynx and trachea connected to each other by an isthmus at the level of the second, third, and fourth tracheal rings. A pyramidal lobe often projects upwards from the isthmus. The gland is highly vascular, enclosed in a fibrous capsule and bound by a sheath of pretracheal fascia to the larynx so it moves up and down during swallowing.

Relations

Superficial Sternothyroid, sternohyoid, omohyoid and their nerves from the ansa hypoglossi. These are overlapped below by the sternomastoid.

Medial Cricoid cartilage, thyroid cartilage, and inferior constrictor above and recurrent laryngeal nerve.

Posterior Carotid sheath, inferior thyroid artery, prevertebral muscles.

Important points

These are the blood supply, lymphatic drainage, the recurrent laryngeal nerves, and the parathyroid glands.

Blood supply

Superior thyroid artery This is the first branch of the external carotid. It passes to the upper pole under the infrahyoid muscles close to the external branch of the superior laryngeal nerve. Damage to the nerve leads to loss of timbre of the voice, which becomes monotone. The superior pedicle should therefore be tied close to its origin.

Inferior thyroid artery Arises from the thyrocervical trunk, a branch of the subclavian artery. It ascends to the carotid sheath and enters the posterior part of the gland. The inferior pedicle should be tied well out to avoid recurrent laryngeal nerve damage.

Superior thyroid vein emerges from the upper pole and drains into the internal jugular vein.

Middle thyroid vein from the lower part of the lateral border of the lobes crosses the common carotid and drains into the internal jugular vein.

Inferior thyroid vein from the isthmus or lower medial lobe passes along the front of the trachea to the left innominate vein.

Lymphatic drainage

To the middle and lower deep jugular, pretracheal, and mediastinal nodes.

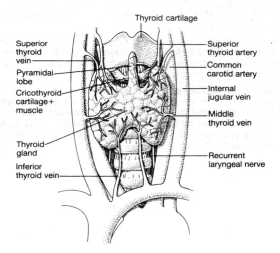

Thyroid cartilage

Superior thyroid vein

Pyramidal lobe

Cricothyroid cartilage+ muscle

Thyroid gland

Inferior thyroid vein

Superior thyroid artery

Common carotid artery

Internal jugular vein

Middle thyroid vein

Recurrent laryngeal nerve

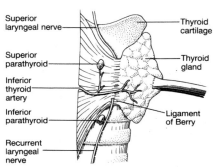

Superior laryngeal nerve

Superior parathyroid

Inferior thyroid artery

Inferior parathyroid

Recurrent laryngeal nerve

Thyroid cartilage

Thyroid gland

Ligament of Berry

Thyroid gland

The thyroid and parathyroid glands (*cont.*)

Recurrent laryngeal nerve

Arises from the vagus and runs upwards in the tracheo-oesophageal groove behind or in front of the inferior thyroid arteries. It supplies all the intrinsic muscles of the larynx. Damage or division causes paralysis of the vocal cords on that side.

The parathyroid glands

These paired reddish-brown glands, each about the size of a small pea, lie behind the upper and lower aspects of the thyroid lobes. They are derived from the third (lower gland) and fourth pharyngeal (upper gland) pouches and lie within the pretracheal fascia often within the thyroid itself. Occasionally there may be more than four.

Superior glands Each is fairly constant in position, lying behind the upper third of the lobe inside the pretracheal fascia lateral to the trachea. Occasionally each may be more anterior and liable to injury during subtotal thyroidectomy.

Inferior glands Their position is more variable. They are usually found behind the lower part of each lobe either above or below the inferior thyroid artery as it enters the substance of the gland.

Unusual positions include retro-oesophageal, either in the neck or posterior mediastinum or the anterior mediastinum in relation to the thymus gland.

The first rib

This rib is important because of its site in the root of the neck, the structures related to it and its attached muscles.

Anatomical features

- Head with single facet for articulation with first thoracic vertebra
- Neck (see below) sloping backwards and upwards
- Shaft sloping forwards at 45°
- Tubercle posteriorly between neck and shaft. It has a medial facet for articulation with the first transverse process. Laterally the costo-transverse ligament and erector spinae muscles are attached.
- Anterior end contains concavity for the first costal cartilage.

Relationships of the neck (of the first rib)

- The anterior primary rami of C_8 and T_1 above and below the medial part
- The sympathetic trunk or stellate ganglion anteriorly near the head
- The supreme intercostal vein and superior intercostal artery laterally
- The cervical dome of the pleura and apex of the lung

Relationships of the upper surface of the shaft

- The lower trunk of brachial plexus
- The subclavian artery with scalenus medius laterally and scalenus anterior medially
- The subclavian vein lateral to scalenus anterior

The undersurface of the shaft is covered by the parietal pleura and crossed by the first intercostal nerve.

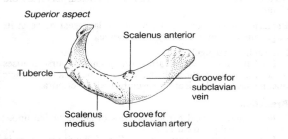

Superior aspect

Scalenus anterior

Tubercle

Groove for subclavian vein

Scalenus medius

Groove for subclavian artery

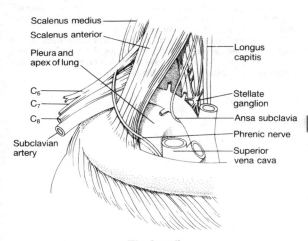

Scalenus medius

Scalenus anterior

Pleura and apex of lung

Longus capitis

C$_6$

C$_7$

C$_8$

Stellate ganglion

Ansa subclavia

Phrenic nerve

Subclavian artery

Superior vena cava

The first rib

633

The carotid arteries

Feel your carotid arteries one at a time by placing your thumb and index on either side of your trachea.

Origins

The right common carotid is one of the two terminal branches of the innominate artery, the other being the right subclavian. It begins at the level of the sternoclavicular joint, runs upwards and slightly laterally in the carotid sheath with the internal jugular vein and vagus nerve. The superior ramus of the ansa cervicalis lies in the anterior wall of the sheath.

Relations

Anterior Skin, fascia, lower third of sternomastoid, deep fascia, superior ramus of ansa cervicalis (descendens hypoglossi).

Medial Trachea, oesophagus, recurrent laryngeal nerve, thyroid gland.

Posterior Sympathetic trunk, recurrent laryngeal nerve at its origin, inferior thyroid artery, vagus nerve.

Lateral Internal jugular vein.

The left common carotid artery runs a similar course. It arises directly from the aortic arch. In the root of the neck the thoracic duct passes lateral to it at the seventh cervical vertebra.

Branches

External and internal carotid arteries originate at the upper border of the thyroid cartilage.

The *external carotid artery* gives off the following branches in the neck: ascending pharyngeal, superior thyroid, lingual, occipital and posterior auricular. Its terminal branches are the maxillary and superficial temporal arteries.

The *internal carotid* gives off no branches in the neck. It lies enclosed in the carotid sheath with the jugular vein lateral and the external carotid anterior and medial. The vagus nerve lies between and behind it and the jugular vein.

The ascending aorta and aortic arch

The ascending aorta

Originates from the left ventricle at the left margin of the sternum at the level of the third costal cartilage. It passes upwards and to the right to end at the level of the junction of the right sternal margin and second costal cartilage where it becomes continuous with the arch of the aorta. It contains three bulges at its origin called *aortic sinuses*. The aortic valve may be auscultated over the aortic area to the right of the sternum in the second intercostal space.

Branches Right and left coronary arteries.

The aortic arch

Lies in the superior mediastinum and crosses its lower part from the right to lie on the left side of the fourth thoracic vertebra. Its concavity is downwards and backwards.

Relations
Inferior
- Bifurcation of pulmonary trunk
- Root of left lung
- Ligamentum arteriosum and superficial cardiac plexus
- Recurrent laryngeal branch of left vagus nerve

Medial
- Trachea and oesophagus

Lateral
- Left pleura and lung

Branches Three: innominate, left common carotid, left subclavian.

The innominate artery arises in front of the trachea in the superior mediastinum and passes upwards to the right. It ends behind the right sternoclavicular joint where it divides into the right common carotid and subclavian arteries. It occasionally gives off a thyroidea ima branch to the thyroid.

The left common carotid artery arises from the aortic arch close to the innominate of its left side. It passes upwards to the left to lie on the left side of the trachea. It enters the neck at the level of the left sternoclavicular joint. It does not give off any branches in the thorax.

The left subclavian artery arises from the aortic arch behind the left common carotid and to its left. It passes upwards in a groove on the medial surface of the left lung and enters the neck at the level of the left sternoclavicular joint. It has no branches in the thorax.

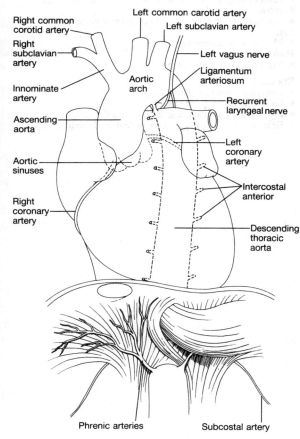

The ascending aorta and aortic arch

The ascending aorta and aortic arch (*cont.*)

The descending thoracic aorta

Is a continuation of the aortic arch at the level of the left side of the fourth thoracic vertebra. It passes downwards and to the right to come to lie in front of the vertebral column. It becomes continuous with the abdominal aorta by passing through the aortic opening of the diaphragm at the level of the twelfth thoracic vertebra.

Relations
Anterior and downwards
- Root of lung, pericardium, oesophagus and diaphragm

Posterior
- Left pleura, left lung, lower and thoracic vertebrae

Left
- Pleura, lung, hemiazygos veins

Right
- Thoracic duct, vein azygos

Branches
- Intercostal arteries (nine pairs)
- Subcostal arteries (one pair)
- Phrenic arteries, right and left bronchial
- Branches to oesophagus, pericardium and mediastinum

The coronary circulation

The blood supply to the heart is from the right and left coronary arteries which arise from the aorta near its origin. The arteries do anastomose with each other, but not sufficiently to establish a collateral circulation.

The right coronary artery

Arises above the anterior cusp of the aortic valve (anterior aortic sinus). It runs forwards between the auricle of the right atrium and the pulmonary trunk to the atrioventricular groove. It descends in the groove and then turns onto the posterior surface of the heart and anastomoses with the left coronary artery at the base.

Branches

Posterior interventricular on the diaphragmatic surface which gives branches to both ventricles.

Marginal gives branches to the right ventricle. It runs from right to left along the inferior border of the heart.

Small branches to aorta, right auricle, pulmonary trunk, right atrium, and right ventricle.

The left coronary artery

Arises from the left posterior sinus of the aorta just above the posterior aortic valve cusp. It runs between the auricle of the left atrium and pulmonary trunk to the atrioventricular groove where it continues its course.

Branches

Anterior interventricular artery gives branches to both ventricles and the septum. It anastomoses with the right coronary artery on the posterior surface of the heart. Small branches are also given off to both atria.

Small branches to aorta, pulmonary trunk, left atrium, and left ventricle.

Veins of the heart are the coronary sinus which drains into the right atrium. It is fed by the great cardiac vein which runs alongside the anterior interventricular artery in the atrioventricular groove, the small cardiac vein (following the marginal artery) and the middle cardiac vein (accompanying the posterior interventricular branch). Small veins, the venae cordis minimae, enter the right atrium directly.

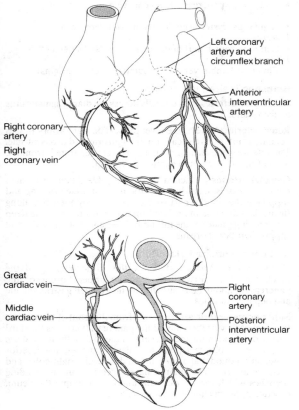

Left coronary
artery and
circumflex branch

Anterior
interventricular
artery

Right coronary
artery

Right
coronary vein

Great
cardiac vein

Middle
cardiac vein

Right
coronary
artery

Posterior
interventricular
artery

Coronary circulation

The abdominal aorta and its branches

At the level of T_{12} the thoracic aorta becomes the abdominal aorta which lies on the left of midline on the bodies of the first four lumbar vertebra and birfurcates into the right and left common iliac arteries at the level of L_4, which corresponds to a point 2 cm below the umbilicus on the anterior abdominal wall. It is retroperitoneal throughout its course.

Relations

Anterior
- Coeliac and aortic plexuses
- Origins of the unpaired visceral vessels (see below)
- Root of the mesentery
- Body of the pancreas and splenic vein

Right side
- Right crus, cysterna chyli, and inferior vena cava

Left side
- Left crus, fourth part of duodenum, coils of small bowel

Posterior
- First four lumbar vertebrae. Lumbar arteries (origins).

Branches (paired visceral)

Suprarenal arteries Pass laterally across the diaphragmatic crus to each gland.

Renal arteries Arise at the level of L_2. The right artery passes behind the inferior vena cava on the right crus and psoas muscle. The inferior vena cava and right renal vein separate it from the head of the pancreas and common bile duct.

Gonadal arteries (testicular or ovarian). Originate lower than the renal vessels, pass down on the psoas muscle crossing and supplying the middle part of the ureter. They then run along (testicular) or cross the pelvic brim (ovarian) to enter the deep inguinal ring (male) or the infundibulo—pelvic fold (female) and supply their target organs.

Branches (unpaired visceral)

Coeliac artery Arises just below the aortic opening of the diaphragm, runs forwards across the upper border of the pancreas for about 1 cm, and divides into the left gastric, splenic and hepatic branches.

Superior mesenteric artery Arises from the front of the aorta at the level of L_1 vertebra behind the pancreas. It crosses the third part of the duodenum and then enters the root of the mesentery. It gives off multiple jejunal and ileal branches, the inferior pancreatico duodenal artery, the middle colic, right colic, and ileocolic arteries. These latter have ascending and descending branches which form a rich anastomosis to supply the ileum, caecum, appendix, and transverse colon.

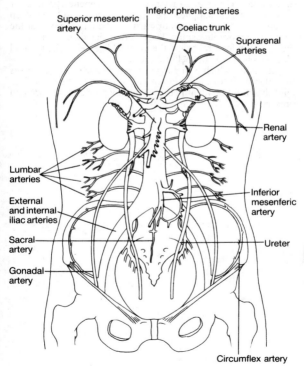

Abdominal aorta and its branches

The abdominal aorta and its branches
(cont.)

Inferior mesenteric artery Arises from the front of the aorta at the level of L_3 vertebra. It gives upper and lower left colic vessels which supply the transverse colon, splenic flexure, descending, and sigmoid colon. The superior rectal artery is a continuation of the inferior mesenteric artery and is the main artery to the rectum.

Parietal

Four lumbar arteries (paired), phrenic artery (paired), median sacral artery (unpaired).

The common iliac arteries

The aorta bifurcates into right and left common iliac arteries to the left of the midline on the body of L_4 vertebra. They bifurcate at the front of the sacro-iliac joint into the external and internal iliac arteries. The external artery gives two branches: the inferior epigastric which supplies the rectus muscles and the deep circumflex iliac which runs upwards to the anastomosis at the anterior superior iliac spine. The external iliac then continues as the common femoral below the inguinal ligament. The internal iliac vessels supply the pelvic organs.

The ureter, which is in danger during pelvic surgery, crosses the bifurcation of the common iliac arteries at the sacro-iliac joint. It should be identified early in any pelvic procedure and preserved.

The oesophagus

The oesophagus is a muscular tube about 25 cm long which extends from the cricopharyngeal sphincter at the level of the sixth cervical vertebra to the cardia of the stomach. It passes through the lower part of the neck, superior, and posterior mediastina and pierces the diaphragm at the level of the tenth thoracic vertebra. The lowermost 4 cm lie below the level of the diaphragm.

Lining

Squamous epithelium except in lower 3–4 cm where this is replaced by gastric type but non-acid-secreting cells.

Musculature

Striated in the upper one third. Smooth muscle in lower two-thirds. There is an external longitudinal layer and an internal circular layer.

Nerve supply

Extrinsic and intrinsic nerve plexuses from the vagus nerve (parasympathetic). Sympathetic supply is from the sympathetic trunks.

Blood supply

This is derived from the inferior thyroid artery, the descending thoracic aorta and the left gastric artery.

Relations

In the neck
- Trachea anteriorly, cervical vertebrae posteriorly
- Carotid artery and thyroid on the right and left
- Subclavian artery and thoracic duct on the left at the root of the neck

In the thorax
- Trachea, left bronchus, left atrium in front
- Right and left pleura and lung
- Aorta mainly behind but on the left at first
- Vagus nerves

Sites of anatomical narrowing

There are three sites of construction at 15 cm, 25 cm, and 40 cm (measured from the incisor teeth). These correspond to the cricopharyngeal sphincter, the aortic arch and bifurcation of the bronchi, and the diaphragmatic hiatus, respectively. Foreign bodies may become arrested at these levels. They are also the sites at which strictures (benign or malignant) are most common.

Barrett's oesophagus

Refers to the presence of columnar epithelium proximal to the gastro-oesophageal junction. This may be due to ectopic mucosa or develop secondary to reflux oesophagitis. Its importance lies in the fact that it is associated with oesophageal ulceration and an increased risk of oesophageal adenocarcinoma of up to 15 per cent.

The stomach

This is the most dilated part of the alimentary canal. It functions as a reservoir for food, as initiator and controller of the digestive processes, and as the site of production of intrinsic factor.

Position

It lies in the upper left quadrant of the abdomen behind the lower ribs and anterior abdominal wall and is separated from the left lung and pleura by the dome of the diaphragm.

Shape

It is described as piriform with anterior and posterior surfaces separated by blunt borders, the greater and lesser curvatures. The greater curvature is convex, the lesser curvature concave. The anterior surface faces upwards, the posterior largely downwards. The oesophagus enters the stomach on its right side about 2.5 cm below its uppermost part.

Divisions

- *Cardiac* at the level of oesophageal orifice
- *Fundus* proximal to oesophageal orifice
- *Body* from fundus to pyloric antrum
- *Pyloric antrum* is the dilated part after the incisura angularis.
- *Pyloric canal* is 2.5 cm long. It passes from the antrum to the pyloric sphincter.
- *Pylorus* is continuous with the first part of the duodenum and has thickened muscle—the pyloric sphincter.

Peritoneal folds

The stomach is covered by peritoneum on each surface. These meet at the curvatures to form the lesser omentum above and the greater omentum below. The greater omentum forms three ligaments—the gastrophrenic from the fundus to the diaphragm, the gastrosplenic to the spleen, and the gastrocolic to the transverse colon.

Relations

- *Anterior*: Liver to the right, diaphragm to the left, anterior-abdominal wall, ribs
- *Posterior*: Body of pancreas, part of left kidney, and left suprarenal, splenic artery and spleen. These form the 'bed' of the stomach. The transverse mesocolon passes from the lower border of the pancreas to the transverse colon.

Blood supply

- *Left gastric* from coeliac axis
- *Right gastric* from hepatic artery
- *Gastroduodenal* from hepatic artery
- *Left gastroepiploic* from splenic artery
- *Short gastrics* (several) from splenic artery
- *Right gastroepiploic* from the gastroduodenal artery

Nerve supply

Anterior and posterior nerves of Latarjet from the vagus. The fundus and body of the stomach receive their nerve supply from the proximal vagus. The antrum is innervated by the anterior and posterior nerves of Latarjet also from the vagus.

Lymph drainage

To glands around the cardia, greater and lesser omenta, pancreas, along the arteries and the first and second parts of the duodenum.

The pancreas

This elongated, lobulated gland lies on the posterior abdominal wall at the level of L_1 in the transpyloric plane. It develops from dorsal and ventral buds of endoderm which initially lie diametrically opposite each other. Subsequent duodenal rotation and asymmetric growth lead to fusion of the buds. Their ducts become the accessory and main pancreatic ducts. Abnormal or incomplete rotation produces an annular pancreas which may cause duodenal obstruction.

Features

It has a head in contact with the duodenum on three sides, a neck which is indented posteriorly by the portal and superior mesenteric veins, and a tail which is contained within the lieno-renal ligament. On cut section the gland is triangular in shape with well-defined superior and inferior borders and a blunt anterior border to which the transverse mesocolon is attached.

Relations

Head
- *Posterior*: Inferior vena cava, right renal vein and artery, terminal portion of common bile duct
- *Anterior*: First part of duodenum, transverse colon, gastro-duodenal artery

Body
- *Posterior*: Aorta, branches of coeliac artery on the upper border, superior mesenteric on the lower border, splenic vein, left renal vein and artery, left suprarenal, hilum of left kidney
- *Anterior*: Transverse mesocolon, stomach, lesser sac

Pancreatic ducts

The main duct arises in the tail, traverses the body and neck, then curves downwards to reach the head. It empties into the ampulla of the bile duct on the medial wall of the second part of the duodenum at the greater duodenal papilla.

The accessory pancreatic duct opens into the duodenum 2 cm above the main duct.

Blood supply

Is from the coeliac and superior mesenteric arteries.

The head and uncinate process are supplied by the superior pancreaticoduodenal artery from the gastroduodenal artery and the inferior pancreaticoduodenal from the superior mesenteric artery. The remainder of the gland is supplied by the splenic artery.

Venous drainage is by corresponding veins to the portal vein.

The liver

This, the largest organ of the body, lies in the right upper quadrant predominantly to the right of the midline and mostly under cover of the ribs. In health it is not normally palpable in the abdomen and extends internally to T_8 and the xiphisternum in the midline and almost level with the nipples on each side.

Lobes

Right and left. The right lobe includes the caudate and quadrate lobes.

Surfaces

There are four: anterior, posterior, superior, and inferior. The inferior surface is related to the viscera, the others to the diaphragm.

Ligaments

- *Falciform* on the anterior and superior surfaces separating the two lobes. It is a fold of peritoneum attached to the peritoneum of the anterior abdominal wall.
- *Ligamentum teres* (vestigial umbilical cord) runs from the umbilicus in the edge of the falciform ligament, to the inferior surface where it enters a deep cleft—the fissure for the ligamentum teres.
- *Ligamentum venosum* (vestigial ductus venosus) lies in a fissure on the posterior surface separating the right and left lobes.
- *Coronary ligaments* are layers of peritoneum reflected on the left side to the diaphragm and enclosing the so-called bare area. Where they join to the right and left bare areas are the right and left triangular ligaments.

The porta hepatis

There is a deep cleft on the inferior surface of the right lobe. The quadrate lobe lies in front, and caudate behind. It transmits the portal vein, the hepatic artery, and the hepatic ducts. The bile ducts lie anterior to the vein and to the right of the artery. Branches of the portal vein, hepatic artery, and bile duct tributaries then run together into the lobes of the liver. These are the portal triads. The lobes of the liver are divided into segments with the portal triads branching to superior and inferior portions. In total there are eight segments, important because their identification permits relatively avascular dissection (considerable experience is essential in hepatic resection) during the operations of segmentectomy and lobectomy.

The vena cava

Lies in a deep groove on the posterior surface 2 cm to the right of midline. There are three main hepatic veins draining approximately equal proportions of the liver. These drain into the vena cava near the diaphragmatic surface of the liver. Accessory smaller hepatic veins are common.

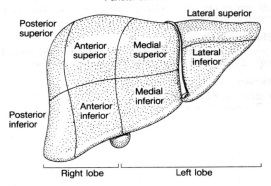

Parietal surface

Lateral superior

Posterior superior

Anterior superior

Medial superior

Lateral inferior

Medial inferior

Posterior inferior

Anterior inferior

Right lobe | Left lobe

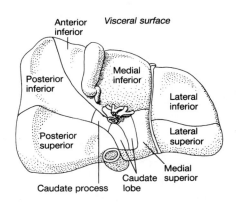

Anterior inferior

Visceral surface

Posterior inferior

Medial inferior

Lateral inferior

Lateral superior

Posterior superior

Medial superior

Caudate process

Caudate lobe

653

The liver (*cont.*)

The lesser omentum

The peritoneal coverings converge inferiorly to form a wide fold under the left lobe of the liver. It is attached to the first 2.5 cm of the duodenum, the lesser curvature of the stomach, and the diaphragm to the left between the liver and oesophagus. Its right edge is free, being the anterior border of the opening into the lesser sac. The common bile duct, the hepatic artery and portal vein lie in its free edge.

The subphrenic spaces

Are common sites of intraabdominal abscesses. These are right and left subphrenic between the liver and diaphragm, the right subphrenic (Rutherford–Morison's pouch), below the liver and related to the gall bladder, duodenum, and right kidney and the left subhepatic space which is the lesser omental sac (see appendix p. 852).

Blood vessels

Hepatic artery and portal vein carry blood to the liver. The hepatic veins drain the liver into the inferior vena cava.

Nerves

- Branches of the vagus nerve
- Branches of the coeliac plexus

The gall bladder and bile ducts

Introduction

The gall bladder is a pear-shaped organ attached to the under surface of the right lobe of the liver and projects slightly beyond its free margin. It lies in the right upper quadrant of the abdomen below the upper ends of the linea semilunaris.

Features

The gall bladder is about 10 cm long and its fundus is completely covered with peritoneum. The body and neck are covered on three sides only, being connected anteriorly to the liver by loose connective tissue from which it is easily separated.

The neck has a dilatation called *Hartmann's pouch* which hangs downwards and may be attached to the duodenum by folds—congenital or inflammatory. The neck becomes the cystic duct which drains into the common hepatic duct to form the common bile duct.

Blood supply

Cystic artery, usually a branch of the hepatic artery.

The bile ducts

The right and left hepatic ducts emerge from the liver through the porta hepatis and unite to form the commmon hepatic duct which continues as the common bile duct after being joined by the cystic duct.

Common bile duct

This duct, 9 cm long, lies in the free edge of the lesser omentum (3 cm) then passes behind the first part of the duodenum (3 cm) and penetrates its second part obliquely (3 cm) behind the head of the pancreas where it lies in a deep groove in front of the right renal vein. Its lower end is dilated to form the ampulla of Vater in common with the main pancreatic duct.

Relations of the common bile duct

- The hepatic ducts and the supraduodenal part of the common bile duct lie in the free edge of the lesser omentum.
- The hepatic artery lies on its left side.
- The portal vein lies posteriorly.
- The right hepatic artery passes behind the common bile duct before giving off its cystic branch.
- The lower one third of the common bile duct may actually lie in a tunnel in the pancreas.

Practical hint

Intermittent compression of the free edge of the lesser omentum can be performed for up to 15 minutes at a time to control hepatic bleeding (Pringle's manoeuvre).

N.B. Variations are common. Some are illustrated. Failure to recognize these can spell *disaster* for the surgeon.

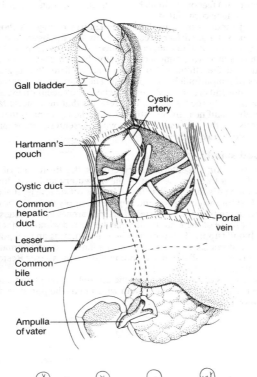

Gall bladder

Cystic artery

Hartmann's pouch

Cystic duct

Common hepatic duct

Portal vein

Lesser omentum

Common bile duct

Ampulla of vater

657

Variations of the cystic artery

Gallbladder and bile ducts

The small intestine

This comprises the duodenum, jejunum, and ileum. The duodenum is retroperitoneal and joins the jejunum at the duodenojejunal flexure, a site which must always be carefully examined in patients with deceleration injuries because disruption can occur.

The jejunum and ileum are both suspended from the posterior abdominal wall by a double fold of peritoneum containing the blood supply and venous drainage—the *mesentery*. Its attachment is about 15 cm long and it lies obliquely from the duodenojejunal flexure to the ileocaecal junction. The 20+ (about 7 m) feet of the small bowel attached to such a small base are therefore convoluted.

As jejunum runs into ileum the features change. Although both have large surface areas internally to aid absorption, the circular folds of mucus membrane decrease in numbers and size in the ileum and are almost completely absent in the lower ileum so the wall feels thinner. Throughout the small intestine aggregations of lymph follicles are present. These are more marked in the ileum and are called Peyer's patches. They occur on the anti-mesenteric border. Another difference is that mesenteric fat is less abundant near the intestinal wall in the jejunum so that the vessels to the gut can be seen with mesenteric transparent windows between them.

Blood supply

The superior mesenteric artery crosses the third part of the duodenum to enter the root of the mesentery and pass downward towards the right iliac fossa. It gives off several jejunal and ileal arteries which divide and reunite to supply the small bowel. In so doing they form a series of arterial arches and arcades, so one or more of the main trunks may be ligated without affecting the blood supply. This is done when constructing an ileal pouch for anastomosis to the rectum in patients with inflammatory bowel disease and familial adenomatous polyposis (see Restorative Proctocolectomy, p. 218).

The veins drain the gut along the arterial pathways and empty into the superior mesenteric vein which joins the portal vein on its way to the liver.

The large bowel

This is approximately 15 m long and comprises the appendix, caecum, ascending colon, hepatic flexure, transverse colon, splenic flexure, descending colon, sigmoid colon, rectum, and anus. The lumen is wide in the caecum and ascending colon but gradually narrows towards the sigmoid colon.

Structure

It has a peritoneal coat complete over the caecum, appendix, transverse colon and sigmoid colon but incomplete elsewhere. There are small pouches of peritoneum filled with fat over all of the colon—*appendices epiploicae*.

The outer muscular coat is disposed in three longitudinal bands—the *taenia coli*. These converge upon the root of the appendix at one end. At the other they become continuous with the longitudinal layer of the rectum. They are shorter than the colon so its wall becomes puckered into sacculations—*colonic haustrations*.

Caecum

The ileum terminates in a slit-like opening—the *ileocaecal valve*. The blind-ending portion below this level is the caecum. It may lie free or be attached to the right iliac fossa by peritoneal folds.

Relations
- *Inferior*: Lateral half of inguinal ligament
- *Anterior*: Greater omentum, coils of ileum, anterior abdominal wall
- *Posterior*: Iliacus and psoas

Colon

The ascending and descending components are partly retro-peritoneal. The transverse and sigmoid colon are suspended on mesenteries.

Transverse colon

The middle portion lies below the umbilicus. Superiorly lies the greater curvature of the stomach, liver, and gall bladder. Posteriorly from right to left are the second part of the duodenum, the pancreas, loops of small bowel, and the spleen. Anteriorly is the greater omentum.

Descending colon

This begins at the splenic flexure and ends at the beginning of the sigmoid colon.

Relations
- *Anterior*: Jejunum, lower part of anterior abdominal wall
- *Posterior*: Left kidney, psoas, quadratus lumborum, iliacus

Sigmoid colon

This begins at the pelvic brim and ends at the level of the third sacral vertebra. It is mobile and variable in length. It is related superiorly to loops of small bowel.

Blood supply

- Caecum and appendix: From the superior mesenteric and ileocolic arteries
- Ascending colon: From ileocolic and right colic arteries
- Transverse colon/flexures: From the right middle and left colic arteries
- Descending colon: Left colic artery
- Sigmoid and upper rectum: Left colic, superior rectal arteries
- Rectum: Three arteries
  —superior rectal from the inferior mesenteric artery
  —middle rectal from the anterior division of the internal iliac artery
  —inferior rectal from the internal pudendal artery

The lymphatic drainage is along the arterial supply. Nodes are situated:

- On the bowel surface
- Between the layers of the mesocolon
- Along the branches of the mesenteric arteries
- Along the trunks of the main vessels

The appendix

Surface markings

The vermiform (worm-like) appendix is an appendage of the caecum lying in the right iliac fossa. When inflamed maximum tenderness is classically felt over McBurney's point (one third of the way along a line drawn from the anterior superior iliac spine to the umbilicus).

Features

It is 8–10 cm long and 6–8 mm in girth. Its base is attached to the caecum at the point of convergence of the three taenia coli on the posteromedial wall (see McBurney's point, above) corresponding to McBurney's point on the surface.

It contains numerous aggregations of lymphoid tissue in its wall and has a small lumen. It is covered by peritoneum and has a well-formed mesoappendix derived from the posterior layer of the mesentery of the lower ileum. It is in the mesoappendix that the appendicular vessels lie.

Blood supply

The appendiceal artery is a branch of the posterior caecal from the ileocolic artery. It gives off 2–3 branches to the appendix during its course in the free crescentic edge of the meso-appendix.

Variations in position

The appendix may be freely movable. Common variations in order of frequency are:
- *Retrocaecal*, sometimes extending to the right lobe of the liver, or in the serous coat of the caecum
- *Pelvic*: rectal examination may produce pain
- *Retroileal*

The inguinal canal

This is a passage about 4 cm long in the anterior abdominal wall through which the spermatic cord or round ligament pass. It lies above the medial half of the inguinal ligament and runs downwards, medially and forwards. It is a potential weak spot for the development of inguinal hernia.

It is a series of three openings in the muscles of the anterior abdominal wall which are staggered so that no opening overlies another. The deepest is the *deep inguinal ring* which is an aperture in the fascia transversalis. Its surface position is about 1.5 cm above the mid inguinal point. The cord or round ligament then traverses an opening in the combined tendons of transversus and internal oblique muscles (*conjoint tendon*) and finally the external oblique aponeurosis which is the *superficial inguinal ring*.

Deep inguinal ring

An indirect hernia enters the canal through this ring. It lies lateral to the inferior epigastric vessels. It transmits the round ligament of the uterus, its vessels and nerve in females. In males the remainder of the processus vaginalis is vestigial superiorly, the cremasteric vessels are medial, and the genital branch of the genitofemoral nerve lies inferiorly. It transmits the spermatic cord and its contents.

Superficial inguinal ring

Is a triangular opening in the external oblique aponeurosis. Its medial and lateral sides are *crura*, the lateral of which is attached to the pubic tubercle. The medial crus is attached to the pubic symphysis. Therefore when an indirect inguinal hernia emerges from the superficial ring it is above and medial to the pubic tubercle.

Walls of the inginal canal

- Anterior: Aponeurosis of external oblique muscle
- Posterior: Fascia transversalis and conjoint tendon
- Superior: Lower edge of internal oblique
- Inferior: Inguinal ligament

The inguinal canal

The femoral canal

A femoral hernia passes through the femoral canal to reach the upper aspect of the thigh below the inguinal ligament.

Features

The femoral artery, vein, and canal are all enclosed in the *femoral sheath* which is a continuation of the transversalis fascia above the inguinal ligament and that covering iliacus and psoas muscles below. The canal lies on pectineus muscle between the femoral vein and the lacunar ligament. It provides space for the femoral vein to expand when venous return from the leg is increased. It is funnel-shaped, about 2 cm long and with an upper opening, the *femoral ring*, which is 1.5–2 cm wide and opens into the abdomen.

The femoral ring The ring contains fat—the *femoral septum*—and a single lymph node (Cloquet's).

Boundaries
- Anterior: Inguinal ligament
- Posterior: Pectineal line, pectineus, and its fascia
- Medial: Lacunar ligament
- Lateral: Femoral vein

The femoral canal Opens below and lateral to the pubic tubercle. The site of appearance of a femoral hernia.

A femoral hernia Descends through the canal and usually hooks round the anterior margin of the femoral ring to come to lie over the inguinal ligament. Its coverings include fat (the femoral septum), lymph glands, the anterior wall of the femoral canal, cribriform fascia, superficial fascia, subcutaneous fat, and skin.

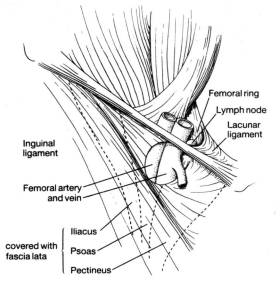

Femoral canal

The popliteal fossa

This is a diamond-shaped space behind the lower femur, the knee joint, and the upper tibia.

Boundaries

Superior Biceps femoris above and laterally, semitendinosus and semi membranosus above and medially + sartorius and gracilis when the knee is bent.

Inferior The two heads (lateral and medial) of gastrocnemius and plantaris below and laterally.

Roof Fascia which is pierced by the short saphenous vein before joining the popliteal vein.

Floor Popliteal surface of femur, posterior capsule of knee joint, popliteus.

Contents

- Fat, lymph nodes
- Popliteal artery + geniculate branches which form an anastomosis around the knee joint + its divisions into anterior and posterior tibial arteries
- Popliteal vein and its tributaries (short saphenous vein)
- The sciatic nerve divides at the upper end into tibial and common peroneal nerves. The common peroneal runs with the tendon of biceps femoris to pass around the neck of the fibula. The tibial nerve runs with the popliteal artery and vein. It is superficial. The artery is deep and lies on bone. It is liable to injury in supracondylar fractures of the femur.
- The posterior cutaneous nerve of the thigh
- The geniculate branch of the obturator nerve

The median nerve

This nerve is often injured because of its exposed situation in the forearm and wrist.

Origin From the medial and lateral cords of the brachial plexus.

Course in the upper arm It descends lateral to the brachial artery then crosses it anteriorly to lie on its medial side in the cubital fossa under the aponeurosis of biceps.

Course in the forearm It leaves the cubital fossa between the heads of pronator teres and descends in the forearm between flexor digitorum sublimis and profundus.

In the lower third of the forearm it becomes superficial at the lateral side of the flexor digitorum sublimis.

2.5 cm above the wrist it lies between palmaris longus and flexor carpi radialis. It enters the palm by passing deep to the flexor retinaculum.

Motor distribution It supplies all the muscles of the forearm *except* flexor carpi ulnaris and the medial half of flexor digitorum profundus.

In the palm it supplies the lateral two lumbricals (L), the opponens pollicis (O), abductor pollicis brevis (A) and flexor pollicis brevis (F) (LOAF).

Sensory distribution It is sensory to the greater part of the palm, the thumb, and the radial 2.5 fingers.

Median nerve palsy

- Causes: injuries at the elbow or wrist (high and low)
- *High injury* (elbow): Paralysis of forearm pronators and flexors of wrist except flexor carpi ulnaris and ulnar aspect of flexor digitorum profundus. There is wasting of the front of the forearm and a pointing index finger.
- *Low injury* (wrist): Paralysis of opponens pollicis. Wasting of radial half of hand. Sensory loss over radial 3.5 fingers.

The ulnar nerve

Injury leads to motor loss of the intrinsic muscles of the hand.

Origin The ulnar nerve is the continuation of the medial cord.

Course in the upper arm In the proximal half of the upper arm it lies medial to the brachial artery. It then pierces the medial inter-muscular septum to enter the back of the arm.

Course in the forearm It lies closely applied to the medial epicondyle and medial ligament at the elbow. It enters the fore-arm between the two heads of flexor carpi ulnaris and passes downwards under cover of this muscle lying on flexor digitorum profundus.

At the wrist it becomes superficial between flexor carpi ulnaris and flexor digitorum profundis. It then pierces the deep fascia to enter the palm and passes superficial to the flexor retinaculum. In the lower third of the forearm and wrist it accompanies the ulnar artery.

Motor distribution
- *Forearm*: Flexor carpi ulnaris medial half of flexor digitorum profundis
- *Hand*: Medial two lumbricals, interossei, adductor pollicis, hypothenar muscles

Sensory distribution Medial 1.5 fingers. Medial border of palm.

Ulnar nerve palsy
- *Causes*: Injuries at the elbow or wrist
- *High*: Paralysis of flexor carpi ulnaris and ulnar aspect of flexor digitorum profundus. Hyperextension of little and ring fingers at the metacarpophalangeal joints. Paralysis of small muscles of the hand *except* the lateral two lumbricals and thenar muscles.
- *Low*: Paralysis of abductor digiti minimi and flexor digiti minimi, claw hand, wasting of intrinsic muscles. Loss of sensation over ulnar 1.5 fingers. Froment's sign (pinching paper between thumb and index leads to flexion of the distal interphalangeal joint).

The radial nerve

Is most commonly injured in fractures of the humerus and its posterior interosseous branch in fractures of the upper third of the radius.

Origin It is the direct continuation of the posterior cord lying posterior to the axillary artery and anterior to subscapularis, latissimus dorsi, and teres major.

Course in the upper arm It passes backwards between the long and medial heads of triceps then winds around the humerus under cover of the lateral head. It then passes forwards through the lateral intermuscular septum, descends between brachialis and brachioradialis to cross the capsule of the elbow joint and enter the forearm giving off its posterior interosseous branch at the level of the lateral epicondyle.

Course in the forearm It descends under brachioradialis to the lower third of the forearm where it passes backwards, piercing the deep fascia, to lie to the lateral side of the distal radius where it gives off branches to the skin of the hand and fingers (see sensory distribution, below).

The posterior interosseous nerve is entirely motor. It reaches the back of the forearm by winding round the lateral side of the upper third of the radius through supinator muscle to descend between the superficial and deep muscle layers.

Motor distribution
- *Main trunk*: Biceps, brachioradialis, anconeus, brachialis, extensor carpi radialis longus
- *Posterior interosseous*: Muscles of the back and radial sides of the forearm—supinator, extensor carpi radialis brevis, extensor digitorum, extensor digiti minimi, extensor carpi ulnaris, abductor pollicis longus, extensor pollicis longus and brevis, and extensor indicis

Sensory distribution The back and lateral side of the upper arm, the back of the forearm, the lateral two-thirds of the dorsum of the hand and dorsal aspects of the proximal phalanges of thumb, forefinger, middle finger, and radial half of the ring finger.

Radial nerve palsy

- *Causes*: Fractures and dislocations of the humerus or around the elbow. Crutch palsy (pressure in the axilla), Saturday night palsy (pressure on the radial groove from prolonged immobility with the arm hanging over the back of chair while asleep).
- *Features*: Wrist drop. Paralysis of triceps and brachio-radialis if the lesion is high. Sensory loss over dorsum of forearm and posterior aspect of first interdigital cleft.

The sciatic nerve

N.B. Beware of injuring the sciatic nerve when giving intra-muscular injections into the buttock. Always inject into the upper outer quadrant of the buttock.

Origin From the sacral plexus (L_4, L_5, S_1, S_2, S_3). It is one of the two terminal branches, the other being the pudendal nerve.

Course It leaves the pelvis through the greater sciatic foramen between its lower margin and piriformis. It lies midway between the greater trochanter and the ischial tuberosity and passes under cover of gluteus maximus into the hamstring compartment of the thigh. It is crossed laterally by the long head of biceps femoris. At the upper border of the popliteal fossa it divides into the tibial and common peroneal nerves.

Motor distribution The hamstrings and adductor magnus.

Sensory distribution There are no cutaneous branches.

The tibial nerve

Crosses the popliteal fossa vertically to enter the posterior aspect of the leg deep to gastrocnemius and soleus but superficial to popliteus.

It then passes downwards deep to soleus accompanied by the posterior tibial vessels. After passing under the flexor retinaculum it divides into lateral and medial plantar nerves.

Distribution It is motor to the flexor compartment of the calf and the sole of the foot and sensory to the sole of the foot, the plantar surfaces of the toes, and the dorsum of the toes as far as the nail beds.

The common peroneal nerve

Leaves the popliteal fossa at its lateral angle and passes on the lateral side of the neck of the fibula where it enters peroneus longus and divides into its terminal branches the superficial and deep peroneal nerves.

Distribution The common peroneal supplies no muscles but its branches supply the extensor and peroneal compartments of the leg and the dorsum of the foot.

20 Operative surgery

Suture materials

The suture materials used in surgery are of two types—absorbable and non-absorbable. They may be constructed of organic or synthetic materials. Their gauge or calibre is expressed in numbers (British or US Pharmacopeia).

Metric gauge	British/US	
0.2	10/0	smallest
0.3	9/0	
0.4	8/0	
0.5	7/0	
0.7	6/0	
1	5/0	
1.5	4/0	
2	3/0	
3	2/0	
3.5	0	
4	1	
5	2	
6	4	largest

Absorbable sutures

Organic Catgut is the commonest example. It consists of twisted collagen (from the intestines of sheep or cattle) and is absorbed by phagocytes over a variable period. When coated in a chromic solution the period of absorption is delayed. With both 'chromic' and 'plain' catgut absorption rates are unpredictable. An inflammatory reaction is instigated and tensile strength is lost early. It is useful in intestinal anastomosis, closure of peritoneum, and as stitch for a fat or subcutaneous tissue. Predictable synthetic absorbables cause less reaction and are superior (see below).

Synthetic (Dexon, Vicryl, PDS) These consist of braided fibres of polyglycolic acid, polygalactin, and polydioxamone. They are stronger than catgut and since they are more slowly phagocytosed induce less tissue reaction. Tensile strength decreases in linear fashion to about 50 per cent in two weeks. These sutures handle and tie better than catgut.

Non-absorbable sutures

Organic (silk, cotton) These braided sutures tie and handle easily but may perpetuate infection by capillary action caused by the braiding. With the exception of silk most have been replaced by synthetic sutures. Linen and cotton are still used in abdominal surgery, but synthetic sutures are superior (see below).

Synthetic (prolene, nylon, Surgilon) These may be mono-filament or braided. They are slightly more difficult to handle and tie than silk, for example, but with experience handling becomes easier. They are relatively inert and therefore induce little tissue reaction. Their use is widespread for skin closure, abdominal wall closure and repair of hernias. They are not absorbed, retain their tensile strength but braided sutures show the same capillary action as silk and are therefore often coated with polytetrafluoroethylene (PTFE) which renders them smooth and reduces the capillary action, e.g. Ethiflex, Ticron.

Sinus formation is a recognized complication of non-absorbable sutures, but only when sepsis prophylaxis is inadequate.

Stainless steel wire

This suture is used for the closure of the sternum after splitting. It is very strong and may be mono- or multifilament. It rarely leads to sinus formation but handles poorly and makes re-exploration of a wound difficult.

Needle types

Cutting needles are usually triangular in cross-section with the apex of the triangle in the concavity of the needle. Reverse-cutting needles have the apex on the convexity and are less traumatic. Both types are used for skin or tendons.

Round-bodied needles (taper point) are oval or round in cross-section. They are used for anastomosis of the GI tract and vascular work.

Needles may be straight, if the tissues are easily accessible, or curved. Half-circle needles are most commonly used. Deeper tissues require a greater circle arc.

Selection of needles used on various types of suture materials

Cutting	Round-bodied
Slim blade curved cutting	Half circle round-bodied
Curved reverse-cutting	Round-bodied heavy Mayo
Half circle taper-cut	Curved round-bodied
Curved super cutting	5/8 circle round-bodied
Taper-cut fish hook heavy	Straight round-bodied
Half circle cutting	Blunt-point curved round-bodied
Precision-point reverse-cutting curved	
Straight cutting	
Trocar point half circle heavy	

Surgical knots and sutures

Mastery of knot-tying by hand and with instruments is essential to good surgical technique and can only be acquired by assiduous practice. These illustrations are examples of common surgical knots and sutures. There are many variations which you can learn from your senior colleagues. Once you can tie knots and suture safely you will be much more use as an assistant and increase your chance of carrying out supervised procedures. Practice the techniques with bare hands at first, then wearing gloves. When you are adept you can perform them during surgical procedures.

The square knot This basic knot can be performed with both hands (slow) or single-handed (fast).

Tying under tension Carry out the first half of a one-handed square knot. Maintain tension by lifting the ends of the ligature. Ask your assistant to apply an artery forceps to the vertex of the ligature and complete the second half. Alternatively keep tension in the ligature and run the second half of the knot down to the first using your right index finger.

Instrument ties Apply an artery forceps to one end of the suture and leave it there throughout the tying manoeuvre. Alternatively the needle can act in place of an artery forceps. It is held in the left hand and the square knot technique performed with an artery forceps or non-toothed dissecting forceps held in the right hand.

Sutures Simple. Vertical mattress. Horizontal mattress. Half-buried horizontal mattress. Subcuticular continuous. Over and over. Continuous locking. Continuous mattress. Purse string. Continuous Lembert. Figure-of-eight. Lembert. Halsted.

Guidelines for obtaining a fine scar

Try to place the incision in contour lines. Use atraumatic techniques and avoid tension. Remove sutures as follows:
- Face and neck: 3–4 days
- Scalp: 5–7 days
- Abdomen and chest: 7–10 days (up to 14 days after aortic surgery)
- Limbs: 5–7 days
- Feet: 10–14 days

Sutures can be removed earlier than these times for cosmetic reasons. Once removed they should be replaced with Steristrip adhesive skin strips. These may also be used as an alternative to skin sutures, especially in children.

Skin clips are a further alternative to skin sutures. They do not penetrate the skin and have the advantage of a lower infection rate. They are, however, more expensive than sutures.

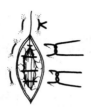

Horizontal mattress

Subcuticular continuous

Continuous overhand

Interrupted Lembert

Halsted

Purse-string

Figure-of-eight

Interrupted

Surgical knots and sutures

Surgical drains

Drains are indicated for established collections of pus, blood or fluid and for prophylaxis to abolish dead space or remove anticipated collections, e.g. after abdominal perineal resection, following splenectomy or mastectomy etc. Most drains consist of latex-based material, or more recently polyvinylchloride, silastic, or polyurethane.

Types
Active
Closed systems These comprise either re-usable or disposable systems. The re-usable system provides high-pressure suction (500 mmHg) with a reservoir capacity of around 300 ml. The bottles can be autoclaved and so exchanged for a sterile system after use.

Disposable drains provide low-pressure suction (100 mmHg) by way of a compressible bottle. The system may be recharged by emptying the bottle, compressing, and reconnecting the drainage tube. Non-return valves should be used to prevent reflux. Since these drainage systems are never entirely empty, colonization of the reservoir contents by bacteria can occur.

Sump drains contain an air inlet lumen which prevents blockage by soft tissue. Although more efficient than closed suction systems, they have the disadvantage of requiring a non-portable suction system and may permit the entry of bacteria, although filters reduce their numbers.

Passive
These may also be open or closed. Closed systems drain into bags by low-pressure suction working on the siphon principle. Open systems drain by capillary action or gravity into dressings or into stoma bags. Their efficiency, then, depends on the position of the patient and the volume and site of the collection.

Complications
- Sepsis (commoner with open drains). This may be reduced by the presence of non-return valves.
- Haemorrhage at the site of insertion of the drain
- Pressure or suction necrosis of bowel leading to leakage of contents

For this reason most drains should be removed within 5 days unless the daily volume is large. If a track is necessary to provide a long period of drainage, inert material like PVC or silastic should not be used. Rubber drains stimulate fibrosis and are better in this situation.

Reference
Broome, A. E., Hanson, L. C., Tyger, J. F. (1983). Efficiency of various types of drainage of the peritoneal cavity—an experimental study in man. *Acta Chirurgica Scandinavica*, **149**, 53–5.

Scrub-up techniques

The aim of preoperative scrub-up is to remove the surface organisms from the hands and forearms. The technique does not remove deeper organisms from the hair follicles or sweat glands, and recolonization usually occurs within 20–30 minutes. However, modern washing agents like povidone-iodine (Betadine) and chlorhexidine (Hibiscrub) will kill organisms for up to 2 hours after scrubbing. There is no advantage to excessive and lengthy scrubbing, and there is some evidence that only the nails need scrubbing. Hand and wrist jewellery should be removed. 3–4 minutes is adequate for the first scrub of the morning. The hands should be washed between cases with a spirit solution or rescrubbed for 1–2 minutes. Staff with boils of the skin or other infective foci should not be in theatre until recovery is complete.

1. Turn on the water and adjust the temperature until it is acceptable. Give the hands and forearms a light wash. Rinse and take a sterile brush from the dispenser. Apply some Betadine or Hibiscrub to the brush and scrub the fingernails carefully. Rinse.
2. Wash the forearms and hands for about 2 minutes. Rinse and allow the water to run downwards towards the elbows.
3. Dry each hand and forearm separately with the disposable towels provided using one half of each of the towels for each hand and forearm. Dry from distal to proximal, e.g. wrist to elbow.
4. Avoid the use of glove powder if possible. (Dry hands do not need powder. Biogel gloves need no powder.) Put your gown on, then put the gloves on left hand first. Pick the folded glove up with the fingers of the right hand on the inside and apply it to the left hand. Now pick the right-hand glove up by passing the fingers of the left hand under the fold (i.e. to the outside) of the right-hand glove. Slip the right hand into the folded glove. Now fold your gown cuffs and complete gloving by straightening the glove folds over your wrists with your gloved fingers on the outside of the folds. An alternative is to have your surgical gloves put on by a nurse (USA) or to use the nurses' method (the 'closed' technique). After donning the gown, the left hand still enclosed in the gown cuff picks up the left glove and applies it over the left cuff and hand. The left cuff is gently pulled down to let the fingers enter the glove and the process is repeated on the right side. This method is usually not acceptable to the surgeon as the cuff ends up on the palms.

Punctured gloves

60–70 per cent of gloves will become punctured during surgery. They and gowns with soiled sleeves should be replaced.

Shaving

The operative field should be shaved, allowing ample room to apply dressings. This is best done by an orderly, a nurse, or the surgeon on the morning of surgery or on the table itself. Do not carry out shaving on the day before because folliculitis may follow leading to postponement of the procedure.

Skin preparation

In theatre the operative field will be prepared with aqueous Betadine or 0.5 per cent chlorhexidine in alcohol. Two applications are usually made. When operating on mucus membranes use normal saline but remove all debris first. In procedures on the head protect the eyes with adhesive tapes to keep them closed.

Vertical abdominal incisions

Abdominal incisions should be large enough to permit the intended procedure to be performed safely and efficiently. They should be closed as cosmetically as possible without compromising their strength.

Upper midline This permits access to the stomach, duodenum, gall bladder, liver, and transverse colon.

The incision extends from the xiphisternum to the umbilicus but can be extended distally if necessary. It is continued through the linea alba to the peritoneum which is held between artery forceps away from the abdominal contents and then incised with a scalpel. Two fingers are then inserted to allow safe division of the peritoneum upwards and downwards. After the procedure the wound is closed in layers with a non-absorbable suture to the linea alba and interrupted nylon, or subcuticular dexon, nylon, or prolene ± beads (easier) to skin.

Mass closure This is closure of all layers of the wound except skin with a single suture which may be an over-and-over technique or a far-and-near figure-of-eight continuous suture. Wide bites greater than 1 cm are taken, with alternate more superficial bites to form the figure-of-eight.

Lower midline This permits access to the pelvic organs.

The incision extends from the umbilicus to the symphysis pubis. Ensure that the bladder is empty because of its relationship when full to the anterior abdominal wall. Close the peritoneum with Vicryl or Dexon, the linea alba with non-absorbable sutures, and the skin with nylon or Dexon.

Full-length midline This incision skirts the umbilicus and is useful when wide exposure is required as in aortic surgery.

Paramedian incision This incision is in the upper or lower abdomen parallel to and 2 cm from the midline. It permits access to the organs of each quadrant of the abdomen according to its site but tends to have been replaced by the faster open–close midlines.

The anterior rectus sheath is divided and the muscle dissected laterally to expose the posterior sheath. This is divided vertically together with the transversalis fascia and peritoneum. The peritoneum and posterior rectus sheath are closed together in one layer of non-absorbable sutures followed by the anterior sheath and skin.

Lateral paramedian incision This is a modification of the paramedian. It is a vertical incision over the junction of the middle and outer one-third of the width of the rectus muscle. The sheath is opened vertically and the muscle reflected laterally. The posterior sheath and peritoneum are opened in the same vertical plane as the anterior sheath. There is a lower risk of wound dehiscence or incisional hernia because of the overlying rectus muscle when the wound is closed.

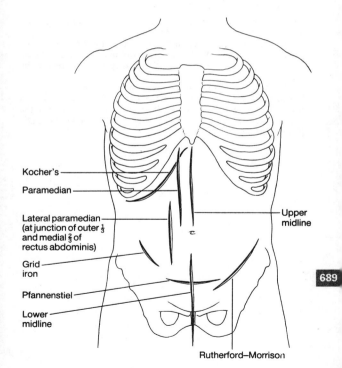

Kocher's

Paramedian

Lateral paramedian
(at junction of outer ⅓
and medial ⅔ of
rectus abdominis)

Grid
iron

Pfannenstiel

Lower
midline

Upper
midline

Rutherford–Morrison

Abdominal incisions

689

Transverse abdominal incisions

Skin crease incisions True skin crease incisions may be made in Langer's lines in the upper or lower abdomen. They may be used to approach the gall bladder, stomach, duodenum, aorta, etc. Exposure and drainage are better. Chest complications and postoperative pain are less common, and incisional herniation is seldom seen.

Upper abdominal transverse incision This exposure allows access to the upper abdomen and may be modified into a 'rooftop' incision (bilateral subcostal incisions confluent transversely below the xiphisternum) to permit access to the pancreas and classically the adrenal glands.

Closure is in layers with vicryl to peritoneum, interrupted or continuous PDS, or nylon to the rectus muscles and nylon (subcuticular with beads) to skin.

The wound offers two advantages:
- A sound scar
- Less postoperative pain

Pfannenstiel incision This suprapubic transverse incision allows access to the pelvic cavity.

The incision is 8–12 cm long, skirts the upper border of the pubis and divides skin, fat, external oblique, internal oblique, and transversalis fascia from one inferior epigastric artery to the other.

The rectus abdominis muscles are split vertically in the midline for several cm to expose the peritoneum which is then opened between artery forceps. If greater access is required, the rectus may also be divided transversely and retracted cranially (Czerny incision).

Oblique abdominal incisions

Kocher's (subcostal) incision This allows access to the contents of the right and left upper quadrants and is particularly indicated for operations on the gall bladder and spleen in patients with a wide costal angle.

The incision extends from the midline to the lateral edge of the rectus muscle. It lies parallel to and about 2–2.5 cm below the costal margin. All tissues are divided in the line of the incision.

Closure is by the mass technique or in layers in the order of peritoneum and posterior sheath together, then anterior sheath with non-absorbable or delayed-absorption sutures, e.g. nylon or PDS, and nylon (preferable) or dexon to skin.

Rutherford–Morrison incision This allows access to the lower ureter, colon, and iliac arteries. The approach can be trans- or extraperitoneal.

The incision extends laterally from just above the anterior superior iliac spine to a point about 2.5 cm above the pubic tubercle medially. All tissues are divided in the same line and the inferior epigastric vessels are ligated. The peritoneum may then be reflected anteriorly to approach the ureter, inferior vena cava, or lumbar sympathetic chain.

Grid-iron incision This is a classic appendicectomy incision extending obliquely from a point two-thirds of the way along the imaginary line drawn from the anterior superior iliac spine to the umbilicus. The incision is made at right angles to this line. Skin, subcutaneous fat, and fascia are divided in the same line but the external oblique, internal oblique, and transversus abdominis are all split in the line of their fibres to expose the peritoneum which is then opened between artery forceps.

Closure is in layers with catgut to peritoneum and loose catgut muscle closure. The external oblique aponeurosis is closed with vicryl. Use nylon or dexon for skin.

The incision has disadvantages in that, although providing strong healing, it is limited medially by the rectus sheath and inferior epigastric vessels. It can, however, be extended laterally and upwards.

If the preoperative diagnosis of appendicectomy is found to be wrong this incision should be closed and a right lower paramedian performed. For this reason it has been condemned as the incision of the 'cocksure'. Its real value is in access to the appendix, but it can be extended to permit a right hemicolectomy. It affords poor access to the gall bladder and duodenum.

Lanz incision This is a right lower quadrant transverse incision which gives a cosmetic scar. It is similar to the grid-iron.

Tracheostomy

This procedure is usually performed under general anaesthetic and may be resorted to after an appropriate period of ventilation with an endotracheal tube. It may also be decided upon for patients undergoing head and neck surgery or managing those with severe head or spinal injuries. It can be performed electively or as an emergency.

Indications

- Severe respiratory or cardiac disease in infants and children
- Long-term care of patients with chest trauma
- Pulmonary insufficiency in the elderly when other treatments are ineffective
- Neurological problems, e.g. coma, stroke, polio, bulbar palsy, etc.

Equipment required

The patient will already have an endotracheal tube in place. If conscious, a full explanation must be given of the procedure. Ensure good lighting and a range of sizes of tracheostomy tubes which may be of silver or polyvinyl chloride (PVC)—ideal as the first tube—ranging from 24 to 42 French gauge (36 to 39 for men, 33 to 36 for women). They should ideally be double-cuffed so that the cuffs can be inflated alternately to reduce the risk of tracheal trauma. Remember too that aseptic technique is required.

Site of incision

Transverse, about 5 cm long, midway between cricoid cartilage and sternal notch.

Procedure

Make the incision and deepen it by dividing the platysma transversely to the pretracheal fascia. Identify the midline. Split and retract the strap muscles in this line to expose the trachea. Secure haemostasis and then identify the cricoid cartilage, then the first tracheal ring. From this point rings 2 to 5 may be covered by thyroid tissue. To expose them it may be necessary to divide the thyroid isthmus.

The removal of large discs of trachea is unnecessary, as is the use of tracheal flaps sutured to the skin. A longitudinal incision from the second to the fourth tracheal ring permits the insertion of an appropriately sized tracheostomy tube. This is done by retracting the edges of the tracheostomy, asking the anaesthetist to remove the endotracheal tube, and inserting the tracheostomy tube. Remember that the endotracheal tube cuff is usually damaged so therefore rapid insertion of the tracheostomy tube is needed. Inflate the balloon, suture the wound loosely, place swabs under the tracheostomy tube and tie the tapes behind the patient's neck using tight knots to avoid inadvertent loosening by well-intentioned staff mistaking them for the patient's nightgown ties. The tube may be changed after 24 hours.

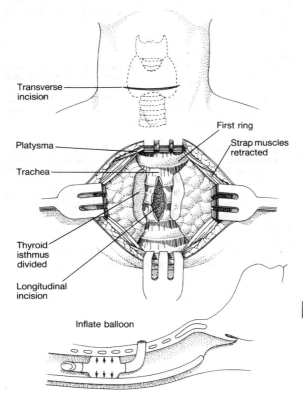

Transverse incision

First ring

Platysma

Strap muscles retracted

Trachea

Thyroid isthmus divided

Longitudinal incision

Inflate balloon

Tracheostomy

Tracheostomy (*cont.*)

Care of the tracheostomy

All procedures should be performed under sterile conditions. Patients require frequent removal of retained secretion by sterile suction every 15–30 minutes. The cuff may also be deflated for 5 minutes in every hour (although with the more recent tracheostomy tubes this may be unnecessary). The inspired gases should be humidified and the patient's hydration well maintained.

Removal of the tracheostomy

As a preliminary the tube may be exchanged for a smaller silver non-cuffed tube so that the patient may breathe through the tube and normally. A speaking tube may also be used, enabling the patient to cough and speak. After a period of a few days the tube may be removed and the wound covered after closing the wound with a few stitches.

Complications of tracheostomy

Early (postoperative)

- *Displacement of tube*: If this occurs the patient should have an endotracheal tube reinserted until a new tracheostomy tube can be reintroduced.
- *Infection* with *Pseudomonas*, *E. coli*, and *Staphylococcus aureus* increases the risk of local necrosis and stenosis. It can be avoided by using sterile technique.

Intermediate

- *Haemorrhage*: From the wound, from erosion of the innominate artery. This is a terminal event and is usually preceded by small bleeds. Further surgery may prevent disaster.
- *Acquired tracheo-oesophageal fistula*: Due to overinflation of the tube. Suspect this complication when gastric juice or feeds appear around the tube. Further surgery is indicated.

Late

- *Stricture formation*
- *Failure of spontaneous closure of tracheostomy*.
  In both situations further surgery is indicated.

Excision of the submandibular gland

Surgical anatomy

The submandibular gland lies under cover of the medial border of the mandible below the mucous membrane of the mouth on the hyoglossus and mylohyoid muscles. It is supported inferiorly by skin, platysma, and fascia. It has superficial and deep lobes continuous behind the posterior edge of mylohyoid. The duct originates from the anterior pole of the deep lobe and passes forward to open near the midline on each side. The facial artery and vein pass deep to the posterior pole of the superficial lobe. The lingual nerve loops around the duct from above and returns to lie medial to it.

Special anatomy

- The mandibular branch of the facial nerve at the angle of the jaw. It may be damaged if the incision is too high.
- The lingual nerve during forward dissection of the duct

Special consent

Warn the patient of potential mandibular nerve damage (drooping angle of lip) and lingual nerve damage (sensory to the anterior two-thirds of the tongue)

Anaesthetic

General anaesthetic.

Position of patient

Use a head ring. Extend the neck. Turn the face away so that the gland presents. Tilt the head up to empty the external jugular vein. Expose a rectangular area of the face with the gland at its centre.

Incision

Transverse, skin crease, 5 cm below the lower border of the mandible. Divide platysma. Insert a self-retainer. Make the skin incision at least 2 finger breadths below the angle of the jaw. Stay as close to the gland as possible. Use retraction widely.

Procedure

Deepen the incision. Expose the lower border of the superficial lobe. Stay close to the gland anteriorly. Identify where the two lobes are continuous. Expose the facial artery and vein above and below. Ligate and divide them at these sites.

Apply tissue forceps to the anterior pole of the superficial lobe. Ask your assistant to retract it laterally and the border of mylohyoid medially. Free the deep lobe from the undersurface of mylohyoid.

Now retract the gland medially. Free the deep surface of the deep lobe from hyoglossus. Dissect along the duct as far as possible. Identify the lingual nerve running forwards and medially. Ligate the duct with Vicryl and divide it.

Closure

Secure haemostasis. Bring a suction drain out via a separate stab wound. Close the skin with interrupted nylon or subcuticular prolene. Apply an Opsite dressing.

Special postoperative care/complications

Observe for haematoma, infection, and nerve injury.

Discharge

2–4 days postoperation.

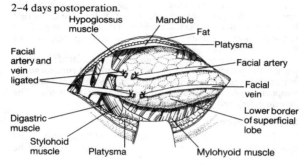

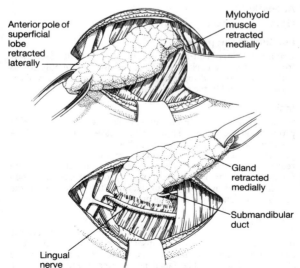

Excision of the submandibular gland

Superficial parotidectomy

Indications

Recurrent parotitis due to inaccessible stone disease and keratoconjunctivitis (Sjogren's syndrome).

Special anatomy

The parotid gland lies in the hollow between sternomastoid and the ramus of the mandible. It extends anteriorly to overlie masseter. Its duct emerges from the anterior extension, runs across masseter, turns medially to pierce buccinator, and opens into the mouth opposite the second molar tooth of the upper jaw. Several structures penetrate the gland: the external carotid artery is at first posteromedial then pierces the gland, emerges level with the neck of the mandible, and divides into superficial temporal and maxillary arteries. It is accompanied by the posterior facial vein.

The facial nerve emerges from the stylomastoid foramen, pierces the posteromedial part of the gland, and runs forwards in its substance where it divides into its five terminal branches: temporal, zygomatic, buccal, mandibular, and cervical. The carotid artery and posterior facial vein lie deep to it.

Special consent

The facial nerve and its five branches must be identified and protected early in the procedure. Informed consent must be obtained.

Anaesthetic

General anaesthetic.

Position of the patient

Use a head ring, head-up tilt, and turn the patient's face away from you with the neck extended. Clean the skin. Apply gowns with the parotid and mandible clearly exposed. Put a swab or cotton wool bud in the external auditory meatus.

Incision

S-shaped in front of pinna extending towards the mastoid process and then into the upper skin crease of the neck. Make the incision in three parts securing haemostasis at each before proceeding to the next. Deepen the cervical part of the incision removing the anterior branch of the greater auricular nerve and retaining in saline until the end of the procedure. Now deepen the facial part until you can identify the junction of the cartilaginous and bony external auditory meatus.

Procedure

Concentrate on the mastoid part of the incision. Dissect the tissue carefully to identify the insertion of sternomastoid. You are approaching the facial nerve. Identify and ligate the stylomastoid artery. The facial nerve runs in much the same direction but a few mm deeper. Identify the main trunk. The aim is now to dissect parotid tissue superficial to the trunk and its branches. Pass fine artery forceps, convex surface down, along the nerve and in contact with it. Allow the tips to protrude through parotid tissue and divide the area so identified with scissors. Repeat this process with each branch until the parotid is reflected. Reflect the skin off the anterior flap and identify the parotid duct at the anterior margin. Dissect and ligate it at the anterior border of the masseter muscle.

Closure

Close the wound with fine interrupted nylon sutures (4/0 or 5/0). Bring a suction drain through a separate stab wound.

N.B. Superficial parotidectomy may also be performed for lumps superficial to the facial nerve. This cannot be established clinically. If the lump is deep to the facial nerve a conservative or radical parotidectomy may be indicated. In conservative parotidectomy the facial nerve is preserved and the lump dissected from below it. In radical parotidectomy the nerve is sacrificed.

Special postoperative care/complications

Observe for haematoma, infection, and nerve injury.

Discharge

2–4 days postoperation.

Excision of branchial cyst, sinus, or fistula

Aim

Excision of all vestigial remnants. Make sure that a large branchial cyst is not a swelling of the lower pole of the parotid or encroaching upon it. If there is doubt proceed as for formal parotidectomy.

Small cyst

Special anatomy
- The external and internal carotid arteries and vagus nerve deep to the cyst
- The hypoglossal and glossopharyngeal nerves deep to the cyst above and below the posterior belly of digastric

Anaesthetic General anaesthetic.

Position of patient Supine. Head turned to the opposite side with upward tilt of the table to collapse the external jugular vein.

Incision Transverse over the lesion about 7 cm long. Raise subplatysmal flaps above and below.

Procedure Divide the fascia over the anterior border of sternomastoid. Retract it laterally. Free the cyst on all sides until its track can be seen. Divide this between two pairs of artery forceps. Often there is no track and the cyst may be removed by blunt dissection.

Closure See below.

Large cyst

It is best to do a superficial parotidectomy if the parotid gland is pushed outwards. Define the facial nerve and proceed.

Branchial sinus and fistula

Incision Elliptical over external opening.

Procedure Deepen the incision through platysma. Apply traction to the ellipse. The track may be felt running upwards between the carotid arteries and above the hypoglossal and glossopharyngeal nerve. It may also proceed towards the ear or parotid region when a superficial parotidectomy may again be necessary.

In most cases it runs towards the mouth. Excise it by dissection, traction, and multiple step-ladder incisions. Divide the posterior belly of digastric and follow the track to the middle constrictor where it blends with the muscle. Excise it with an ellipse of muscle.

Closure Secure haemostasis. Bring out a single suction drain through a separate stab wound. Close skin with interrupted nylon.

Special postoperative care

Remove the drain in 3–5 days, sutures in 7 days.

Discharge

2–4 days postoperation.

Excision of pharyngeal pouch

Aim
- To deal with the pouch—excise or reduce
- Cricopharyngeal myotomy

Special anatomy
The pouch herniates between the fibres of the inferior constrictor. Avoid recurrent laryngeal nerve damage and causing an iatrogenic postcricoid oesophageal stricture.

Special consent
Warn the patient of the risks. Obtain informed consent.

Anaesthetic
General anaesthetic.

Position of the patient
Supine with the head rotated to the right, left arm by the side, sandbag or pillow between the shoulder blades.

Incision
10–15 cm long parallel to the anterior border of left sternomastoid centred on the cricoid cartilage.

Procedure
Incise platysma and deep fascia. Retract sternomastoid, internal jugular vein, and the carotid sheath posteriorly (*carefully*). Retract the thyroid gland and the larynx anteriorly. Use blunt (pledget) dissection to identify the sac. Clean it down to its neck grasping its fundus with tissue forceps to provide gentle traction.

The pouch Small pouches may be left alone. Larger sacs should be excised. Ask the anaesthetist to pass a 28–30 French gauge bougie into the upper oesophagus. This procedure avoids narrowing. Apply a vascular clamp across the neck of the sac and excise it. Suture the sac with interrupted catgut sutures and tie them after releasing the clamp.

Crico-pharyngeal myotomy This is done to prevent postoperative fistula or late recurrence. Gently insert curved artery forceps between the cricopharyngeus muscle and the oesophageal mucosa. Open and close them to free the two layers. Incise the muscle between the open forceps for a length of 2–4 cm distally. When this is done small or moderate sacs may be suspended to the prevertebral fascia and their contents emptied into the gullet.

Closure
Suction or corrugated drain to site. Close platysma and skin in separate layers.

Special postoperative care

Limit oral fluids on the first 2 days to 30 ml/60 ml/hour respectively. Start free fluids on the third day and graduate thereafter to diet.

Complications

Fistula formation, mediastinitis, inhalational pneumonia.

Discharge

Between 7 and 10 days postoperation.

Heller's oesophageal cardiomyotomy (thoracic approach)

Aim

To restore normal swallowing by division of the lower oesophageal musculature without breaching the mucosa.

Special consent

Informed.

Anaesthetic

General anaesthetic.

Position of patient

Right lateral position with left side uppermost.

Incision

In the line of the seventh rib to the costal margin.

Procedure

Approach the oesophagus through the bed of the seventh rib. Divide the pleura, retract the lung forwards and diaphragm downwards. Free the oesophagus from hiatus to inferior pulmonary vein. Pass a rubber sling around it. Ask your assistant to apply gentle traction. Pass the index around the lower oesophagus and gently incise the muscle layer longitudinally. Separate the divided muscle with artery forceps or right-angled vascular clamps until mucosa pouts outwards.

Apply tissue forceps to the edge of the divided muscles and continue this process until the lower 10–12 cm of the muscle has been divided to the cardia. Take extreme care not to breach the mucosa. Ligate small vessels or apply gentle pressure until haemostasis is secure.

If perforation does occur repair it with fine sutures (vicryl). If this is unsuccessful it may be necessary to resuture the muscle layer to avoid fistula formation. In such cases consider oesophagectomy with either oesophagogastric anastomosis at the aortic arch or short segment colon replacement.

Closure

Bring a chest drain through a separate stab wound. Approximate the ribs with rib approximator. Close the rib bed with continuous nylon and the muscle layers separately with continuous dexon. Close the skin with continuous nylon or dexon.

Complications

- Leaking from unnoticed mucosal perforation. If this happens re-exploration should be carried out with repair if possible. Oesophagogastrectomy may have to be considered.
- Dysphagia due to failure of the procedure. A further myotomy is necessary.

- Peptic stricture due to reflux. Prevent by the use of oral H_2 antagonists or an antireflux procedure at the time of myotomy.
- Carcinoma of oesophagus. Patients should be screened on a regular basis by endoscopy or barium swallow. Achalasia is a premalignant condition.

Heller's oesophageal cardiomyotomy (abdominal approach)

Aim

To restore normal swallowing by longitudinal division of the lower oesophageal musculature.

Anaesthetic

General anaesthetic. Ask the anaesthetist to pass a nasogastric tube.

Position of patient

Supine.

Incision

Left upper paramedian.

Procedure

Mobilize the left lobe of the liver by division of the left triangular ligament. Fold the lobe to the right. Explore the abdomen and confirm the diagnosis.

Identify the oesophagus by palpation of the nasogastric tube. Divide the peritoneum and phreno-oesophageal ligament in front of the oesophagus. Preserve the vagus nerves. Now encircle the oesophagus with the index and thumb so that they meet behind. Pass a rubber tube around the gullet to provide traction.

Divide the muscle on the anterior wall of the stomach 1 cm below the cardia until *intact** mucosa bulges through. Extend this upwards for about 8–10 cm so that tissue forceps can be applied to the edges of the muscle.

*N.B. The oesophageal mucosa is like wet blotting paper when isolated. Do take care not to perforate it. It takes sutures poorly.

Identify the plane between muscle and mucosa with artery or vascular forceps alternately opening and closing them to split the muscle longitudinally and allowing mucosa to bulge through.

Check that there are no leaks. Stop bleeding.

Closure

Close the wound in layers with a suction drain brought out through a separate stab wound. Suture skin with interrupted sutures.

Discharge

7–10 days after operation for both thoracic and abdominal approaches.

Exploratory laparotomy

This operation is indicated in elective patients when investigations have failed to confirm the diagnosis or when the operability of a condition is in doubt. In emergency patients laparotomy may be indicated to determine the diagnosis and subsequent treatment in the patient with an acute abdomen. All abdominal procedures should be combined with a laparotomy.

Preoperative assessment

Appropriate haematology and biochemistry results should be at hand. Resuscitation should be as complete as possible in emergency cases. Cefuroxime (1500 mg) and metronidazole (500 mg) IV should be given on induction if septic peritonitis is suspected. Be prepared to start intravenous antibiotics by ensuring that an IV line has been established.

Anaesthetic

General anaesthetic.

Incision

Midline, skirting the umbilicus, or right paramedian centred on umbilicus.

Procedure

Divide the anterior rectus sheath then reflect the rectus muscle laterally. Open the posterior sheath and peritoneum (paramedian). Is there free fluid or pus? Take a bacteriology swab if there is. Now examine the abdominal contents systematically. Begin by examining the lobes of the liver, then gall bladder and spleen. Examine the oesophageal hiatus, stomach, and pylorus. Is there a perforated duodenal ulcer? Are the bile ducts normal?

Palpate the right kidney and head of pancreas. Then lift the greater omentum out of the wound and pass the right hand down behind it to examine first the body and tail of pancreas, the left kidney, then all of the transverse colon, both ascending colon and caecum, then descending, sigmoid colon, and pelvic colon as far as possible. Is there diverticular disease or a carcinoma? Now inspect the root of the mesentery with its vessels and proceeding from the duodenojejunal flexure examine all of the small bowel to the appendix. Is there inflammation (?Crohn's), a Meckel's diverticulum, or appendicitis? (If obvious pathology is found early, e.g. appendicitis, it is best dealt with first then followed by a laparotomy.)

Finally examine the pelvic organs and hernial orifices. Note whether there is dilatation of the ureters (hydronephrosis). Is the uterus fibroid? Are the tubes and ovaries inflamed? (salpingitis). Is there tubal bleeding (ruptured ectopic pregnancy)? Note should also be made of the retroperitoneal structures. Is there an aortic aneurysm (usually extremely obvious)? Are there abnormalities of the iliac vessels? Is there a retroperitoneal mass? Once the diagnosis has been confirmed carry out the definitive procedure indicated in the circumstances.

Closure

Close the peritoneum and posterior rectus sheath with vicryl. Close the anterior rectus sheath with non-absorbable sutures. Close skin with interrupted or subcuticular nylon. Suture any drains in position, bringing them through a separate stab wound.

Special postoperative care

Maintain the IV infusion until bowel sounds return, then start oral fluids. Monitor temperature, respiration, and urine output. Continue antibiotics if indicated.

Complications

See chapter 4.

Discharge

7–10 days postoperation in uncomplicated cases.

Ramstedt's pyloromyotomy

Aim

To restore normal gastric emptying by dividing the hypertrophic pyloric muscle.

Special consent

Explain the procedure and the necessity for it to the parents. Obtain their informed consent.

Anaesthetic

General anaesthetic.

Position of patient

Supine.

Incision

Transverse incision 3–4 cm long placed 1 cm above the liver's edge (which must be felt before proceeding).

Procedure

Divide the right rectus muscle. Open peritoneum and retract the liver's edge in a cephalad direction. Use the left index to hold down colon and grasp the 'tumour' with Babcock's forceps. Do not close the blades.

Confirm the diagnosis. Make a longitudinal incision over the anterior aspect of the pylorus dividing the peritoneum only, beginning distal to the tumour and continuing proximally to the anterior wall of the stomach. Gently deepen the incision over the constricting ring for 1–2 mm.

When the incision has been deepened a little insert the tips of a pair of artery forceps into it, opening them alternately until the mucosa bulges out. Continue this procedure for the whole length of the incision ensuring that all pyloric muscle is divided and that the mucosa is intact. This can be checked by gently squeezing gastric air into the pylorus. Repair any defects with catgut.

Closure

Return the duodenum to the abdomen. Close peritoneum and muscle sheath separately with continuous vicryl. Close skin with subcuticular dexon or prolene (leaves a neater scar).

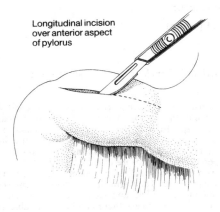

Longitudinal incision over anterior aspect of pylorus

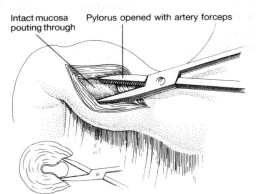

Intact mucosa pouting through

Pylorus opened with artery forceps

Cross-section of pylorus with mucosa pouting through

713

Inguinal hernia in infants and children

Aim

To excise the hernial sac (herniotomy). Repair of the defect is rarely necessary in children.

Special anatomy

The internal and external rings in infants and children are superimposed upon each other. It is unnecessary to split the external oblique.

Anaesthetic

General anaesthetic.

Position of patient

Supine.

Incision

Inguinal incision 3–5 cm long over the external ring.

Procedure

Isolate the cord. Now remove its covering dividing each one in turn in the line of the cord by opening it between two pairs of fine artery forceps. Identify the sac and gently dissect it free using two pairs of non-toothed forceps, one gently retracting the sac, the other dissecting the surrounding structures carefully from it. Open the sac and ensure that the tips of the forceps can be passed into the peritoneal cavity. Dissect the sac proximally taking great care to avoid damaging the vas deferens and vessels. Transfix and ligate the sac and excise it. Usually no other procedure is necessary unless there is a large palpable gap in which case the external ring should be gently plicated with one or two interrupted catgut sutures.

Closure

Catgut to subcuticular tissue, subcuticular dexon.

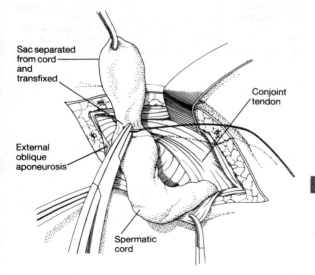

Sac separated from cord and transfixed

Conjoint tendon

External oblique aponeurosis

Spermatic cord

Strangulated inguinal hernia in children

Aim

Operate without delay. Resection and anastomosis of bowel increase the risk of mortality.

Special consent

Advise the parents of the urgency of the procedure and the implications of not operating.

Anaesthetic

General anaesthetic.

Position of patient

Supine.

Incision

As for the elective procedure.

Procedure

If there is bowel in the sac gently draw it downwards and examine the viability of the bowel which has been caught in the neck of the sac. If it appears to be non-viable then it should be resected with an end-to-end anastomosis. If it is viable the bowel should be returned to the abdomen. (Sometimes anaesthesia allows the hernia to reduce spontaneously. In such cases operation should then be carried out along the same lines as an elective procedure.) After the contents of the sac have been dealt with, repair the hernia as previously described.

If for some reason gangrenous bowel is returned to the abdomen, repair the hernia first, close the hernia wound, and carry out a laparotomy with resection of bowel and end-to-end anastomosis.

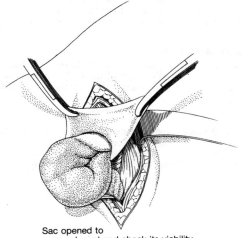

Sac opened to
expose bowel and check its viability

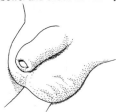

Strangulated inguinal hernia in children

Circumcision

This is the commonest operation performed on children.

Indications
- Religious reasons
- Phimosis (check blood sugar in adults). Diabetes mellitus is frequently associated.
- Recurrent balanitis
- Cancer of penis affecting the prepuce

Procedure
General anaesthetic. Skin preparation. Separate all preputial adhesions. Attach two small artery forceps to the dorsum of the prepuce at 11 and 1 o'clock. Make a scissor cut between the clips towards the corona. Stop 2 mm from the edge to leave mucosa for suture. Apply a clip to hold the mucosa and dorsal skin together. Now apply a clip to the frenulum and excise each side leaving a 3 mm fringe of mucosa. Secure haemostasis with ligatures. Then suture the skin and mucosa together with chromic catgut or vicryl-type sutures and apply a paraffin gauze dressing.

Postoperative analgesia
Use plain marcain 0.5 per cent. Inject 5–10 ml in total under the symphysis pubis at either side of the bulbo-spongiosus. This lasts 12 hours. An alternative is to infiltrate the remaining fringe of foreskin with marcain.

Complications
- External urethral meatus stenosis. Do not carry out circumcision in the presence of inflammation.
- Bleeding. Stop with pressure or suturing.
- Infection. Usually superficial. Use topical cicatrin.

Truncal vagotomy and drainage

Aim

To divide the vagus nerves to the stomach and so reduce the output of hydrochloric acid. It leads to a considerable degree of gastric stasis and so a drainage procedure such as a gastro-enterosotomy or pyloroplasty is carried out to prevent gastric stasis.

Special consent

Warn the patient of the possible postoperative complications, especially dumping and diarrhoea, thus obtaining informed consent.

Preoperative preparation

The procedure is most commonly performed today for perforated or bleeding duodenal or gastric ulcer. However, it is still done in some hospitals for patients who fail to respond to H_2 antagonist therapy. The diagnosis will have been established by endoscopy or barium studies. Patients with bleeding duodenal or gastric ulcer may need preoperative resuscitation with blood and IV fluids.

Routine haematological and biochemical parameters should be established before theatre, and antibiotic prophylaxis should be given in emergencies especially after H_2 receptor antagonist therapy which leads to changes in the gastric/duodenal flora.

Anaesthetic

General anaesthetic.

Position of patient

Supine.

Operation

Incision Upper midline.

Procedure The peritoneal cavity is opened between artery forceps. The contents of the abdomen are examined. The size of the diaphragmatic hiatus is established and the presence or absence of gallstones or diverticular disease is especially noted. Insert a self-retaining retractor. Instruct your assistant to hold the retractor under the xiphisternum so that the left coronary ligament of the liver can be easily identified. Divide this ligament and fold the left lobe of the liver towards the midline. The peritoneum overlying the gastro-oesophageal junction can now be seen and identified as a fine white line. Gently incise this peritoneum and enlarge the opening so that the index finger of the right hand can be gently insinuated around the oesophagus to encircle it so that a length of red tubing can be passed around the lower end of the oesophagus to control it and so perform the vagotomy. Apply an artery forceps to the two ends of the rubber tubing.

Vagotomy technique Apply gentle traction to the rubber tubing and using the right index finger carefully identify the anterior (which may be in multiple strands) and posterior vagus nerves. Clamp and divide them. They can be felt as tight bands on the anterior and posterior surfaces of the oesophagus. Ensure that haemostasis is secured. When the vagotomy has been performed the oesophagus will appear to descend for some distance into the abdomen.

Drainage procedure Either a pyloroplasty (preferable) or a gastroenterostomy may be performed.

Pyloroplasty (Heineke–Mikulicz) Identify the pylorus. Insert non-absorbable stay sutures at its upper and lower aspects. With a knife divide the circular muscle longitudinally in line with the duodenum for a distance of only about 1 cm. Aspirate any contents, suture the pylorus transversely to increase the size of its lumen using interrupted fine dexon or vicryl (more popular) sutures. If necessary place an omental tag over the pyloroplasty.

Gastroenterostomy This procedure involves a gastrojejunal anastomosis between the most dependent part of the stomach and the proximal jejunum. This may be performed in antecolic or retrocolic (better because it drains more easily).

Retrocolic isoperistaltic gastroenterostomy Identify the duodenojejunal flexure and select a portion of jejunum which is mobile enough to reach the most dependent part of the stomach. Ask your assistant to lift the transverse colon vertically. Make an incision on the lower aspect of the transverse mesocolon well clear of the middle colic artery. Enlarge this opening so that the most dependent part of the stomach can be pulled through it using Babcock's forceps. Apply soft gastrointestinal clamps or gastroenterostomy clamps (Lane's) to the stomach border and the portion of jejunum selected for anastomosis. Carry out a gastrojejunal anastomosis in two layers with continuous absorbable sutures. Fix the edges of the gastroenterostomy to the mesocolon with interrupted absorbable sutures to prevent intussusception or herniation by a loop of small bowel.

Closure Unless mass closure is used close the wound in layers with absorbable suture to the peritoneum, non-absorbable to the midline and interrupted nylon or subcuticular prolene to skin.

721

Special postoperative care

Aspirate the nasogastric tube hourly for the first 12 hours then remove it. (Some surgeons will remove it at the end of the operation.) Give ice on the first postoperative day then 30 ml hourly by mouth on the second day. Encourage the patient to sit up to improve drainage. Increase fluids on the third day and graduate to light diet.

Highly selective vagotomy

(parietal cell vagotomy, proximal gastric vagotomy)

Aim

To denervate the acid secreting portion of the stomach. The anterior and posterior nerves of Latarjet are preserved but branches which they give to the parietal cell area are divided, thus the vagus nerve remains intact to supply the alkali-secreting cells of the gastric antrum and the vagal branches to the other abdominal viscera are preserved.

Special anatomy

The anterior and posterior nerves of Latarjet pass down the lesser curvature giving off branches to the parietal cell area of the stomach. At the angulus they spread out in a 'crow's foot' to supply the antrum.

Preoperative preparation

As for truncal vagotomy and drainage. Cefuroxime prophylaxis is warranted in case of inadvertent gastrostomy and should always be used in emergencies.

Anaesthetic

General anaesthetic.

Position of patient

Supine.

Operation

Incision Upper midline or left upper paramedian.

Procedure After carrying out a laparotomy open the gastrocolic omentum through an avascular area. Lift the stomach forwards and carefully separate its attachments posteriorly from the pancreas. Pass the right index finger through the defect in the gastrocolic omentum and grasp the gastric antrum pulling it downwards. Stretch the lesser curvature of the stomach and so identify the anterior nerve of Latarjet. This nerve runs parallel to the lesser curve of the stomach and separates into branches over the lesser curve. When the nerve has been identified make a hole through the lesser omentum. Pass a tape around the angulus of the stomach to keep it on the stretch. Now keeping close to the gastric wall proceed proximally along the lesser curve doubly clamping, dividing, and ligating the vessels and accompanying nerve filaments so as to strip the lesser omentum from the lesser curve. Continue this process along the anterior surface of the stomach and the gastro-oesophageal junction. Identify the angle of His between the fundus and the left edge of the lower oesophagus. Carefully incise the peritoneum and the angle and use finger dissection to completely encircle the cardia. Divide all loose tissue overlying the oesophagus and anteriorly and posteriorly but keeping the oesophageal muscle wall intact.

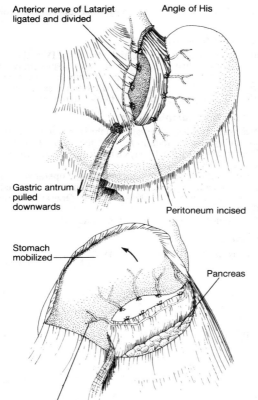

Anterior nerve of Latarjet ligated and divided

Angle of His

Gastric antrum pulled downwards

Peritoneum incised

Stomach mobilized

Pancreas

Terminal fork of nerve is preserved

Highly selective vagotomy

Highly selective vagotomy (*cont.*)

Now complete the mobilization of the greater curve of the stomach with division of the gastrocolic omentum to show the posterior nerve of Latarjet which terminates in the antrum distal to the incisura. Preserve the terminal fork of the nerve, divide all other branches to the stomach. The objective of this operation is to separate the lesser omentum from the lesser curvature from incisura to cardia by dividing all blood vessels and nerves that enter the lesser curvature from the lesser omentum. Vessels and nerves should be divided individually and it is important to have early access to the posterior aspect of the stomach since the blood vessels run in two distinct bands, one to the anterior aspect and one to the posterior aspect of the stomach.

Closure Layers with interrupted nylon or subcuticular prolene to skin.

Special postoperative care

Withdraw the nasogastric tube and remove the IV at the end of the procedure. Give ice only on the first postoperative day, graduating to 30 ml and 60–90 ml of water by mouth on the second and third days respectively. Free fluids can commence on the fourth day, graduating to light diet thereafter.

Postoperative complications

- Necrosis of the lesser curvature of the stomach can occur due to ischaemia or trauma to the gastric wall by diathermy or instrumental damage. It may cause death in 1 in 1000 highly selective vagotomy operations.
- Splenectomy may be necessary because of operative damage to the splenic capsule. If the organ cannot be repaired by suture or patched, it should be removed. Operative mortality 0.3 per cent.
- Recurrence in up to 15 per cent of patients

Discharge

7–10 days postoperation.

Perforated duodenal ulcer

The gastric contents spill through the perforation leading to chemical peritonitis. Patients who undergo early surgery have an excellent prognosis. Conservative treatment should only be considered in patients who have had a long-standing perforation or who are too aged or infirm to survive surgery.

Preoperative preparation

Establish urea and electrolyte status and start an IV infusion. Exclude the possibility of acute pancreatitis by serum amylase estimation. In elderly patients or those with chronic obstructive airways disease take arterial samples for blood gas estimation.

Anaesthetic

General anaesthetic.

Position of patient

Supine.

Incision

Upper midline (preferable) or right paramedian.

Procedure

On opening the peritoneal cavity a variable quantity of gastric fluid intermingled with bile will be encountered. Aspirate this with suction and retain a specimen for bacteriology. Identify the perforation. It is usually on the first part of the duodenum on its anterior wall. If not, check the posterior and superior aspects and the stomach. Perforations can also occur into the lesser sac. Oversew the perforation with one, two, or three vicryl sutures (1/0 or 2/0) placed some way (0.5 cm) proximal and distal to the ulcer. Tie them to close the perforation. Leave the ends long and bring an omental tag over the perforation. Lightly tie it in place. If it necroses it falls off.

Now aspirate all four quadrants of the abdomen, the paracolic gutters, and the pelvis. There is usually no need to drain the abdomen unless the perforation is long-standing, but lavage with warm saline ± antiseptics is valuable.

Closure

Layers with non-absorbable or absorbable to rectus sheath and nylon to skin.

Complications

Infection, abscess formation. 25 per cent will relapse and may need definitive surgery which should always be considered in fit patients, those with a proven chronic ulcer, those with a history of previous complications, and those with perforations induced by non-steroidal anti-inflammatory drugs.

Discharge

7–10 days postoperatively.

Insertion of Angelchik prosthesis

The Angelchik prosthesis is an antireflux device which is indicated in patients who have severe unremitting gastro-oesophageal reflux symptoms due most commonly to hiatus hernia.

Preoperative preparation

As for vagotomy procedures. Ask the anaesthetist to pass a nasogastric tube to facilitate identification of the oesophagus once the patient has been anaesthetized. Give prophylactic antibiotics (cefuroxime/metronidazole) on induction and 8 and 16 hours postoperatively.

Anaesthetic

General anaesthetic.

Position of the patient

Supine.

Incision

Upper midline or left upper paramedian.

Procedure

Insert a self-retaining retractor. Mobilize the left lobe of the liver by dividing the left triangular ligament as in truncal vagotomy procedures. Divide the peritoneum overlying the oesophagus and encircle it. At this stage assess the extent of any hiatus hernia and attempt to reduce it gently into the abdomen.

When the oesophagus has been encircled, pass the C-shaped Angelchik prosthesis around the gullet so that it sits at the cardio-oesophageal junction. Use Ligaclips to clamp its tapes together at two or three sites. Cut the redundant tape.

Closure

Mass or close in layers with interrupted nylon to skin. No drain is required.

Special postoperative care

As for vagotomy and drainage.

Complications

- Infection. When this occurs the prosthesis must be removed, therefore operative technique must be as aseptic as possible and prophylactic antibiotics must be used.
- Migration of the device proximally. This may occur when there is a large diaphragmatic defect. Often it produces no symptoms.
- Ulceration of the prosthesis through the oesophagus or stomach.

- Oesophageal stricture. If stricture or ulceration of the prosthesis occurs it may be necessary to remove it. The patient should be reassessed both endoscopically and radiologically and full account taken of symptoms.

Discharge

7–10 days postoperation.

Celestin intubation of inoperable oesophageal carcinoma

This is a palliative procedure which should be considered in patients who have advanced disease or who are too frail to undergo major oesophageal resection. It may also be performed as a preliminary to radiotherapy in patients with squamous carcinoma to ensure patency and maintain swallowing.

Preoperative preparation

The patient may need a short period of IV feeding or IV fluid and electrolyte replacement, but there is no evidence that prolonged attempts at nutrition change the outcome (2–3 weeks). If the obstruction is complete, nasogastric aspiration of the oesophageal contents will prevent inhalation.

Anaesthetic

General anaesthetic.

Position of patient

Supine.

Incision

Upper midline.

Procedure

Identify the tumour. It usually lies at the oesophagogastric junction, the lower oesophageal third, or the proximal stomach. Confirm that it is not operable.

Make a gastrostomy distal to the tumour. Give the patient cefuroxime (500 mg) IV at induction and 8 and 16 hours post-operatively. Aspirate the stomach contents. Insert the index finger proximally. Can you pass it into the oesophageal lumen? Is the nasogastric tube in the stomach? If so, ask the anaesthetist to attach the Celestin introducer to the nasogastric tube by passing a suture through both after removing the funnel of the nasogastric tube. Railroad the introducer into the stomach. The anaesthetist now ties the Celestin tube to the introducer using the special groove provided and the introducer is gently pulled downwards by the surgeon, the tube being guided into the gullet at the mouth by the anaesthetist. Continue traction on the tube until its 'thistle-funnel' impacts against the shoulders of the tumour. Suture the tube in place with two 2/0 black silk sutures tied loosely and passing from outside the stomach through the lumen of the tube. Remove the redundant tubing by bevelling the end of the tube. Close the gastrostomy with two layers of continuous 2/0 catgut or vicryl.

Closure

Peritoneum and linea alba may be closed together with continuous non-absorbable sutures. The skin may be closed with interrupted nylon.

Special postoperative care

Do a chest X-ray to exclude perforation. Give ice on the first postoperative day and then graded fluids on the second and third days (as in vagotomy and drainage).

Complications

Since the procedure is not completely aseptic and the patient is often ill, wound infection is a potential risk. Be prepared to remove one or two sutures to allow drainage of pus, if necessary.

Difficulties

Sometimes the nasogastric tube cannot be passed into the stomach. In such cases a guide wire may be passed from above endoscopically or the introducer may be passed upwards through the gastrostomy, using the index finger to guide it into the lumen of the oesophagus.

Discharge

7–10 days postoperation.

Operative procedures for hiatus hernia

Aims

- Closure of the hiatus after reducing the hernia
- Maintaining the oesophagogastric junction beneath the diaphragm and establishing a length of intra-abdominal oesophagus
- Fashioning of an acute angle between the oesophagus and gastric fundus

Procedures

Nissen fundoplication In this procedure the hernia is reduced at laparotomy and a wide-bore tube passed into the stomach. The limbs of the right crus are then approximated behind the oesophagus by sutures (one or two usually suffices). The gastric fundus is then carefully mobilized by dividing the gastrosplenic omentum and the fundus is then literally wrapped around the lower oesophagus to form a cuff of stomach behind, in front, and on each side. The stomach is held in place by non-absorbable sutures.

Belsey Mark IV operation This is carried out through a left sixth interspace thoracotomy. The entire oesophagus in its lower part is freed up to the level of the aortic arch. The cardia is then mobilized and pulled up through the hiatus. The acute angle of entry of the oesophagus into the stomach is then re-established by a series of mattress sutures passing from the oesophagus to the fundus and back up through the diaphragm. At the end of the procedure the right crus is approximated behind the oesophagus.

Other procedures may be performed such as the Collis operation in which the right crus is sutured in front of the oesophagus and the fundus sutured to the undersurface of the diaphragm to re-establish the acute angle of entry or gastropexy in which the anterior surface of the stomach is sutured to the linea alba (Boerema anterior gastropexy) or the lesser curve fixed to the median arcuate ligament (Hill posterior gastropexy). If increased oesophageal length is required a gastroplasty (Collis) may be performed. In this a tube is fashioned from the lesser curvature to lengthen the gullet and the gastric fundus is sutured to the arcuate ligament (see also Angelchik prosthesis, p. 728).

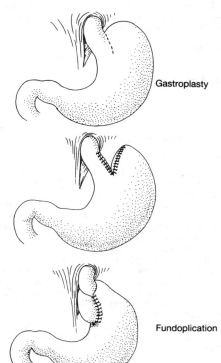

Gastroplasty

Fundoplication

Gastrectomy

Indications

Partial gastrectomy carcinoma of distal stomach, benign gastric and duodenal ulcers, benign tumours.

Total gastrectomy carcinoma of proximal stomach, multifocal carcinomas, Zollinger–Ellison syndrome. Intestinal continuity is re-established by gastroduodenal (Billroth I) or gastrojejunal anastomosis (Billroth II).

Preoperative preparation

Check haemoglobin, electrolyte, and nutritional status. Correct if necessary by replacement therapy or parenteral nutrition. Treat shock if present. Wash out the stomach if it contains food or debris. Assess cardiorespiratory status. Pre- and post-operative physiotherapy is of value. Check liver and renal function. Use prophylactic antibiotics especially in emergency, in patients with carcinoma, and in those who have received prolonged H_2 receptor antagonists.

Anaesthetic

General anaesthetic.

Position of patient

Supine.

Incision

Upper midline or paramedian. Oblique with thoracic extension if oesophagogastrectomy is indicated.

Procedure

Identify the pathology. Assess the extent of spread of a carcinoma. Are there metastases locally or in the liver? If so consider palliative or no surgery.

Mobilize the stomach along its curves as required. If there is a carcinoma excise the omentum and lymph nodes. At the pyloric end of the lesser curve identify, ligate and divide the right gastric artery; on the greater curve the right gastroepiploic artery. Divide the short gastric arteries and the left gastroepiploic arteries at the cardiac end of the greater curve.

Mobilize the duodenum by dividing the peritoneum along its lateral border. Divide between clamps and oversew it. Use the stomach now as its own retractor by lifting it upwards by its distal clamp. Identify and ligate the left gastric artery at the proximal end of the lesser curve.

Divide the stomach at its midpoint between clamps, partially close the lesser curve and perform an antecolic gastrojejunal anastomosis in two layers using an absorbable suture.

Closure

In layers. Nylon to skin.

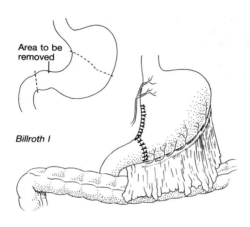

Area to be removed

Billroth I

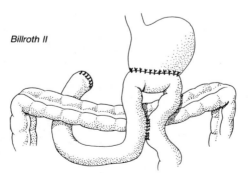

Billroth II

Polya   As Billroth II but with retrocolic anastomosis

Gastrectomy

735

Gastrectomy (*cont.*)

Special postoperative care

Insert a nasogastric tube after the operation. Aspirate it hourly for the first 12 hours to observe for haemorrhage. Begin oral fluids on the third postoperative day. Observe for evidence of gastric bloating or anastomotic leakage.

Complications

Early:
- Haemorrhage
- Anastomotic leakage (up to 3 days postoperatively)
- Gastric bloating

Late:
- Recurrent disease
- Metabolic complications
- Postgastrectomy syndromes

Discharge

7–10 days postoperatively.

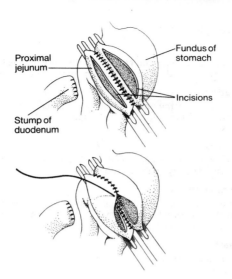

Proximal jejunum

Stump of duodenum

Fundus of stomach

Incisions

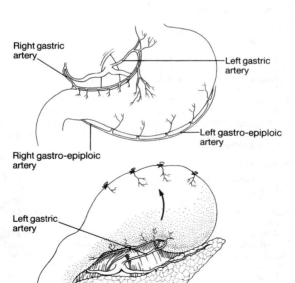

Right gastric artery

Left gastric artery

Right gastro-epiploic artery

Left gastro-epiploic artery

Left gastric artery

Cholecystectomy

Aim

This is an elective operation (usually) to remove the gall bladder and its contents and retain the anatomy of the biliary system.

Indications

Cholelithiasis, carcinoma, cholangiohepatitis, empyema or perforation of the gall bladder, traumatic rupture, bile duct cancers.

Diagnosis

This is suspected by clinical signs and symptoms and confirmed by ultrasonography or cholecystography.

Special anatomy

There is great variability in the anatomy of the gall bladder and bile ducts. You should not attempt this operation until you are fully conversant with these.

Anaesthetic

General anaesthetic. Give IV cefuroxime (1500 mg) on induction.

Position of patient

Supine.

Incisions

Several approaches may be used:
- Right paramedian incision (commonest)
- Kocher's (subcostal) incision (in obese patients)
- Transverse
- Midline

The rectus muscle is reflected in a paramedian incision or divided with diathermy in a Kocher's or transverse incision.

The peritoneum is opened between artery forceps. A self-retaining retractor is inserted.

Procedure

A full laparotomy is performed. There is an incidence of colorectal carcinoma of about 5 per cent. If this is found it should be dealt with first. If a Kocher's incision has been made it may be necessary to close it and make a fresh incision to obtain access.

The gall bladder is identified and the small bowel packed downwards to expose the biliary tree. The assistant's left hand maintains exposure. The liver is retracted with a Deaver retractor held in the assistant's right hand.

Sponge-holding forceps are applied to the gall bladder near its junction with the cystic duct and gentle traction applied by the operator. The peritoneum overlying the neck of the gall bladder is then carefully incised and the junction of the cystic duct with the common bile duct carefully exposed by sweeping

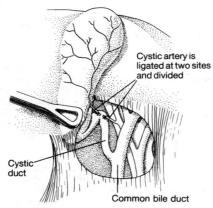

Cystic artery is ligated at two sites and divided

Cystic duct

Common bile duct

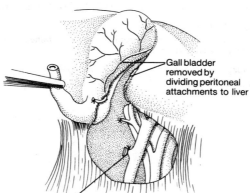

Gall bladder removed by dividing peritoneal attachments to liver

Cystic duct ligated almost flush with common bile duct

Cholecystectomy

739

away the peritoneum with gauze. The cystic artery is exposed
close to the cystic duct by the same method and its destination to
the gall bladder confirmed. It is then ligated at two sites and
divided between the ligatures. Now the gall bladder is attached
to the common bile duct by the cystic duct only. At this stage
carry out on-the-table cholangiography either by passing a
cholangiogram catheter into the common bile duct via the cystic
duct or by the direct puncture technique. Take two films (or use
image intensification) after injection of 2–4 ml and 6–8 ml of
contrast. If this excludes stones and shows good flow into the
duodenum ligate the cystic duct almost flush with the common
bile duct with vicryl ensuring that there is no narrowing.

The gall bladder can now be removed by dividing its
peritoneal attachments to the liver.

Closure

Secure haemostasis in the gall bladder bed either by diathermy
or by oversewing the peritoneum with catgut. Bring a suction
drain through a separate stab wound from the gall bladder bed
(Redivac). Close the wound in layers with dexon/vicryl to
peritoneum and non-absorbable suture to the rectus sheath.
Insert interrupted or continuous sutures to the skin. Stitch the
drain in position. Check that the swab count is correct.

Specific postoperative care

As with all upper-abdominal surgery, physiotherapy should be
prescribed for at-risk patients to prevent pulmonary atelectasis.
Revacuum the drain daily and measure the drainage. As it
reduces, the drain may be removed between 3 and 5 days postoperatively. Remove the sutures after 7–10 days (7 is the ideal).

Complications

Pulmonary atelectasis, subphrenic abscess, retained stones,
damage to the common bile duct.

Discharge

7–10 days postoperation.

Exploration of the common bile duct

This is usually performed in the course of a cholecystectomy or as a secondary procedure for retained stones. A single dose of prophylactic antibiotics is usually given unless there is cholangitis when antibiotic treatment is indicated. Prophylactic antibiotics are routinely given.

Indications

- Stones or block seen on operative cholangiography performed as a routine during cholecystectomy
- Stones palpated in the common bile duct
- Common bile duct thick-walled or dilated
- Obstructive jaundice
- Retained stones

The exploration may be supraduodenal (commonest), retroduodenal or transduodenal.

Procedure

Divide the hepaticocolic ligament if not already done at cholecystectomy. Insert two stay sutures into the wall of the supraduodenal duct parallel to each other. Place a swab in the hepatorenal pouch to collect bile leakage and stones. Open the duct with a knife for about 1 cm. Note and aspirate the bile. If it is a thick sludge consider transduodenal sphincterotomy or a choledochoduodenostomy in older patients to prevent further problems. Fortunately most stones can be flushed out of the common bile duct with sterile saline injected through a soft rubber catheter inserted into the opening in the duct. Difficult stones may be removed by passing a Fogarty catheter past the stones, inflating the balloon and gently withdrawing the catheter. Prevent proximal migration of stones above the choledochostomy with gentle digital pressure or a bulldog clamp. Explore the proximal duct in similar manner.

Insert Desjardin's forceps upwards and downwards, and feel for stones. Wash out the ducts with saline passed through a catheter. If stones are still felt pass the Desjardin's forceps and try to remove them. If they cannot be removed use the choledochoscope to visualize and remove them either complete or after crushing. The use of the choledochoscope requires considerable experience. Flush the ducts proximally and distally several times to remove clot and debris.

Prepare as small a T-tube as possible (Maingot 14) by fashioning its short limb into a gutter. Bring the long limb through a stab wound on the abdominal wall and insert the short limb into the choledochostomy. Ensure that bile flows freely up the long limb. Close the choledochostomy above the T-tube with 2/0 catgut.

Impacted stones?

If a stone is impacted at the lower end of the duct, mobilize and open the duodenum along its convex border to expose the ampulla. Pass a Bakes' dilator from above. If the stone passes through all is well. If not slit the ampullary orifice gently until the stone passes through followed by the probe. Close the duodenostomy transversely in two layers of 2/0 catgut. An alternative is endoscopic sphincterotomy in selected patients.

Stricture?

If there is a stricture distally perform a duodenotomy and pass a probe or fine Bakes' dilator to mark the sphincteric orifice. Cut down on it for 1 cm or so with a knife or diathermy and divide the sphincter. Stop bleeding. Suture the duodenal mucosa to that of the common bile duct on each side. Close the duodenotomy and insert T-tube in the common bile duct. This procedure is a sphincteroplasty. Choledochoduodenostomy is an alternative to sphincteroplasty. Longitudinal incisions are made in the supraduodenal portion of the common bile duct and the duodenal bulb. The two are anastomosed with interrupted vicryl inserting all the sutures first and tying them at the end of the procedure bringing the two openings together to form a diamond-shaped anastomosis. Ascending cholangitis is a complication of the procedure.

Splenectomy

The spleen plays a vital role in normal immunological function. It may be removed electively in cases of hypersplenism and massive splenomegaly. It may also be removed as an emergency after trauma. Every effort should be made to preserve the normal spleen.

Anaesthetic

General anaesthetic.

Position of the patient

Supine.

Elective procedure

Incision Midline incision extending to umbilicus or beyond. This may be extended transversely or into the chest for very large spleens.

Procedure Identify the splenic artery arising from the coelic axis and running along the superior border of the pancreas. Tie it first in continuity. This is done by opening the lesser sac of the peritoneum and isolating the artery before passing a ligature around it.

Now place the left hand, palm downwards, over the top of the spleen and draw it medially. Divide the peritoneum lateral to the spleen proximally and distally. Draw the spleen forwards lifting the splenic flexure of the colon and the tail of the pancreas. Identify the splenic pedicle, tie the vein and the artery separately and divide them. Now divide the other attachments of the spleen: phrenicosplenic ligament and lienorenal ligaments. Identify and divide the short gastric arteries and remove the spleen.

Emergency procedure

The assistant retracts the left side of the abdominal wall and the left hand of the operator is passed over the spleen which is then drawn forwards and the peritoneum lateral to and above it divided. The splenic pedicle is then identified and the artery and the vein ligated and divided in turn. The procedure is then carried out as for an elective splenectomy.

Closure

The wound is closed in layers with a suction drain brought out through a separate stab wound and interrupted silk or nylon to skin.

Complications

- Haemorrhage
- Acute pancreatitis from damage to the tail of the pancreas
- Subphrenic abscess
- Overwhelming postsplenectomy infection

Salvaging the spleen

Overwhelming postsplenectomy infection is a major complication and may cause death in around 5 per cent of splenectomized patients. Some patients (25 per cent) may be managed conservatively if they are stable with no other intra abdominal injury and if the extent of splenic damage is minimal on CT scan, radionuclide scanning, or laparoscopy.

At laparotomy splenic artery ligation is valuable for preservation especially for multiple tears of the hilar region. Partial splenectomy may also be effected by ligation of individual branches of the splenic artery. Other methods include ligation of the short gastrics to control bleeding, the insertion of mattress sutures, staples, or the application of collagen-impregnated packs.

If the spleen has to be removed, autotransplantation of sliced spleen into the omentum may provide protection. Otherwise long-term antibiotics or vaccination are necessary.

Reference

Buyukunal, C., Danismend, D., and Yeker, D. (1987). Spleen saving procedures in paediatric splenic trauma. *British Journal of Surgery*, **74**, 350–2.

Appendicectomy

The clinical presentation varies according to the anatomical position. Classically, tenderness is maximal over McBurney's point.

Preoperative preparation

Give the patient one suppository of metronidazole per rectum when the diagnosis is made. If there is peritonitis, antibiotics will be required. Nasogastric intubation, antibiotics, and IV fluids may be indicated if there is severe peritonitis.

Anaesthetic

General anaesthetic.

Position of patient

Supine.

Incision

Lanz or right grid-iron centred on McBurney's point. The muscles are split in the line of their fibres. The incision may be extended upwards and laterally or downwards and medially to gain access to a retrocaecal or pelvic appendix respectively (muscle cutting if necessary).

Procedure

Identify the peritoneum and open it between artery forceps. Extend the opening to gain access. Insert the index finger. Identify the caecum by grasping a taenium and deliver it. The appendix is at its base. Alternatively the appendix may be delivered directly and a Babcock's applied to its tip. Secure the appendiceal artery with forceps. Clamp, divide, and ligate it. The appendix mesentery may then be divided. Clamp the base of the appendix with two artery forceps and divide it between them. Place the appendix and the instruments in a dirty dish, taking care at division and after not to soil the wound or peritoneum. Ligate the appendix stump with vicryl and bury it with a purse-string suture.

Closure

Close the peritoneum with continuous vicryl. Then close the muscle layers with loosely inserted vicryl sutures. Insert interrupted nylon or subcuticular prolene to skin.

Special postoperative care

Encourage early oral fluids. If there has been peritonitis proceed cautiously, introducing fluids orally only when bowel activity returns.

Complications

- Wound infection
- Pelvic or abdominal abscess formation
- Paralytic ileus. Treat with IV fluids and nasogastric suction.
- Incisional herniation (rare)

Discharge

3–5 days postoperation.

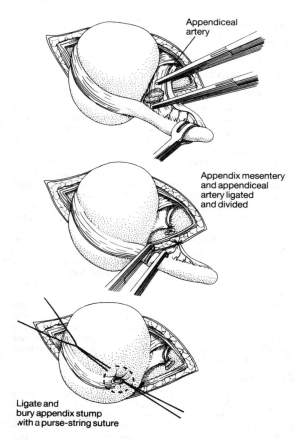

Appendiceal
artery

Appendix mesentery
and appendiceal
artery ligated
and divided

Ligate and
bury appendix stump
with a purse-string suture

Appendicectomy

Right hemicolectomy

This procedure is most often performed for malignant disease affecting the caecum, ascending colon, and hepatic flexure.

Preoperative preparation

For elective procedures the patient should have a full bowel preparation carried out. Prophylactic broad-spectrum antibiotics are given on induction of anaesthesia and at 8 and 16 hours postoperatively.

Anaesthetic

General anaesthetic.

Position of patient

Supine.

Incision

Midline or transverse.

Procedure

Insert a self-retaining rectractor and carry out a laparotomy. Is the tumour resectable? Does the liver contain metastases? Use large packs to pack the small bowel to the medial side of the abdomen away from the right hemicolon. Incise the lateral peritoneal attachment of the right colon and mobilize it from the posterior abdominal wall. Identify and preserve the spermatic or ovarian vessels, the ureter, and the second part of the duodenum. Mobilize the hepatic flexure by dividing the greater omentum from the right extremity of the transverse colon. If the omentum is involved with tumour it should also be removed. Avoid handling the carcinoma itself so as not to disseminate the cancer cells.

When the right hemicolon is mobilized note the ileocolic and right colic vessels. Ligate and divide then near their origins. The branches of the middle colic can now be identified and preserved as required. Now divide the mesocolon from its side on the distal ileum (to be used for the anastomosis) to the mid transverse colon. Ensure that there is at least 5–10 cm clearance on either side of the tumour and divide the transverse colon and the terminal ileum between clamps. The transverse colon may now be closed using loose over-and-over vicryl sutures with the clamp as a retractor. Remove the clamp and pull the suture line tight. Bury the first suture line by inserting a second line of linen sutures picking up serosa only.

Gastrointestinal continuity

This is established by an ileotransverse end-to-end or end-to-side anastomosis (less conventional) in two layers of absorbable or non-absorbable sutures. Control of leakage of the gastrointestinal contents is maintained by using non-crushing intestinal clamps. Bring a suction drain out through a separate stab wound in the abdominal wall and lay it near the anastomosis.

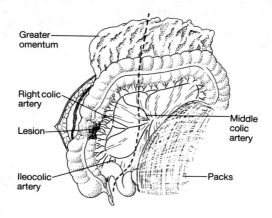

End-to-End anastomosis

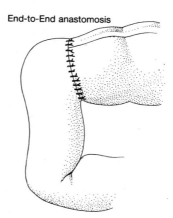

Right hemicolectomy

Right hemicolectomy (*cont.*)

Closure

Close the abdomen in layers or by mass closure with interrupted nylon to skin.

Postoperative complications

As for other major abdominal surgery, the most important disaster is leakage from the anastomosis. This may occur due to poor surgical technique or as a result of intestinal ischaemia or poor general perfusion. It should be suspected when there is abdominal pain and distension with a rise in temperature accompanied by guarding or rebound. Early re-operation is the best management. Those patients with peritonitis need drainage, peritoneal lavage with noxythiolin, and broad-spectrum antibiotics.

Transverse colectomy

This operation is indicated for resection of malignant disease of the transverse colon. Benign lesions may also be resected. If resection is for malignant disease the greater omentum, transverse colon, transverse mesocolon are included in the resection specimen.

Preoperative preparation

As for right hemicolectomy.

Incision

Upper midline.

Procedure

Detach the stomach from the transverse colon by serial division of the gastrocolic omentum. Identify and ligate the origin of the middle colic artery and its vein. Choose the sites for division of bowel and divide the mesocolon in a V-shape up to and including these sites. Apply occlusion clamps to the portions of colon for anastomosis and crushing clamps to the transverse colon which is to be resected. Divide the bowel between these clamps. Resect the transverse colon, greater omentum, and mesocolon. Anastomose the ascending colon to the descending colon with two layers of sutures. It may be necessary to mobilize the ascending or descending colon by dividiing the peritoneum lateral to them in order to establish an anastomosis without tension.

Closure

Bring a suction drain out through a separate stab wound. Close the wound with continuous vicryl or PDS and nylon to skin.

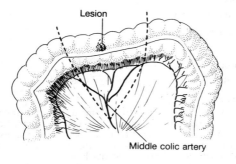

Lesion

Middle colic artery

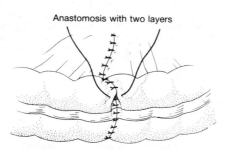

Anastomosis with two layers

Transverse colectomy

Left hemicolectomy

This procedure is indicated for malignant disease of the left hemicolon. It may also be performed in benign disease such as severe and extensive diverticulitis.

Preoperative preparation

Standard bowel preparation. Ensure preoperative catheterization of the patient and that the table can be tilted into the Trendelenburg position if necessary. If the lesion is low in the sigmoid colon or rectum, place the patient in the lithotomy position.

Special consent

All patients undergoing large-bowel surgery should be advised whether or not a stoma will be necessary and counselling provided.

Anaesthetic

General anaesthetic.

Position of patient

Supine.

Incision

Lower midline extended proximally if required.

Procedure

Mobilize the left hemicolon by dividing the peritoneum lateral to it and extending the mobilization around the splenic flexure. This is often best mobilized by first mobilizing the transverse and descending colons drawing them downwards into the wound and dividing the phrenicocolic ligament.

The extent of resection depends on the size and site of the tumour with the possible involvement of regional lymph nodes. If the procedure is a radical left hemicolectomy then anastomosis is established between the transverse colon (which must then be mobilized) and the rectosigmoid colon. Less extensive resections (segmental resections) involve resection of the affected portion of colon with its mesentery and affected lymph nodes with end-to-end anastomosis of the descending colon to the rectosigmoid junction. Such lesser procedures are carried out for small tumours or in the elderly and unfit patient and the inferior mesenteric artery can be left intact, the vessels being ligated and divided close to the bowel wall.

Radical left hemicolectomy involves ligation of the inferior mesenteric pedicle, sigmoid vessels, and middle colic vessels at their left extremity.

Closure

The wound is closed in layers, a suction drain being brought out through a separate stab wound and interrupted nylon sutures are applied to the skin.

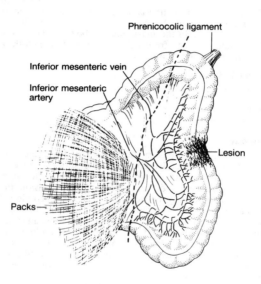

Left hemicolectomy

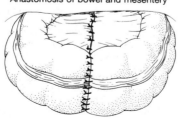

Anastomosis of bowel and mesentery

Left hemicolectomy

Anterior resection

This procedure is carried out for malignant disease affecting the rectum. Intestinal continuity is re-established by end-to-end anastomosis either by hand-sewn anastomosis or by use of a stapling gun.

Preoperative preparation

Standard large-bowel preparation should be carried out.

Position on table

The patient should lie in the lithotomy Trendelenburg position, for if an anterior resection is not possible then an abdomino-perineal resection may be carried out. The use of the lithotomy position also permits the anastomosis to be carried out with the EEA stapling device from below or permits a pull-through technique to be carried out.

Anaesthetic

General anaesthetic.

Incision

Lower midline.

Procedure

A careful laparotomy is carried out and the presence of any metastases in the liver, lymph nodes, or peritoneum noted. The sigmoid colon is now mobilized by dividing the sigmoid mesentery on each side and extending the incision in the peritoneum anteriorly between the bladder and colon. The rectum is now fully mobilized as for abdominoperineal resection, and right-angled occlusion and crushing clamps applied immediately distal to the palpable margins of the tumour allowing at least 2 cm of clearance. The rectum is divided at this level. The proximal colon is now divided between a crushing and an occlusion clamp and the resection specimen removed (the vascular supply to the resected specimen is divided immediately prior to resection once the feasibility of the procedure has been established). The proximal colonic stump is cleaned using swabs soaked in chlorhexidine solution, as is the distal stump, and an assistant carries out rectal irrigation with Betadine from below. Gastrointestinal continuity is re-established with either one- or two-layer anastomosis or by use of a mechanical stapling device. When a two-layer anastomosis is used stay sutures are placed on each side of the rectum and interrupted 3/0 silk sutures placed on the posterior wall of the anastomosis and left untied. The proximal colon is now parachuted down and the sutures tied. The posterior wall is oversewn with continuous all-layer sutures which is continued around the anterior wall. The anastomosis is completed by inserting several anterior wall interrupted vicryl sutures.

Single-layer anastomosis

This is carried out using horizontal mattress sutures. The posterior wall is completed first and the knots are tied on the inside. When the posterior wall is completed the anterior wall anastomosis is carried out also using interrupted mattress sutures.

Completion of the anastomosis with the stapling gun

Most surgeons use a stapling gun (EEA) passed from below by an assistant. The bowel ends are closed over the open instrument with purse-string sutures. The gun is then fired. A circular knife within the instrument removes two 'doughnuts' of bowel which should be inspected for completeness after the gun has been gently removed.

Closure

A suction drain is brought out through a separate stab wound. The gap between the mesocolon and parietal peritoneum is closed to prevent herniation. The abdominal wound is closed in layers with interrupted nylon to skin.

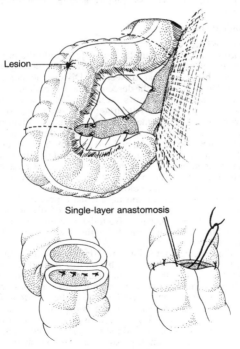

Lesion

Single-layer anastomosis

Anterior resection

Excision of the rectum

This procedure is indicated for rectal carcinomas which are too close to the anal margin to permit adequate clearance by the anterior resection technique. It is also carried out during panproctocolectomy for ulcerative colitis.

Preoperative preparation

Full bowel preparation. Biochemical and haematological parameters.

Anaesthetic

General anaesthetic.

Position of the patient

Modified lithotomy position with the thighs flexed at about 45° and the sacrum raised upon a sandbag to allow exposure for the perineal operator. The table should have a slight head-down tilt.

Operation

The procedure involves two phases, abdominal and perineal dissection.

Abdominal dissection

Incision Long midline.

Procedure Carry out a full laparotomy. Identify the presence and extent of metastases. If the tumour is deemed operable, complete the abdominal dissection. Divide the congenital peritoneal folds lateral to the descending and sigmoid colon with light curved dissecting scissors. At the level of the left common iliac vessels expose the left ureter and sweep it from the posterior aspect of the mesocolon to protect it. Identify the vessels to the pelvic mesocolon and select the site for their division which should be carried out early in the procedure. If possible divide the inferior mesenteric artery at its origin and the inferior mesenteric vein at the inferior border of the pancreas. If there is doubt about the viability of the colon, however, divide these vessels nearer to the colonic wall.

Divide the peritoneum and the medial aspect of the mesocolon from sacral promontory to the site of division of the vessels. Elevate the mesentery from the anterior surface of the aorta with the index finger and divide the vessels between heavy ligatures (0/0).

Lift the rectosigmoid colon upwards away from the sacrum, insert scissors downwards and backwards behind the mesorectum, and identify the presacral plane or cleavage. Sweep the hand from side to side to free the rectum completely posteriorly.

Divide the peritoneum on either side of the rectum as far as the peritoneal reflection to join the two peritoneal incisions anteriorly and deepen this incision by dividing the fascia of Denonvilliers in a transverse direction. There is now a distinct plane of cleavage extending downwards to the apex of the prostate. Develop this with the fingers. Displace the rectum

758

from one side to the other to render the lateral ligaments taut and divide them on each side with scissors as far downwards as possible. Ligate the middle haemorrhoidal vessels if necessary. Now identify the site of division of the colon proximally, carefully preserving its marginal vessels and dividing it between De Martell clamps so that 5 cm of viable bowel can project through the abdominal wall.

Perineal dissection Insert two heavy (0/0, 1/0) black silk sutures around the anus and tighten and tie them in purse-string fashion. Apply an artery forceps to the ends to permit gentle traction.

Incision Elliptical longitudinal, 5 cm long between the anus and the bulb of the urethra. Extend this posteriorly on each side to meet at the tip of the coccyx.

Procedure Deepen the incision posteriorly to expose the coccyx. Make small lateral incisions on either side, dividing the coccygeus muscle. Ligate any vessels. Insert a self-retaining perineal rectractor. Apply traction anteriorly to the rectum to identify the fascia of Waldeyer posteriorly. Divide this in semi-circular fashion to expose the mesorectum. Insert the fingers to separate mesorectum from the hollow of the sacrum as far upwards as the promontory. Free all attachments of the rectum posteriorly to meet the abdominal operator in the same plane.

Anteriorly divide the subcutaneous fat to expose the superficial transverse perineal muscles and then the deep transverse perineal muscles. Divide the external sphincter behind them and deepen the plane. Divide the pubococcygeus muscles on each side. Split rectourethralis in the midline, puborectalis laterally as well as the lower parts of each lateral ligament.

The rectum is now fully mobilized and should be passed downwards and removed by the perineal operator.

Closure of the perineum

Secure haemostasis of the deep layers with interrupted absorbable sutures (vicryl) and subcuticular prolene to skin. Bring a suction drain out through a separate stab wound.

Closure of the abdomen

Fashion a left iliac colostomy but complete and open it only when the wound is closed and sealed by Opsite or other dressing. Suture the pelvic peritoneum with continuous catgut. Close the wound in layers with interrupted nylon to skin. Apply the wound dressing. When the colostomy is completed apply a colostomy bag.

Types of colostomy

End colostomy

This is established following abdominoperineal resection. The proposed site is first marked in the left iliac fossa with indelible ink by the surgeon or stoma therapist on the day before surgery. Cardinal rules are that the colostomy must be under no tension, the left lateral space must be closed to prevent internal herniation, and the viability of the colon must be established by visible arterial flow. Stitch the colostomy to the skin at the time of surgery to prevent postoperative stricture.

Procedure Apply Babcock's tissue forceps to the colostomy site and exert light traction. Excise a circular disc of skin, deepen the incision through the muscle layers, and open the peritoneum. Now draw the mobilized and clamped colon through the colostomy site. Close the left gutter with interrupted catgut and close by abdominal wound. Apply a wound dressing. Apply Babcock's tissue forceps to the sides of the colostomy. Excise its distal portion just below the clamp. Ensure adequate arterial flow. Suture the edges of the colostomy to the skin edges with interrupted dexon. Apply a colostomy bag at the end of the procedure.

Transverse colostomy

This may be carried out as a decompressive colostomy for intestinal obstruction or as a protective colostomy for a more distal anastomosis, but it should be avoided if possible as management is difficult.

Incision Vertical, 4 cm lateral to the midline through the right rectus muscle.

Procedure Open the peritoneum, identify the transverse colon, encircle it with the fingers and draw a segment of colon through the wound as far to the right as possible through the greater omentum as required by serial application of arterial clamps, division, and ligation. Hold the segment of colon to the light so that the mesocolic vessels can be seen and choose as avascular a site as possible. Make a hole through this close to the bowel wall with large artery forceps. Pass a glass rod or plastic colostomy bridge (Hollister) through this window. Draw the colon out of the wound and gently suture the serosal surfaces of the afferent and efferent limbs together for a distance of about 3–4 cm. Allow the colon, except for the glass rod or bridge, to fall back into the abdomen. Close the wound in layers above and below the colostomy loop. Apply a dressing.

Opening the colostomy If the bowel is obstructed, open the colostomy at the time of surgery and apply a colostomy bag. If there is no obstruction you may wait for 24–48 hours before opening the colostomy on the ward with the diathermy by making a small cross-shaped incision at its most prominent surface. Many patients do not like this procedure, so colostomies are probably best opened in theatre.

Caecostomy

This is performed to protect a distal anastomosis or to decompress the bowel in certain obstructive situations.

Incision Grid-iron in the right iliac fossa.

Procedure Bring out the caecum as for appendicectomy. Apply a catgut purse-string suture to the serosa. Suture the lateral margin of caecum to the peritoneal stab wound before caecotomy. Make a small stab wound in the caecum and pass a large-bore, self-retaining catheter into it. Tighten the purse-string suture and apply a further purse-string to ensure complete envagination. Blow up the balloon of the catheter and bring it out through a separate stab wound. Suture the margins of the caecostomy to the peritoneal stab wound. Close the wound in layers leaving the catheter in place.

The base of the appendix may also be used after appendicectomy to perform decompression. The caecostomy tube should be irrigated at intervals to prevent blockage and may also be used for lavage.

Intestinal resection and anastomosis

Learn this technique under supervision until you become competent. It is indicated in congenital lesions, trauma, inflammatory disease, mesenteric infarction, the resection of infarcted bowel as a result of herniation or intussusception, the resection of some fistulas, benign and malignant tumours.

Ideals for successful anastomosis

- Exposure must be adequate.
- The bowel should be as empty as possible.
- There should be good and even approximation of the two ends.
- There should be no tension.
- There should be no peritoneal soiling.
- Defects in the mesentery must be closed to prevent internal herniation.
- There should be no complicating factors, e.g. poor blood supply to cut ends, anaemia, malnutrition, radiotherapy.

Procedure

Select the section of bowel to be resected and gently lift it out of the wound. Hold the bowel to the light so that the vascular pattern can be seen, and choose the line for section. If the resection is for benign disease the vessels may be divided and ligated close to the bowel. If for malignant disease, a generous V-shaped portion of mesentery should be resected with the affected segment. Divide the near side layer of mesentery with scissors in a V-shape and gently ligate and divide the vessels between artery forceps. Do this up to the bowel wall. Apply light crushing clamps (Kocher's, Stevenson's) from the anti-mesenteric border and angle them to include more of the anti-mesenteric border than the mesenteric border. Apply a second pair of crushing clamps at the same angle parallel to these and again at the same angle apply soft occlusion clamps about 8 cm off the double clamps after first emptying the bowel by milking the contents in either direction. Now divide between the crushing clamps and remove the resection specimen. In selected patients resection may be carried out without clamps.

End-to-end anastomosis

Ask your assistant to hold the crushing clamps so as to bring the two ends of the bowel together. Now gently turn the clamps back on themselves and join the adjacent walls of the small bowel with a seromuscular Lembert suture. When this has been completed cut across the bowel beneath the crushing clamps and remove them. Now join the bowel together with an all-coats over-and-over suture using 2/0 chromic catgut or vicryl. The stitches should be less than 0.5 cm apart picking up about 0.5 cm of bowel wall. Continue this layer round the corner and on to the anterior wall and tie it at the other corner. Remove the non-crushing clamps and complete the seromuscular Lembert stitch on the other wall, tie it when completed. Now place interrupted

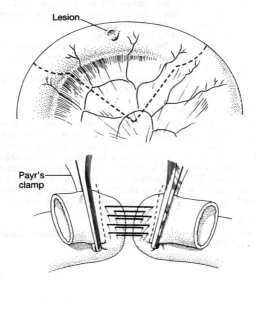

Anastomosis in two continuous layers

Intestinal resection and anastomosis

Intestinal resection and anastamosis
(*cont.*)

sutures in the mesenteric defect to close it, taking care not to pick up mesenteric vessels. The use of a double-ended suture makes end-to-end anastomosis even easier.

N.B. To turn the corner pass the needle from mucosa outwards on one corner to serosa inwards on the other followed by mucosa outwards on the same side to serosa inwards on the other, thus forming a series of loops on the mucosal surface (The Connell suture, which is not haemostatic at the edges). Once the corner is passed, continue with the over-and-over suture.

Disparity of the lumina

When the lumina of the bowel are of different sizes the anastomosis may be effected by dividing longitudinally the anti-mesenteric border of the bowel with the smaller lumen. Anastomosis is then carried out as above.

This procedure, the *Cheatlke split*, is commonly indicated in cases of obstruction where the bowel distal to the obstructive lesion is collapsed.

Linear and side-to-side stapling instruments can also be used for anastomosis.

Colonic anastomosis

Blood supply to the cut ends of the bowel is critical in colonic surgery. Blood flow measurements in the colon indicate that if flow is reduced the mesenteric side is preferentially affected. A simple method of lessening the risk of leakage is to rotate one of the free edges through 90° so that the two mesenteric edges do not lie together. The anastomosis is effected either by one or two layers of sutures or by using the circular stapling gun which frequently spares the patient from a permanent colostomy.

Reference

Gillespie, I. E. (1983). Intestinal anastomosis. *British Medical Journal*, **286**, 1002.

Ileostomy

The terminal ileum is exteriorized as a spout in the right iliac fossa. Effluent is collected in a collecting bag (ileostomy bag). The procedure is an option in the treatment of inflammatory bowel disease, familial adenomatous polyposis, and proximal colonic obstruction.

Loop ileostomy

This is exteriorization of a loop of terminal ileum usually as a temporary procedure to decompress obstruction or protect a distal anastomosis (e.g. following restorative proctocolectomy).

Preparation

Counsel the patient. As for colostomies, you or the stoma nurse should site the ileostomy in the right iliac fossa clear of the umbilicus and anterior superior iliac spine. The patient should wear an ileostomy bag for 24 hours before siting to determine the best spot.

Technique

Divide the terminal ileum between clamps. Remove the distal part with the colon if this is a total colectomy. If not, close the distal end.

Pick up the skin marked for the ileostomy with tissue forceps and excise a circular area. Repeat the procedure through aponeurosis and muscle. Do not divide peritoneum if the ileostomy is to be extraperitoneal. Do if it is going to traverse the peritoneal cavity.

For an extraperitoneal ileostomy make a tunnel under the peritoneum from the caecal reflexion to the anterior abdominal wall (preferred method).

For transperitoneal ileostomy draw the ileum through the opening on the anterior abdominal wall so that it protrudes for several cm. Insert a few interrupted stitches between aponeurosis and ileum to anchor it. Close the lateral peritoneal space to prevent herniation then close the main abdominal wound. Now trim the bowel ends if required and gently evert the edges with Babcocks forceps to draw them down to skin level thus forming a spout which should be at least 5 cm from the skin to permit close application of the ileostomy bag with discharge of effluent well away from the skin. Stitch the mucosa to the skin with interrupted catgut sutures.

Procedures for dealing with colonic obstructive lesions

There are several procedures less extensive than abdomino-perineal resection which can be used to deal with cancers of the colon, pericolic infection, intussusception, or volvulus.

Hartmann's procedure

This is ideal in the frail elderly patient. Through a left lower paramedian incision the affected colon is excised with full protocol. The proximal colon is brought through the abdominal wall as a colostomy. The distal rectal stump is either oversewn or covered with peritoneum. The pelvic cavity may be drained through the anus.

Exteriorization resection method (Paul Mikulicz procedure)

This can be used for gangrenous small bowel, colonic cancers with obstruction and proximal distension, volvulus, or localized diverticular infection. The affected segment of bowel is exteriorized at laparotomy then the abdominal wound is closed. The bowel is then resected between Kocher clamps after anchoring the two loops to skin. This procedure reduces the risk of contamination of the peritoneal cavity and allows decompression of the bowel. The spur of tissue between the double-barrelled colostomy can be crushed with an enterotome and the colostomy closed after 10 days of the procedure, although it is more common to wait for 6–12 weeks then close it formally.

Palliative/decompressive colostomy

This should be done only if senior help is available. If primary resection is not possible at the time of surgery a transverse colostomy should be fashioned in the right upper quadrant. This permits decompression of the bowel and the formal resection can be carried out later. The colostomy is closed at a third operation (the three-stage procedure). Equally if disease is advanced and the patient obstructed a palliative transverse or left iliac colostomy will relieve obstruction.

N.B. The blood supply is poor to the left sigmoid colon, making later closure risky.

Local procedures

Polyps and small tumours can be resected transanally using a resectoscope similar to that used for transurethral resection of the prostate. The instrument makes use of a cutting diathermy loop, Frankenfeldt snares for polypoidal lesions, and electro-coagulation.

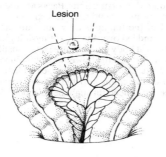

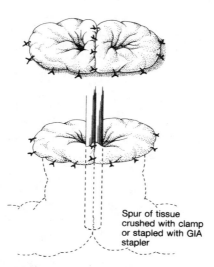

Procedures for dealing with colonic obstructive lesions (exteriorization resection method—Mikulicz procedure)

Procedures for dealing with colonic obstructive lesions (*cont.*)

'On-table' bowel lavage

Following removal of an obstructive lesion the proximal bowel should be washed out before fashioning the anastomosis. A large Foley catheter is passed into the caecum from the distal ileum and held in place with a purse-string suture. The proximal bowel is mobilized to bring it out of the wound, a length of anaesthetic scavenging tube inserted into it and secured with nylon tapes tied onto the corrugations. Packs are then applied to the abdominal cavity and wound to collect spillage.

The distal end of the tubing is now tied into the neck of a double plastic bag and led over the table. Physiological saline or Hartmann's solution at body temperature is then instilled, faeces being broken down through the bowel wall with finger pressure. Washouts are continued until the fluid appears clear on entry to the bag. Suction of the colon into the tube is prevented by a large-bore needle introduced through the colonic wall to act as a vent.

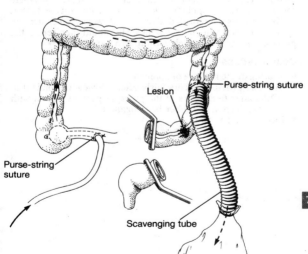

'On-table' lavage and colectomy

Lord's procedure for haemorrhoids

This is suitable for some patients with haemorrhoids. Do not carry out the procedure in those with laxity of the anal sphincter or a history of incontinence.

Anaesthetic
General anaesthetic.

Position of patient
Sim's (left lateral decubitus).

Preoperative assessment
Perform a rigid sigmoidoscopy to exclude proximal disease.

Procedure
Insert the index fingers of each hand. Feel for and break down any fibrous constrictions. Gently dilate the anus and lower rectum exerting the force laterally and not anteroposteriorly (to avoid sphincter damage). Proceed until 6–8 fingers can be inserted.
Do too little rather than too much.

Postoperative care
Relieve postoperative pain. Give stool softeners and an anal dilator. Teach the patient how to insert this using K-Y jelly. He should feel the anal sphincter slot into the groove on the dilator.
The process is repeated daily for 2 weeks and then on alternate days for the next 2 weeks. He should then repeat the process once or twice weekly for up to 6 months.

Complications
- Tearing of the mucosa or sphincter
- Mucosal prolapse occurs fairly often. Review the patient at SOPD after 6–8 weeks. If the prolapse persists treat it with phenol injection or rubber-band ligation.
- Incontinence is rare

Haemorrhoidectomy

Indications

- Patients whose haemorrhoids do not resolve with conservative measures
- Patients with large prolapsing haemorrhoids associated with skin tags
- Prolapsed thrombosed haemorrhoids
  All are increasingly rare.

Preparation

Ensure the bowel has been emptied preoperatively. Give Picolax 1 sachet to be taken orally at 6 p.m. on the two days before surgery.

Position of the patient

Lithotomy position.

Preoperative assessment

Identify the haemorrhoids which usually lie at the 3, 7, and 11 o'clock positions. Carry out rigid sigmoidoscopy and barium enema, especially in older patients, to exclude proximal disease.

Procedure

Start with the 7 o'clock haemorrhoid. Apply tissue forceps to it. Hold the forceps in the palm of the right hand whilst inserting the index finger into the anus to place traction on the haemorrhoid with scissors in the left hand. Make a semicircular incision in the skin at the lower end of the haemorrhoid. Using a gauze swab gently push the skin lateral to this incision away so that the circumferential fibres of the anal sphincter can be identified. Do not damage this muscle. Continue the dissection in the submucosa for about 2 cm above the lower edge of the muscle. The haemorrhoid is now well mobilized and should be transfixed and ligated at its base with vicryl.

Repeat the process at each haemorrhoidal mass. Sometimes only two masses need to be excised, sometimes four. Remember to leave a bridge of anal skin between each mass to allow regeneration and prevent stricture. Apply paraffin gauze dressing, dressing swabs, and T-bandage.

Postoperative care

- Relieve pain.
- Observe for haemorrhage.
- Give faecal softeners and encourage daily bathing.
- Allow home after the bowels have moved, usually on the third or fourth day.
- Arrange for follow-up examination in 6–8 weeks.

Complications
- Posthaemorrhoidectomy bleeding
- Faecal impaction
- Urinary retention
- Fissure and stricture formation
- Incontinence

Rubber-band ligation of haemorrhoids (Barron's ligation)

This is an outpatient procedure which can be repeated if required. It is an effective treatment for haemorrhoids which can be combined with injections and may avoid the need for surgery.

Position of the patient

Left lateral (Sim's).

Procedure

Pass a sigmoidoscope first to exclude proximal disease. Note the position of the piles. The banding gun is loaded by placing its cone on the banding ring. The band is then pushed up the cone until it lies on the groove near the edge of the ring. The cone is then removed.

A proctoscope is then passed into the anus well above the anorectal junction. The tip is directed upwards and forwards towards the base of the right anterior pile. At this stage the assistant or patient holds it steady. Pile grasping forceps (held in the left hand) are then passed through the ring of the banding machine (held in the right hand). The base of the pile is then grasped and pulled down into the ring. If this hurts the patient do not release the band for you are too low. Push the proctoscope in further and repeat the procedure. Once satisfied, shoot the ring over the base of the pile and repeat the procedure for other piles.

Band two haemorrhoids at each session until treatment is complete.

Postoperative care

Warn the patient that there may be minor bleeding for several days. Provide mild analgesia if needed and also a mild aperient.

Review

Often no review is necessary but a 6-week assessment at the clinic permits further examination and banding if there are still symptoms.

Ingrowing toenail

This procedure can be performed under local or general anaesthesia. If there is cellulitis, give the patient perioperative broad-spectrum antibiotics.

Procedures

Simple nail avulsion This is indicated for the relief of suppuration, but regrowth and recurrence of the problem is common.

Wedge excision Lateral or medial nail and nailbed are removed with granulation tissue and a wedge of the nail fold. The tissue is excised down to periosteum of the lateral or medial side of the phalanx distal to the joint.

Zadik's procedure (complete excision of the nailbed) The nail is avulsed, incisions are then made on either side of the skin covering the matrix of the nail. A flap is thus fashioned and drawn back and the germinal matrix so exposed is completely excised ensuring adequate lateral and medial dissection. The skin flaps are loosely sutured with interrupted nylon sutures.

Complications

- *Recurrence*: Common after simple avulsion. Uncommon if adequate lateral and medial dissection of the germinal matrix is carried out. Nail spikes are difficult to manage.
- *Wound infection*: This responds to antiseptic dressings rather than antibiotics unless there is cellulitis.
- *Osteomyelitis and septic arthritis* can also occur occasionally and need treatment with antibiotics.

Inguinal hernia

Indications

- *Elective*: all symptomatic hernias, especially if indirect, need operation
- *Emergency*: irreducible or strangulated hernias

Preoperative preparation

Check for cardiorespiratory disease and renal function. In emergency situations, associated with intestinal obstruction, intravenous fluids, antibiotic prophylaxis and nasogastric suction are required.

Anaesthetic

General anaesthetic.

Position of patient

Supine.

Incision

Curved inguinal incision in skin crease approximately 2 cm above the medial two-thirds of the inguinal ligament. Incise fat and fascia ligating the two or three veins which cross the line of the incision.

Procedure

Identify the external ring. Divide the external oblique aponeurosis along the line of its fibres over the inguinal canal. Apply artery forceps to the edges. Turn back the edges and identify the internal oblique and conjoint tendon above the cord and the inguinal ligament below. Steps now include identification and clarification of the sac after picking up the cord at its medial end, splitting its investing fibres in the line of the cord, and dissection of the sac to its origin. Excise the sac after transfixing its base in indirect hernias except when a sliding hernia is suspected. Repair the transversalis fascia margins of the internal ring and repair the posterior wall of the inguinal canal with non-absorbable sutures.

In direct hernias push the sac inwards and repair the posterior wall. If the sac is very large it may be necessary to divide the sac near its origin and leave the distal portion open. In sliding hernias reduce the sac and viscus into the abdomen and repair the posterior wall.

Repair

Use non-absorbable sutures in darn fashion. Bassini, Halsted, and Shouldice are common methods of repair.

In cases of recurrence the testis may be removed in elderly men to achieve repair.

Closure

Approximate the external oblique and close the wound in layers with clips or adhesive strips to skin.

Complications

- Urinary retention
- Haematoma
- Infection
- Exacerbation of prostatic symptoms
- Recurrence

Bassini repair

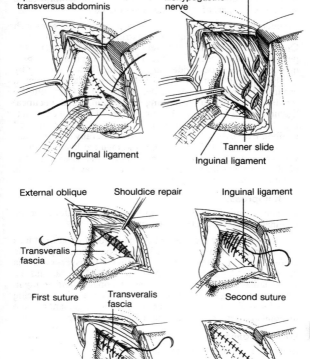

Aponeurosis of transversus abdominis

Iliohypogastric nerve

Rectus muscle

Inguinal ligament

Tanner slide

Inguinal ligament

External oblique

Shouldice repair

Inguinal ligament

Transveralis fascia

First suture

Second suture

Transveralis fascia

Third suture

Inguinal ligament

Closure of external oblique aponeurosis

Repair of femoral hernia

There are several approaches to the repair of femoral hernia which are all safe, but the operator must witness all of them and decide which procedure suits his needs most.

Anaesthetic

General anaesthetic.

Position of patient

Supine.

McEvedy approach

Incision Oblique vertical or transverse over the femoral canal with the lower part of the incision over the sac.

Procedure First isolate the sac by a preperitoneal approach through the external oblique aponeurosis and conjoint tendon. Open the transversalis fascia and dissect between it and the peritoneum to the neck of the sac. The inferior epigastric vessels may now have to be divided in this plane. Reduce the sac by manipulating it from above and below then isolate it, open it, ensure it is empty, transfix, ligate, and divide it. Carry out the repair of the hernia by uniting the inguinal and pectineal ligaments for a distance of 0.5–1 cm laterally without constricting the femoral vein. Use monofilament nylon, or braided Ethibond or Ethiflex. Elevate the vein laterally with the left index finger, insert the stitch into the inguinal ligament, have your assistant retract the ligament upwards, and take a large bite of the pectineal ligament. A single stitch may be all that is required; if not, insert a further one.

High approach

Incision Inguinal incision above the medial two-thirds of the inguinal ligament and about 1 cm above it.

Procedure Identify the external oblique aponeurosis. Split it as for the approach to an inguinal hernia repair then displace the cord or round ligament upwards. Incise the transversalis fascia in line with the incision to identify the neck of the sac and the external iliac vein. Gently isolate the sac, open it, and empty it. Then transfix, ligate, and divide it. Effect the repair by placing non-absorbable sutures between the pectineal ligament and the inguinal ligament. Do not constrict the femoral vein.

Low approach

(not recommended for strangulation)

Incision 6–12 cm long in the line of the groin crease centred between the anterior superior iliac spine and the symphysis pubis. Deepen the incision to identify the hernial sac. Ligate veins as necessary. Clean the sac, open it, empty its contents, transfix, ligate, and divide it and carry out the repair as above.

Closure

The skin wounds are closed in layers with interrupted nylon to skin.

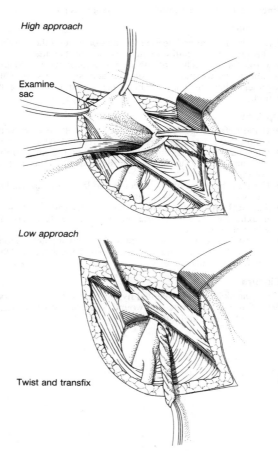

High approach

Examine sac

Low approach

Twist and transfix

Repair of a femoral hernia

Repair of incisional hernia

Aim

To close the abdominal wall defect and prevent recurrence by side-to-side suture of healthy aponeurosis, overlap of the defect edges or prosthetic patch or mesh interposition.

Anaesthetic

General anaesthetic.

Position of the patient

Since most hernias are midline the patient is usually supine.

Procedure

Excise the old scar and deepen the incision around the margins of the hernia until healthy aponeurosis is identified then dissect towards the free edges of the defect. Return the sac to the peritoneal cavity intact is possible and suture its margins together. 'Clean' the edges of the defect so that they are free all round.

Options

- If the defect edges come together easily suture them with continuous (mass-type closure) or interrupted nylon sutures.
- Create a two-layer closure by incising along the edges of the muscular defect. Suture the opposing sides together in two layers.
- Overlap the edges of the defect in two layers of sutures. Do this only if no tension is created.
- The 'keel' repair involves inversion of the sac with suture of its margins followed by several layers of sutures to invert the edges of the defect.
- If the defect is very large use a mesh (marlex) or patch (teflon) to close the defect. Suture the prosthesis all the way round the palpable margins of the defect.

Closure

Close the skin with interrupted nylon sutures. Bring a suction drain through a separate stab wound. When using a prosthesis give prophylactic antibiotics (p. 112).

Simple mastectomy

This operation is still commonly performed for carcinoma of the breast. The aim of the procedure is to excise and control the disease locally and establish a cosmetic scar.

Preoperative assessment

Ideally the diagnosis should be made preoperatively by fine needle aspiration, Trucut biopsy, drill biopsy, or suspected on mammography, and the woman should be informed what to expect. If this is not possible, then frozen section biopsy with subsequent mastectomy if the lump is malignant is performed (becoming increasingly rare). Determine whether the patient would wish a postmastectomy reconstruction.

Anaesthetic

General anaesthetic.

Position of patient

Supine.

Incision

Use a marking pen to indicate the site of the skin flaps. These should be above and below the palpable margins of the tumour. Make the incisions (2) above and below the tumour to form a transversely lying (if possible) ellipse. If it is not possible to make a transverse ellipse, fashion it obliquely.

Procedure

Instruct your assistant to control the upper flap with an abdominal pack and deepen the flap to the upper limit of the breast ensuring that no 'button-holing' of the skin occurs. If there is bleeding control this by diathermy or ligation as you go. Beware of burning the skin.

Repeat the technique with the lower flap.

Return to the medial part of the upper flap and dissect down until you see the pectoral fascia. Insert a finger or gently dissect laterally with the knife until the plane between the breast and pectoralis major is found. Using a combination of traction on the breast and blunt and sharp dissection (with the scalpel blade held obliquely) dissect the breast laterally and downwards until it and the axillary tail are removed. Blood vessels perforating the fascia should be controlled as you proceed.

Closure

Ensure that haemostasis is complete. Bring a suction drain through a separate stab wound and arrange it so that the upper and lower flaps are drained. This may involve introducing the drain from the outside through the inframammary fold and bringing it out again via a separate stab. The loop so formed is then divided to from two drains.

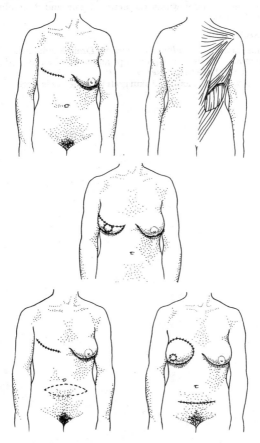

Methods of reconstruction of the breast after mastectomy

Simple mastectomy (*cont.*)

Close the wound by inserting nylon or silk sutures at two to three sites along the wound to obtain accurate skin flap approximation and avoid dog ears. This done, complete the closure with interrupted sutures or cosmetically with subcuticular nylon.

Apply dressings. Swabs and chest bodice for 24 hours to prevent haematoma. Alternatively, adhesive dressings may be applied.

Postoperative care

Ensure adequate postoperative analgesia. Begin shoulder mobilizing exercises within the limits of pain and discomfort after 48 hours.

Complications

- *Haematoma formation and infection*: Drain if indicated
- *Seroma formation*: Frequent aspirations may be required. A drain may have to be inserted
- *Flap necrosis*: Excise and skin graft if the defect is large

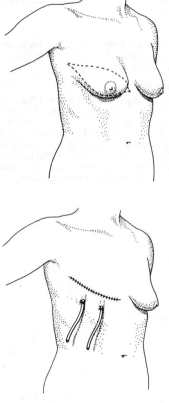

Simple mastectomy

Removal of axillary nodes

The presence of axillary lymph nodes may indicate spread of carcinoma. If there are nodes which are mobile they should be removed for histological inspection.

Procedure

The breast and axillary nodes are usually removed *en bloc* by opening the subcutaneous tissue of the axilla using gentle finger or scissor dissection (alternately opening the blades to dissect the tissue).

Divide the tendon of pectoralis minor, insert a retractor, and remove the nodes from above downwards. Send them to the laboratory indicating to the pathologist which are the apical nodes.

Do not attempt to remove fixed nodes because you may damage axillary structures.

Closure

The axilla should be drained by a suction drain brought out through the lateral edge of the wound. The mastectomy wound is then closed in the usual manner.

Patey's operation

This variation of simple mastectomy has become increasingly popular to strip the axilla of all nodes. The pectoralis minor muscle is routinely removed from its origin on the coracoid process to its insertion on the ribs. The contents of the axilla are removed from above and below the axillary vein, ligating venous tributaries. The lateral pectoral nerve and the acromiothoracic trunk are preserved but it is necessary to sacrifice the medial pectoral nerve (supplying the pectoralis minor).

Arterial anastomosis

In arterial surgery the aim of anastomosis is to contain a water-tight closure without tension on the vessels concerned whilst maintaining the normal lumen. This is best achieved by first excising the adventitia. It is vital that the intimal surface remains undamaged.

Materials

Vascular sutures of non-absorbable material are used. These are usually produced with a needle at each end. In general 3/0 sutures are used for large vessels such as the aorta, with 4/0 and 5/0 being reserved for smaller vessels. Microsurgical anastomosis may be performed with 8/0 to 10/0 sutures. Synthetic sutures such as prolene, braided polyester (Ticron), and PTFE (Gortex) are used. Taper-pointed needles permit penetration of atheromatous vessels.

Techniques

Anastomosis may be end-to-end or end-to-side.

Insertion of sutures inside to outside the lower vessel of the anastomosis avoids raising intimal flaps.

End-to-end—1 Insert one of the double-ended needles from the inside to the outside of the proximal end of the artery. Insert the other from inside to outside of the distal end. Tie the sutures. Now take each needle and suture round the vessel using an over-and-over stitch. Avoid tension and intimal damage. Tie the sutures at the front of the vessel.

End-to-End—2 Where it is possible to rotate the vessel, insert a double-ended suture inside to outside at each corner of the vessel. Tie the knots. Now complete the anterior and posterior walls with over and over sutures. Complete the anterior wall first, tie the knot, then pass the opposite suture under the vessel maintaining traction on the other to turn the vessel over. Complete the posterior wall and tie the stitch.

End-to-side This technique is useful for femoropopliteal grafting and onlay aortic grafting.

Use a double-ended suture. Tailor the two free vessel edges (or vessel and graft) to suit each other. The anastomosis may be fashioned with the vessels at right angles to each other or at an angle. Insert one needle into the apex of the arteriotomy from inside to outside. Insert the other into the graft from inside to outside. Now tie the knot.

Proceed down each side of the anastomosis passing the needles from outside (graft) to the inside of the artery. Take one suture around the lowermost part of the anastomosis and tie the knot away from the lower angle.

Variations

- The first passes of the needles can both be made from outside the graft to inside of the apex of the arteriotomy. Now without tying the knot complete the anastomosis as above. T

permits the operator to place the sutures more accurately in tricky anastomoses. The graft can be gently snuggled onto the arteriotomy at the end and the knot tied.
- Two double-ended sutures can be used one at the apex and one at the lower end. The anastomosis is completed as in the rotation technique.

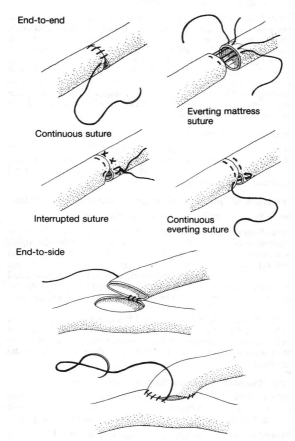

Arterial anastomosis

Access to arteries

In vascular surgical practice it is essential to be able to expose arteries for anastomosis or repair. Complete control of the vessel and any branches should be achieved before the procedure is performed.

Carotid bifurcation

The patient lies supine with the head turned so that the side to be operated on is uppermost.

Incision 2 cm below the mandible along the line of sternomastoid to 2–3 cm above its insertion.

Procedure Deepen through platysma. Incise the deep fascia along the anterior border of sternomastoid. Identify the internal jugular vein and free it from the carotid sheath. Ligate the anterior facial vein. Now feel the common carotid artery and divide the fascia overlying it with scissors. Pass tape around the artery. Proceed now to expose the external (the superior thyroid is its first branch) and internal carotid arteries by carefully dividing the fascia with scissors. Pass tapes around them. The vagus nerve is lateral to the artery. The hypoglossal and carotid sinus nerves are above the artery.

Subclavian artery

Proceed as for cervical sympathectomy (p. 798). After division of the scalenus anterior muscle (preserving the phrenic nerve) the artery is visible. Pass tapes around it and its internal mammary and thyrocervical branches.

Exposure of the abdominal aorta and iliac bifurcation

Incision Long midline, paramedian, or transverse.

Procedure Place self-retaining retractors in the wound. Lift the transverse colon out of the wound or pack it under the xiphisternum and lower thorax. Lift the small bowel out of the wound and retract it within moist warm green towels. Identify the peritoneum over the aorta and divide it over the bifurcation and upwards to the right of the inferior mesenteric artery. At the upper end displace the duodenum to the right to avoid damaging it.

To expose each common iliac, lift the overlying peritoneum with forceps and carefully divide it. Diathermize bleeding points and beware of damage to the ureters (which cross the iliac bifurcation) and iliac veins (which are often adherent to the arteries).

The common femoral artery and its bifurcation

Incision Make a 10–12 cm vertical incision two finger breadths lateral to the pubic tubercle, one third above, two thirds below the inguinal ligament. Feel for the pulsation with your fingers.

Procedure Incise fat and fascia between the muscles of the femoral triangle. Expose the artery using careful sharp dissection. Some minor veins may require ligation and division.

Access to arteries (*cont.*)

Continue sharp dissection to expose the common femoral artery, profunda femoris, and superficial femoral. Pass slings around them. The common femoral often appears to narrow at its bifurcation because a posterior muscular branch is also given off.

The popliteal artery

Above the knee

Incision Vertical, 10–15 cm long extending from the medial femoral condyle along the posterior border of vastus medialis.

Procedure Incise fat and deepen the incision through the fascia. Retract sartorius downwards to expose the neurovascular bundle below the hiatus in adductor magnus. Dissect the venae comitantes and the popliteal vein from the artery whilst preserving the geniculate vessels. Expose the popliteal artery for about 5 cm. Pass slings around it.

Below the knee

Incision 0.5–1 cm behind the medial tibial condyle extended distally parallel to the medial tibial border for 10–12 cm. Preserve the saphenous vein.

Procedure Incise the deep fascia and retract the medial head of gastrocnemius downwards. This exposes the fascia over the neurovascular bundle which can then be incised and the vessel exposed and slings passed round it.

Tibioperoneal trunk

Through the above incision the tibial vein is divided and the tibioperoneal trunk dissected by incising the soleus muscle.

Peroneal artery

Incision Vertical from the head of the fibula for 6–10 cm downwards along the fibula.

Procedure Dissect the muscle from the fibula, incise its periosteum throughout the length of the incision, and strip it from the bone. Divide the fibula above and below. Remove the segment. The peroneal artery lies in the bed of the fibula. The fascia is incised and the vessel isolated from its accompanying veins.

Anterior tibial artery

Incision Vertical, 15–20 cm long, lateral to anterior tibial border.

Procedure Divide the deep fascia. Identify the plane between tibialis anterior medially and extensor digitorum longus laterally. The vessels lie deeply and should be carefully dissected until clearly seen. Good retraction is essential.

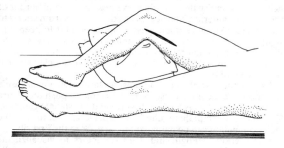

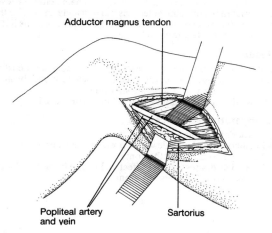

Adductor magnus tendon

Popliteal artery
and vein

Sartorius

Approach and exposure to the popliteal artery

Cervical sympathectomy

This technique is indicated for patients with refractory Raynaud's syndrome, Buerger's disease, causalgia, and hyperhidosis. It is also of value in patients with obliterative arterial disease of the upper limb where reconstruction is not possible.

Preoperative preparation

The patient is anaesthetized and lies supine on the table with feet-down tilt. The head is turned to the opposite side.

Operation

Incision 10 cm long, above the medial half or two-thirds of the clavicle.

Procedure Deepen the incision by dividing the clavicular head of the sternomastoid and the inferior belly of the omohyoid if necessary. Identify the phrenic nerve near the medial border of the scalenus anterior. The nerve runs downwards from lateral to medial side. Mobilize it and pass a tape around it to keep it in sight at all times. The scalenus anterior muscle may now be divided from the lateral to medial side, close to the 1st rib to expose the underlying subclavian artery. The subclavian vein lies in front of the muscle.

Palpate the first rib and clear Sibson's fascia. Divide the fascia to expose the pleura and use blunt dissection to push the pleura down. Identify the neck of the first rib and the stellate ganglion. The cervical sympathetic chain is palpable as a cord and the stellate ganglion can be palpated on the neck of the first rib.

Use a good spotlight and deep retractors to visualize the ganglion and chain. Pick up the chain with a hook and carefully free it as far as possible distally. Divide the chain below the level of the first rami comminicantes. Dissect it distally and excise the chain for at least two interspaces distally. Apply Ligaclips if it is necessary to divide intercostal veins.

Closure

Ensure there is no pneumothorax by asking the anaesthetist to inflate the lungs. Check for excessive bleeding and bring a suction drain through a separate stab wound.

Close the wound in layers with nylon or clips to skin.

Postoperative complications

- Horner's syndrome
- Pneumothorax (exclude with a postoperative chest X-ray)
- Neuralgia—postsympathectomy
- Excessive dryness of the affected hand. The use of handcreams may relieve this symptom.

Transaxillary sympathectomy

Position of patient

Lateral position with the arm abducted at right angles and maintained so with a sling.

Incision

Transverse/oblique in line of third rib about 15–20 cm long.

Procedure

Deepen through fat of axilla. Identify and retract the long thoracic nerve and muscle to expose the periosteum of the rib. Remove a section of rib. Incise pleura and open the chest with a rib spreading retractor. Displace the lung downwards to expose the sympathetic chain which lies under the mediastinal pleura. Expose the chain and excise the fourth ganglion (the first is not easily accessible by this route).

Closure

Haemostasis. Chest drain. Allow the lung to reinflate. Approximate the ribs. Close periosteum and muscle with continuous nylon. Subcuticular nylon or vicryl to skin.

Postoperative care

Postoperative chest X-ray. Remove the drain at 24 hours if the lung is inflated.

Complications

- Surgical emphysema
- Pneumo- or haemothorax
- Temporary (sometimes permanent) Horner's syndrome

Lumbar sympathectomy

Complete sympathectomy of the calf and foot is achieved by excising the second and third lumbar ganglia. Excision of the first, second, and third lumbar ganglia will sympathectomize the thigh, leg, and foot. The technique is used most often for limb salvage, especially in patients with rest pain, digital gangrene, and ischaemic ulceration.

Phenol block technique

(chemical blockade of the sympathetic chain)

This procedure, which is valuable in elderly patients who are anaesthetic risks, should be performed under image-intensification. The patient lies on his side with the theatre table 'broken' and the side for injection uppermost. The skin is prepared with antiseptic solution and the area between the twelfth rib and the iliac crest draped. Two sites between these landmarks are infiltrated with local anaesthetic a few cm apart at the upper edge of quadratus lumborum. After raising a skin bleb, infiltrate 5 ml deep to each. Now insert a spinal needle into one of the blebs (125 mm/18 G). Direct it towards the spinal column. When this is felt, 'march' the needle forwards by angling it slightly until it meets the vertebral body tangentially. The needle is now close to the sympathetic chain in the retroperitoneal space. Remove the stylet a little and place a drop of 6 per cent phenol in water on the open end. Now observe the drop of phenol. If it is in the correct space the phenol will drop. If it is in a vessel it will rise and the manoeuvre should be repeated. Aspirate the needle. If there is no blood inject 3–5 ml of 6 per cent phenol in water into the space stopping for a few minutes after each ml injected. Check that the patient feels all right. Repeat this process at the other site. The successful procedure will result in a warm, dry limb which is observed almost immediately.

Complications The commonest is groin and upper thigh neuralgia. Serious dangers can be avoided by careful technique. The aortic pulsation can often be felt on the right side if the needle is misplaced. Aspiration of blood on the left may indicate that the needle is in the vena cava. It should be withdrawn a little and aspirated again.

Surgical lumbar sympathectomy

This procedure can be performed transabdominally at the time of aorto-iliac surgery. It may also be performed as a separate procedure by the extraperitoneal approach.

Position of the patient The anaesthetized patient lies supine with a sandbag tucked under the side to be operated upon.

Operation

Incision Transverse from lateral rectus margin to a point about halfway between the anterior superior iliac spine and the twelfth rib.

Procedure Split the muscle fibres in grid-iron fashion or divide them if the patient is well-muscled. Identify the transversalis fascia, open it, and develop the extraperitoneal space deep to it sweeping the peritoneum forwards until you feel the definite step which is the psoas major muscle. Stay in front of the muscle. Insert two deep retractors anteriorly and superiorly, the anterior retractor protecting the ureter.

Now feel for the sympathetic chain which lies on the anterior longitudinal ligament medial to the psoas major. It can be rolled with a pledget. Pick it up with a nerve hook and remove about 3–5 cm of chain. Control bleeding by diathermy. If lumbar vessels are in the way of your dissection ligate and divide them.

On the left the chain lies lateral to the aorta. On the right it is below the vena cava. Take extreme care to identify the correct structure, for the edge of vena cava can also 'roll' to the un-initiated finger.

Closure Drain the wound with a suction drain and close it in layers with nylon to skin.

Complications
- Wound haematoma and infection
- Injury to the ureter, vena cava, or aorta
- Postsympathectomy neuralgia affecting the groin and antero-lateral aspect of the thigh. Treat with simple analgesics. Symptoms resolve spontaneously.
- Genitofemoral nerve injury
- Sexual dysfunction occurs in about 50 per cent of males who have bilateral removal of the first lumbar ganglia.

Resection of an abdominal aortic aneurysm

This procedure is best carried out electively, when the operative mortality is in the order of 5 per cent. When the aneurysm ruptures the operative mortality is in excess of 50 per cent. If left untreated, half of all patients with abdominal aortic aneurysms will die within five years. Even small (< 6 mm) aortic aneurysms have a 20 per cent rate of rupture. Elective resection and grafting is therefore indicated.

Preoperative assessment

Ultrasonography is accurate in assessing the size of the aneurysm and confirming the diagnosis. The patient should be as fit as possible and well advised about the implications of the procedure. Preoperative chest physiotherapy and medical control of diabetes, bronchitis and cardiac disease should be carried out. Haematological status, biochemistry, and 4–6 units (10 units for emergencies) of blood should be available.

Anaesthetic

General anaesthetic.

Position of patient

Supine.

Incision

Long midline from xiphisternum to pubis, skirting umbilicus. Insert a large self-retaining retractor (or two).

Procedure

Divide the peritoneum over the aneurysm. Mobilize the small bowel and pack it between moist packs either on the chest or inside the right upper quadrant.

Palpate the neck of the aneurysm (usually below the left renal vein). Identify and retract the inferior mesenteric vein proximally or divide it if necessary. Dissect the common iliac vessels distally after dividing the peritoneum overlying them. Beware of damaging the ureters. Beware of the iliac veins. They may be adherent to the deep surface of the artery.

When these steps are complete, choose a suitable knitted graft (either tube or bifurcate—USCI, Meadox, Bard) aspirate 40 ml of blood from the inferior vena cava. Completely immerse the graft in a kidney dish in the blood until there is clot. Then heparinize the patient with IV heparin 1 unit/kg body weight. Apply pressure to the puncture site in the vena cava and cross-clamp the aorta proximally and both common iliac arteries. Incise the aneurysm sac longitudinally. Remove debris and thrombus. Oversew bleeding lumbar or the inferior mesenteric arteries using 3/0 vascular sutures. Identify the aneurysm neck (usually protruding forwards) and incise the sac on each side to get a clear view of the proximal anastomosis site. Is the bifurcation suitable as the site for the distal anastomosis? If so, use a tube graft. If not, a bifurcated graft is required.

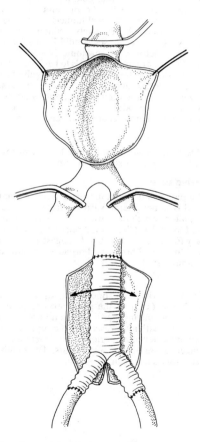

805

**End to end anastomosis for an aortic bifurcation graft
following resection of an aortic aneurysm**

Resection of an abdominal aortic aneurysm (*cont.*)

Select a suitable graft (18–22 mm are most common) and make sure that it is of suitable length to extend from the neck of the aneurysm to both groins, if necessary. Suture it in position proximally using 2/0, 3/0, or 4/0 prolene or Ticron. When this proximal anastomosis is complete, release the aortic clamp to flush the graft then cross-clamp the graft close to the suture line.

Now complete the lower aortic anastomosis (if a tube graft) or anastomose the limbs of the bifurcate into the junction of the common femoral arteries with the profunda and superficial femoral arteries via retroperitoneal tunnels after tying off both common iliac vessels. Alternatively, bilateral bifurcate to common iliac vessel anastomoses should be carried out using 4/0 or 5/0 sutures. Before each anastomosis is complete release the clamp on the iliac or femoral vessels to back-bleed the vessel and reclamp it. Also release the clamp proximally to flush out clot. Then complete the anastomosis. Suture the wall of the aneurysm over the graft and close the peritoneum to prevent formation of an aorto-duodenal fistula. Always ensure that the anaesthetist knows when the clamps are being released.

Closure

The wound is closed in layers with buried non-absorbable deep tension sutures and clips or sutures to skin.

Postoperative care

Maintain the patient's temperature using a space blanket. This prevents acidosis, peripheral vasoconstriction, and cardiac arrhythmias, all of which can result from hypothermia. Close monitoring of the pulse, CVP, blood pressure, and urine output are necessary for the first 12 hours. Nasogastric suction should be monitored for 48–72 hours, and broad-spectrum antibiotics should be administered.

Prognosis

Most patients survive the operation, even for ruptured aneurysms. Deaths occur in the postoperative period due to cardiac arrhythmias and myocardial infarction. Renal failure is a common cause of death after resection of a ruptured aneurysm. Preoperative hypertension, perioperative metabolic acidosis and hypokalaemia are aggravating factors. The death rate for elective aneurysms is less than 5 per cent.

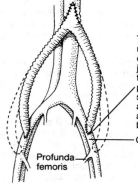

This anastomosis is most commonly constructed at the junction of the common femoral with the profunda and superficial femoral arteries. Aorto femoral anastomosis gives best results.

Common femoral

Profunda femoris

Aorto femoral bypass graft
(onlay type) following atheroma

807

Femoropopliteal bypass grafting

This procedure is indicated for atheromatous occlusion of the superficial femoral artery in patients with crippling intermittent claudication (< 100 metres) which shows no sign of improving spontaneously. It is also a limb salvage procedure.

Preoperative preparation

Attention is given to myocardial, cerebral, and renal status and the possibility of diabetes. The patient should stop smoking. Serum should be grouped and retained if needed. The leg, pubis, and lower abdomen are shaved and if a reversed saphenous vein is to be used its course in the thigh is marked out. Details of the extent of disease and the feasibility of the procedure may be determined from the preoperative arteriogram. 'On-table' arteriograpy may be indicated.

Anaesthetic

General anaesthetic.

Position of patient

Supine, with the knee of the affected leg flexed to 45°. The hip is flexed and externally rotated and the skin is prepared and draped to expose an area from groin to below the knee. This area is then covered with a transparent adhesive incise-drape.

Incisions

Expose the femoral artery in the groin and the popliteal artery above the knee through a medially placed incision between vastus medialis and sartorius tendons. Extend the incision for about 10 cm. Deepen it between the anterior border of sartorius and adductor magnus and carefully free the popliteal vessel from the vein. Pass tapes around the artery. Is it suitable for distal anastomosis? If not, expose the vessel below the knee by extending the incision along the medial border of the tibia and reflect the gastrocnemius backwards. This exposes the neuro-vascular bundle.

Subcutaneous tunnel This can also be made subsartorially. Start with finger dissection in the groin and distal incision. The subsartorial tunnel is begun distally. Then a finger is passed under the deep fascia in front of the femoral artery. The tunnel may also be completed by a device.

Procedure Reversed saphenous vein or 5 or 6 mm Gortex may be used (conduit). Pass the conduit through the tunnel ensuring there are no kinks. Give 5000 units of heparin IV. Control the popliteal vessel and open it. Fashion an end-to-side anastomosis with continuous 5/0 arterial sutures. Clamp the conduit close to the anastomosis and release the popliteal clamps to check for excessive leaks.

Now proceed to control the femoral artery and its branches. Fashion an end-to-side anastomosis to the common or superficial femoral artery with continuous 4/0 sutures. Release the clamps. You may check the flow with arteriography or flow meter.

In situ bypass grafting is performed by exposing the long saphenous vein and anastomosing its proximal end to the common femoral artery near its bifurcation. As blood flows down the vein, puckering will appear at the stie of the valves. A Hall valve stripper is then passed from below to obliterate them. Major tributaries and perforators are identified either by direct exposure or by use of Doppler ultrasound and are ligated. When this has been done the distal end is anastomosed to a suitable site on the popliteal artery or one of its three branches below the knee.

Closure

Layers. Suction drain. Clips to skin.

Postoperative care

Check the pulses either by palpation or Doppler. If there is evidence on arteriography of internal damage or clotting, repair or embolectomy may be necessary.

Femoral embolectomy

Most emboli affect the lower limb in the femoropopliteal segment and (90 per cent) are due to myocardial disease, especially atrial fibrillation, myocardial infarction, and rheumatic heart disease. Debris from atheromatous plaques or aortic thrombus is also common. This latter group has a poorer prognosis, and bypass grafting may be necessary.

Preoperative action

Control of the primary cause includes cardiac support in the form of antiarrhythmic drugs, diuretics, and dopamine as indicated. Careful monitoring of the patient with titration of drugs to the response, the use of analgesics, and oxygen are essential.

Anaesthetic

General or local anaesthetic. General anaesthetic is preferable in confused or unco-operative patients.

Position of patient

Supine.

Incision

Make a 12–15 cm longitudinal incision over the femoral artery from the level of the inguinal ligament.

Carefully expose the common femoral artery and its bifurcation. Use sharp dissection with scissors. Pass silastic loops round the common femoral and superficial femoral using curved forceps. Retract these tapes to expose the profunda femoris. Ligate and divide the large vein which crosses this vessel. Pass looped ligatures round small branches (especially the posterior muscular branches).

Procedure

Heparinize the patient with 10 000 units of heparin intravenously. Control the vessels with the loops and make a transverse arteriotomy over the bifurcation of the common femoral artery. Remove obvious clot with dissecting forceps. Pass a number 3 or 4 embolectomy catheter with its stilette removed into the profunda femoris and remove any clot by inflating the balloon. Is there good backflow? If so, instil 15–20 ml of heparinized saline (5000 units in 500 ml) into the vessel and control it either by clamping or by means of the silastic loops. Make several passes to ensure that all clot is removed.

Repeat the technique in the superficial femoral artery.

Now use a number 5 or 6 embolectomy catheter proximally into the aorta until good proximal flow is achieved. Control the vessel and close the arteriotomy with 5/0 or 6/0 monofilament nylon passing the suture from the inside to the outside of the vessel to prevent intimal damage.

Closure

Bring a suction drain through a separate stab wound. Approximate the fat with interrupted vicryl and close the skin with interrupted clips or nylon sutures.

Similar technique may be applied to embolectomy of the vessels of the upper limb. The use of perioperative angiography ensures patency. 20 ml of Conray are injected distally and the limb screened. If the artery is patent, close the wound. If not, repeat the embolectomy or consider bypass grafting.

Technical points

- Be gentle. The catheter can produce intimal damage. Allow the balloon to deflate a little as resistance is felt.
- If you cannot pass the catheter there may be extensive occlusive disease. Most patients are elderly.

Lower limb amputations

Peripheral vascular disease accounts for 90 per cent of amputations performed in England and Wales.

Indications

Acute ischaemia When this fails to respond to surgical or medical measures, amputation should be performed to prevent septicaemia resulting from gross tissue necrosis.

Chronic ischaemia Patients with severe rest pain associated with ulceration, gangrene, and infection may require amputation but only after angiography when treatable lesions can be identified. Lumbar sympathectomy may improve skin blood flow and so the viability of the flaps.

Gangrene When 'dry', amputation can be carried out once the line of demarcation is established. If the gangrene is 'wet', i.e. associated with putrefaction and sepsis, the risk of proximal spread and systemic sepsis is high. Preoperative control of the situation should be aimed for by the use of antibiotics, e.g. cephalosporins, aminoglycosides plus metronidazole (there are often combined aerobic and anaerobic organisms). If control is not possible amputation should be performed anyway.

Level of amputation

Skin flaps on all amputations must be free from tension to ensure good healing.

Above-knee amputation

The stump should be as long as possible, with the bone being divided at least 15 cm above the knee-joint line.

Incision 'Fish-mouth' or circumferential

Procedure The quadriceps is divided anteriorly and the hamstrings posteriorly. Divide the sciatic nerve as high as possible to prevent postoperative entrapment. Ligate and divide the popliteal artery and vein then elevate the muscles from the femur so that it can be divided at a higher level than the flap. This is done with a saw and the ends smoothed with a file.

Closure Anterior and posterior muscles are sutured with vicryl. The anterior and posterior layers of the deep fascia are then closed and the skin sutured with interrupted nylon. A vacuum drain should be brought out through healthy skin well away from the amputation site.

Below-knee amputation

This is the best amputation for mobility but is contraindicated when there is extensive lower limb sepsis or gangrene close to the tibial tuberosity.

Incision Hemicircumferential 8 cm distal to the tibial tuberosity and extending distally as a posterior flap for a further 16 cm.

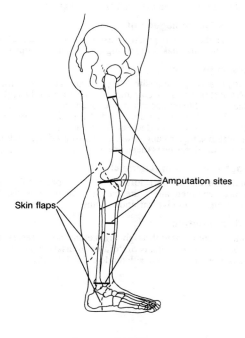

Amputation sites

Skin flaps

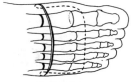

Plantar flap

Amputation sites

813

Lower limb amputations (*cont.*)

Procedure Divide tibialis anterior and its vessels at the level of the incision.

The nerve is divided at a higher level. Elevate the tibial periosteum and divide the bone with a saw 1 cm proximal to the skin incision. Bevel the tibia anteriorly. Use a Gigli saw or bone-cutting forceps to divide the fibula 2 cm proximal to the wound.

Now expose the posterior tibial and peroneal vessels. Ligate and divide them. Taper gastrocnemius and soleus distally.

Closure Smooth the tibial stump and approximate the flaps fascia first using interrupted catgut. The skin is closed with interrupted nylon sutures and the flap bandaged lightly.

Postoperative management

Ensure that the patient practises extension of the knee joint. Flexion of the knee makes the fitting of a prosthesis impossible.

Syme's amputation

A healthy heel is a prerequisite. The posterior flap is constructed from the heel pad and the amputation is performed through the talotibial joint. Subcutaneous tissue is dissected from the os calcis and the Achilles tendon divided. The foot is removed, then the malleoli are divided above the line of the joint. The posterior flap is then swung upwards and sutured with nylon. A plaster of Paris dressing is then carefully applied.

Forefoot amputation (Lisfranc)

This makes use of a short dorsal and a long plantar flap. The bones are sectioned along the base of the metatarsals. The midtarsal amputation is a variation of this.

Digital amputation

Toes may be amputated through a racket incision. Preservation of the base of the phalanx is preferable to disarticulation. This maintains the joint capsule with tendon attachments.

Surgery for varicose veins

Varicose veins occur in the long and short saphenous systems due to valvular incompetence. They may also result in these systems due to incompetence of the perforating veins, usually after deep venous thrombosis. Surgical procedures aim to eradicate incompetence at these levels.

Preoperative preparation

The veins are marked on the skin with a marking pen before surgery. Points of control, sites of perforators, and proposed incisions should also be marked.

Trendelenburg's procedure

Anaesthetic General anaesthetic.

Position of patient Supine, with the legs abducted. The skin of the legs, groin, and lower abdomen is prepared.

Incision Medial to the femoral pulse, 5 cm long, and below and parallel to the inguinal ligament.

Procedure The long saphenous vein is identified and all its tributaries are ligated and divided (the superficial circumflex iliac, superficial, and deep external pudendal and superficial circumflex iliac veins). The long saphenous vein is ligated and divided after identifying its junction with the femoral vein at the cribiform fascia. Use a double tie proximally and clamp the vein distally with artery forceps. A vein stripper is then passed down it distally to below the knee, cut down upon, and brought out through the incision. The vein is clamped distally, ligated, and divided.

An olive is now placed over the upper end of the stripper which is gently pulled down until it is flush with the upper end of the saphenous vein which is then ligated. The artery forcep is now removed and the passage of the stripper controlled with the fingers.

Any other marked veins can now be identified through small incisions and delivered with fine mosquito forceps. Divide these veins between forceps, then gently apply traction until a segment of vein can be delivered with each forcep. Ligate and remove the veins and repeat the procedure at other sites. Close each wound with single fine nylon sutures or Steristrips.

Close the groin wound with interrupted or cosmetic sub-cuticular sutures. Close the distal wound but leave the stitches untied so that the stripper can be removed. Apply crepe bandaging to the leg. Pull the stripper down (with the long saphenous vein), tie the distal sutures, apply a dressing to the wounds, and reapply further crepe bandages.

Postoperative care Walking should begin on the first post-operative day. Supportive bandages should be worn for 2–3 weeks.

Cockett's procedure

(ligation of calf perforators)

Incision Parallel to the subcutaneous porterior border of tibia, 1 cm behind it. The incision is vertical and extended distally as required.

Procedure Deepen the incision and divide the deep fascia in the same vertical plane. Reflect the flaps so formed until the perforators can be seen. Ligate and divide their tributaries and the perforators themselves flush with the fascia.

Closure Close the skin with nylon sutures. Apply a compression bandage.

Multiple ligations/multiple stab avulsions

When varicose veins are extensive they do not always communicate with the saphenous systems. In such cases they can be dealt with by making multiple incisions, delivering as much of each vein as possible, and ligating it proximally and distally. When there are perforators they should be ligated too. This may be combined with sapheno-femoral disconnection if appropriate.

An alternative to this is to make multiple incisions, raise the leg to prevent bleeding, and avulse each vein in turn applying a pressure pad to each puncture site.

After both procedures, compression bandaging should be applied ensuring that it is not too tightly applied.

Cystoscopy

This is the visual examination of the interior of the bladder using a cystoscope or fibrescope.

Instruments

The cystoscope is a rigid instrument relying on rod lenses for vision and a fibre-optic source for light. Passage of the instrument is painful, and requires general or regional anaesthesia. The patient should have reasonable hip abduction to allow introduction. The clarity of vision is excellent and the large calibre allows a wide range of manoeuvres to be carried out through the instrument.

The fibrescope is flexible, using fibre-optics for vision and lighting. Passage of this instrument is relatively pain-free and can be done under local anaesthetic. The clarity of vision is limited, and because of the small calibre and 'bed' of the instrument limited operative manoeuvres are possible.

Uses

Diagnosis of urethral stricture, bladder stones and carcinoma of urethra and bladder, assessment of prostatic size, and normality of bladder mucosa, muscle, and ureteric orifices.

Anaesthetic

General anaesthetic (see above).

Position of patient

Lithotomy.

Procedure

After cleansing the genitalia the urethra is lubricated (antiseptic in the lubricant). In the male urethroscopy should always be performed, using a 0, 25, or 30° telescope passed with the irrigating fluid running, otherwise urethral tumours, false passages, and strictures may be missed. A forward-viewing telescope is required for assessment of prostatic size, but change to a 70° telescope to examine the bladder. Orientation in the bladder can be difficult; the air bubble is always at 12 o'clock. The ureteric orifices should be at 5 and 7 o'clock on the interureteric bar but may be deviated pathologically (e.g. reflux). Examine the mucosa for tumours, abnormalities of blood vessel pattern, and appearance. Check the muscle for trabeculation and ureteric orifices for normality of position. If there are saccules or a diverticulum record their position and check there is not a hidden tumour. Record the size and number of stones. If in doubt about anything, take a biopsy. Always biopsy a 'first-time' tumour, never just cauterize with diathermy.

Fibreoscopy is very similar but the instrument is forward-viewing, narrower, flexible, and has a steerable tip which makes introduction often easier than the rigid instrument especially if the patient has a deformity which interferes with abduction of the hips.

Cystoscopic manoeuvres

Biopsy of suspicious lesions, diathermy of small bladder tumours or bleeding points, bladder washout (e.g. clot retention, small stones), catheterization of the ureter either to allow injection of radio-opaque contrast for X-ray studies or bypass obstruction, passage of stone baskets (e.g. Dormia, Segura) to remove small stones from the lower ureter.

Pyeloplasty

Indication

Obstruction at the pelviureteric junction leads to renal tissue destruction with increased risk of stones, infection (pyonephrosis), traumatic damage, and hypertension.

Causes

- Neuromuscular incoordination at the pelviureteric junction (cf. Hirschsprung's)
- Muscular hyperplasia (cf. pyloric stenosis)
- Kinking by lower pole vessels (30 per cent of cases)

Aims of treatment

- To improve drainage of urine
- To reduce the pelvic deadspace

Anaesthetic

General anaesthetic.

Position of patient

Lateral with affected side uppermost.

Procedures

There are various operations: Anderson–Hynes; Foley V–Y plasty, Culp. The latter two procedures are only suitable for a small renal pelvis and no lower pole vessels.

Approach

This may be via a loin or Anderson–Hynes incision. This incision is horizontal in a line from the tip of the twelfth rib towards the umbilicus. The approach is extraperitoneal on to the ureter and pelviureteric junction.

Advantages

Quick. Heals well.

Disadvantages

It is only suitable for straightforward cases. If there is doubt use the loin approach.

Method

Carry out an Anderson–Hynes pyeloplasty if the renal pelvis is large or there are lower pole vessels only. This procedure allows the vessels to be placed behind the anastomosis after reduction in the pelvic size. If there is doubt about the viability of the suture line use a nephrostomy ('Cummings' with a 'tail' down the ureter or a ureteric catheter down the ureter) to prevent obstruction while doing the anastomosis. Use plain catgut and suction drainage.

A–H pyeloplasty

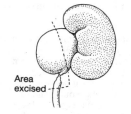

Area
excised

Foley Y–V

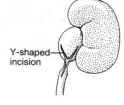

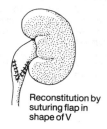

Y-shaped
incision

Reconstitution by
suturing flap in
shape of V

Culp

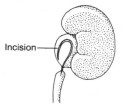

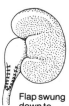

Incision

Flap swung
down to
reconstitute
ureter

821

Pyeloplasty

Cystoplasty

Essence
This is augmentation or reduction in the size of the bladder.

Indications
- *Augmentation*: TB, radiotherapy, Hunner's ulcer, chronic infection, irritable detrusor
- *Reduction*: Diverticulum, chronic obstruction, atony

Methods
- *Augmentation*: An isolated piece of bowel (stomach, small bowel, or colon) is sutured to the bladder
- *Reduction*: Part of the bladder (or a diverticulum) is excised

Details
Augmentation Expose the bladder through a lower midline incision and open it between stay sutures. Divert a segment of bowel (large or small) on its mesentery and restore continuity. Open it along its antimesenteric border and suture it to the bladder as a patch. Use plain catgut for the inner (mucosal) layer and chromic catgut for muscle. Close the wound and drain with a tube drain for 5 days. Restrict oral fluids for 2 days. Solids should not be taken until about 4 days in view of the bowel anastomosis.

Reduction The bladder is exposed and opened as above. If a simple reduction is required free the bladder from the peritoneum and other organs and excise the excess. As long as the bladder is healthy it is amazing how little can be left and still retain good capacity.

If there is a diverticulum dissect it free and excise it with a two-layer closure of the defect. If the ureter is close pass a catheter up it to protect it.

Types of urinary diversion

This may be temporary (to relieve obstruction) or permanent (following excision of the bladder).

Indications

Distal urinary obstruction, loss of bladder sphincter control (spina bifida, spinal injuries), vesico-vaginal fistula, pelvic exenteration, ectopia vesicae.

Methods

Temporary
- Pyelostomy/nephrostomy
- Suprapubic cystostomy

Permanent
- Ureterosigmoidoscopy
- Ileal conduit
- Cutaneous ureterostomy

The ileal conduit is the favoured method for permanent diversion. Although it has the disadvantage of a stoma there is a lower risk of urinary tract infection which is inherent in colonic diversions. There is also a decreased risk of hyperchloraemic acidosis associated with reabsorption of the urine because the urine–mucosa contact time is short. Cutaneous ureterostomies have the serious disadvantages of stricture formation and the presence of two stomas.

Procedure

Following cystectomy and mobilization of the ureters a segment of lower ileum 20 cm long is isolated about 20–25 cm from the ileocaecal junction. The remaining ends are anastomosed. The mobilized left ureter is brought through a tunnel in the peritoneum. The ureters are then joined along their adjacent borders with catgut to form a single opening. A stent is passed up each ureter and, using a long artery forceps passed down the isolated loop from its distal end, the ureters are drawn with their stents into the proximal end where the anastomosis is completed with fine catgut sutures. The distal end is brought out as an ileostomy in the right iliac fossa. The stents are anchored to the tip of the ileostomy spout. Urine is then collected in an ileostomy bag. The stents can be removed (they usually fall out) at 10 days.

Nephrectomy

This is removal of a kidney.

Indications

Malignancy, chronic infection, stone, obstruction with irreparable damage, hypertension (rare).

Approach

Loin.

Anaesthetic

General anaesthetic.

Position of patient

Lateral with the table 'broken' to increase access between ribs and pelvic brim. The approach is usually through the bed of twelfth rib but may be through the eleventh if a higher approach is required, e.g. upper pole problems.
• *Advantages*: Quick, well tolerated, good access for most procedures
• *Disadvantages*: Postoperative costal nerve neuralgia may occur, mobilization of the kidney is required before vascular control obtained. Incisional hernias are very difficult to repair.

Transabdominal

Extends from costal margin across both rectus muscles.
• *Advantages*: Reflection of the hepatic or splenic flexure allows immediate control of the vascular pedicle with unlimited room, good access to major vessels and related lymphatics (if node dissection is required).
• *Disadvantages*: More traumatic, longer period of ileus, dehiscence, and long-term adhesions

Lumbotomy The patient is prone and the table partially broken. A vertical incision is made in the lumbar triangle through skin, fat, and the lumbar fascia directly to the renal pelvis and pedicle.
• *Advantages*: very quick and atraumatic, avoids previous operative fibrosis, is relatively painfree and heals easily
• *Disadvantages*: limited access (between two nerves), nephrectomy only possible for small kidneys

Closure

Each muscle layer should be closed individually. A small suction drain will remove any blood and can be removed at 48 hours. For pus leave 5 days. No nasogastric tube is required. Oral fluids can be started next day and normal diet as soon as flatus is passed (usually second day).

Nephroureterectomy

Removal of kidney and ureter in one piece.

Indications Transitional cell carcinoma of renal pelvis or ureter (not adenocarcinoma or squamous carcinoma), associated gross hydroureter.

Method Having mobilized the kidney and divided the pedicle the kidney is pushed down towards the iliac fossa. The loin wound is then closed in the usual way and the patient placed supine. A lower midline incision is made and the lower ureter approached by an extraperitoneal route. The ureter is mobilized upwards and the kidney delivered into the wound. The ureter is then freed towards the bladder. If the indication is transitional cell carcinoma the ureter must be mobilized through the bladder muscle and the ureteric orifice included in the specimen. For a large ureter this is not so important. Close the wound with a small suction drain which is removed on the fifth day.

Orchidopexy

Bringing the testicle down into the scrotum.

Indication

Undescended or ectopic testicle.

Incidence

Diminishes with age. Descent occurs after birth. (Descent should occur in eighth month. 6 per cent of term infants have undescended testes, but only 1–2 per cent at 1 year.)

Reason

To allow normal development of the testicle, diminish the risk of neoplasia in later life, diminish the risk of trauma, cosmetic.

Background

It is not known whether the undescended testicle is abnormal because of maldescent, or whether the abnormality causes maldescent. Biopsies of the abnormally descended testicle often show abnormalities and the risk of malignancy is higher than average in the contralateral 'normally' descended testicle. Cancer in an undescended testicle, whether brought down or not, is increased by 30 times. However, enlargement of the testicle will be easier to detect in the scrotum (rather than in the groin) so that orchidopexy is always worth while regardless of age. Hormone production is preserved despite maldescent.

Examination

The child should be relaxed. Hands should be warm. Sprinkle powder or oil over the pubic area. Gently slide the fingers of the left hand along the line of the inguinal canal, 'pushing' the testicle towards the scrotum. If descent appears normal, confirm by palpating the 'descended' testicle with the right fingers to confirm its position in the scrotum. A final check under anaesthetic should be performed since some testicles will be normally descended once the stress of the outpatient clinic is removed.

Method

Incise medial to the pubic tubercle. Use blunt dissection because the testicle may be very close the surface. Having found the cord and testicle, make sure there is no indirect sac. Separation of an indirect sac will usually provide sufficient mobility to place the testicle in the scrotum. If still short, separate the peritoneum as much as possible. Feel for a 'band of tissue' on the dorsum of the cord. Division (making sure this is not the vas or artery) may give another cm of length.

Dartos pouch

A finger is placed through the groin incision into the scrotum. An incision is made on the finger but only skin deep. Using artery forceps develop a pocket under the skin large enough to accept the testicle. Incise the tissue under the skin onto the finger and pick up the rubber of the glove with a clip. Withdraw the finger (plus clip) into the groin and attach the clip onto loose tissue of the testicle. By withdrawing the clip the testicle is brought down and out through the scrotal skin into the pouch prepared for it. The scrotal skin is closed with dexon. The groin is closed with subcutaneous dexon or vicryl and the skin with silk.

If a unilateral testicle Bring down as low as possible and reoperate at a later age. If the other testicle is normal and the undescended testicle cannot be brought down, remove it.

Transurethral resection

This is excision of tissue from the bladder or urethra using an electrical cutting current. Blood vessels are cauterized by a coagulation current. The technique is commonly used for prostatic resection and resection of bladder tumours.

Physics

An oscillating current is produced by a diathermy machine. Depending on whether the setting on the oscilloscope is 'parallel' or 'sinusoidal', cutting or coagulation can be carried out. These currents are directed to a metal loop which cuts cleanly with minimal tissue charring (cutting) or leads to local temperature rise (coagulation).

The resectoscope

Resection is carried out via a resectoscope inserted into the urethra or bladder. The instrument may be irrigating or non-irrigating. The advantages of the irrigating instrument lie in good vision and drainage reducing the risk of circulatory absorption of irrigation fluid with subsequent overload. Its disadvantage is its smaller working loop which reduces the amount of tissue removed with each 'cut'.

Anaesthetic

General anaesthetic.

Position of patient

Lithotomy.

Transurethral resection of prostate

Assess the size of the prostate at cystoscopy. Exclude bladder stones and tumour. Stones should be crushed and removed. A bladder tumour carries a risk of implantation in the raw prostatic bed and so may necessitate a different approach to the prostate, e.g. retropubic instead of transurethral.

Orientation is vital and depends on sighting the veru montanum. This is the guide to the external sphincter. It should not be resected nor its position forgotten. It is customary to resect the middle lobe then the lateral lobes from 2 o'clock down to the veru and 11 o'clock to the veru by moving the loop backwards and forwards excising the tissue in 'chips'. Always witness these stages at demonstration by a competent urologist before carrying out supervised resection.

When the prostate has been resected to its capsule (circular fibres) wash the 'chips' out with Ellik's evacuator. Ensure that the washout is complete to prevent blockage of the catheter which is passed at the end of the procedure. Coagulate the bleeding vessels.

The catheter passed should be no wider than size 20 and preferably three-way to permit drainage and irrigation in the immediate postoperative period. The irrigation is stopped 24 hours postoperatively and the catheter removed on day 2 (or when there are no more clots). The patient is allowed home when the urine is clear (usually day 4). Instruct him to take a high fluid intake for 2 weeks. 20–30 per cent bleed around the tenth day due to secondary infective haemorrhage. Instruct the patient to keep drinking and the bleeding usually clears. Review should be at 6 weeks.

Complications

- Blockage of the catheter due to blood clot or fragment of prostate. Dislodge with a bladder syringe. If unsuccessful change the catheter.
- Haemorrhage may occur postoperatively. If persistent take the patient back to theatre.
- Pyrexia due to bacteraemia. Give prophylactic antibiotics. (p. 112)

Thyroidectomy

Indications

Absolute
- Suspicion of malignant change
- Compression of the trachea by a benign lesion

Relative
- Treatment of thyrotoxicosis
- Unsightly goitre

Choice of operation

- *Thyroid nodule*: thyroid lobectomy
- *Thyroid cancer*: total thyroidectomy
- *Nodular colloid goitre* (with hyperthyroidism): subtotal thyroidectomy

Anaesthesia

General anaesthesia with endotracheal intubation.

Position of patient

Supine with neck extended to 20–25° and the table foot down, the head placed on a padded ring with a pillow or pad beneath the shoulders. Place packs on either side of the neck. Use a double towel as a head set. Stand on the side opposite the lobe to be removed.

Incision

Look for skin creases. Mark it beforehand with ink or an 2/0 thread pressed on the skin. Make a transverse incision as near to midway between the thyroid cartilage and manubrial notch, in a crease if possible. The incision should reach the sternomastoid on each side.

Procedure

Use a size 15 blade. Deepen the incision through fat, platysma, and cervical fascia to expose the strap muscles. Reflect skin flaps proximally and distally. Begin with the knife then use gauze dissection. Stop bleeding. Insert a Joll's retractor.

Divide the strap muscles in the midline. The deep layer (sternothyroids) may be stretched or adherent to the gland. When they are split the gland is exposed.

Thyroid lobectomy Palpate both lobes to assess the extent of disease. Ligate the middle thyroid veins. Now draw down the upper pole. Identify the superior thyroid vessels. Ligate them with care using 2/0 catgut. Draw the upper pole down with a Kocher forceps and divide the vessels below the ligature with a scalpel. The upper pole can now be peeled downwards under gentle traction.

Apply gentle traction on the forcep to rotate the pole to the opposite side. Wipe away the fascia on its undersurface to identify the superior parathyroid. Divide any vessels between thyroid and parathyroid, thus keeping it safe.

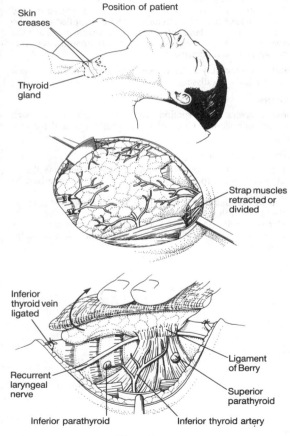

Position of patient

Skin creases

Thyroid gland

Strap muscles retracted or divided

Inferior thyroid vein ligated

Recurrent laryngeal nerve

Inferior parathyroid

Ligament of Berry

Superior parathyroid

Inferior thyroid artery

Thyroidectomy

Thyroidectomy (*cont.*)

Now pay attention to the recurrent laryngeal nerve and the inferior thyroid artery. Move the assistant's retractor to exert a pull laterally. Apply a small swab to the gland and exert a medial pull with the left hand to expose the fascia overlying the inferior thyroid artery, recurrent laryngeal nerve and the inferior parathyroid. Gently divide the fascia longitudinally to identify the nerve close to the ligament of Berry. Do not disturb the inferior parathyroid. Simply locate it. Ligate the artery in continuity then the inferior thyroid veins. The lobe is now free and may be removed by dividing the isthmus, dissecting it off the trachea, and oversewing the isthmus with catgut.

In a subtotal lobectomy the gland is not dissected off the trachea. Instead several small haemostats are applied along the lateral aspect of the gland anterior to the recurrent nerve and the tissue anterior to the forceps excised, the remnant being oversewn with catgut if possible.

Total thyroidectomy Total lobectomies are performed on each side ± radical neck dissection.

Closure

Close in layers. Bring suction drains (one each side) through separate stab wound. Close platysma with catgut and the skin with clips.

Postoperative care

- Check the vocal cords postoperatively. The anaesthetist usually does this (but the appearance of the cords may be misleading).
- Ensure an adequate airway. If haematoma develops open all the layers of the wound.
- Remove drains on day 2. Check serum calcium daily.
- Home day 3–5. SOPD and thyroid function tests after 6 weeks.

Parathyroidectomy

Indications

Parathyroid adenomas, carcinoma, 2- or 3-hyperparathyroid-ism (all four glands).

Preoperative arrangements

Frozen section facilities should be available. The laboratory should be informed.

Anaesthesia

General anaesthetic.

Position of patient

Supine with the head on a neck ring, the neck extended and a 20–25° downward tilt.

Incision and access

The same approach as for subtotal thyroidectomy is used (p. 832).

Procedure

Mobilize the thyroid gland and retract its upper and lower lobes forward. Also assess its areolar bed. The normal parathyroids are difficult to identify but lie posterior to each of the four poles of the thyroid. Each gland is yellowish and is flattened with a small vascular pedicle.

If there is an adenoma carefully remove it. If the patient has 2- or 3-hyperparathyroidism, remove all four glands.

Apply tissue or dissecting forceps to the gland and dissect it free by a combination of sharp and blunt dissection. Diathermize bleeding vessels. If you are in doubt about the nature of the tissue send biopsies for frozen section. Sometimes an adenoma can be found in the mediastinal region or in relation to the superior or inferior thyroid arteries.

Closure

Layers. Clips to skin. Suction drainage.

Postoperative care

- Observe for signs of hypocalcaemia and haematoma.
- Remove the drain at 2–3 days.
- Repeat serum calcium daily to observe the trend.
- Most patients can have their sutures removed on the fourth or fifth day and go home.
- Arrange SOPD appointment in 6–8 weeks.

Complications

- Haemorrhage
- Nerve injury
  - Unilateral recurrent laryngeal nerve injury: hoarseness
  - Bilateral: stridor
  - Superior laryngeal nerve: change in 'timbre' of voice
  - Cervical sympathetic chain: Horner's syndrome

Adrenalectomy

Indications
- Cushing's disease
- Phaeochromocytoma
- Aldosteronoma (Conn's)
- Advanced breast cancer

Preoperative care
- Locate tumours by CT scan or arteriography.
- Check serum potassium (Conn's syndrome).
- Prevent alkalosis due to hypokalaemia which leads to tetany.
- Check and correct serum sodium which contributes to hypertension.
- Estimate serum catecholamine levels.
- Rogitine (phentolamine) test for phaeochromocytoma. Blood pressure falls for a short time in response to drug.
- Use beta-blocking agents to control blood pressure for several days before surgery to allow adequate expansion of plasma volume. Discontinue for 24 hours presurgery.

Anaesthetic
General anaesthetic. In cases of phaeochromocytoma tranquillizers are given as premedication because opiates stimulate catecholamine release. IV lignocaine, propranolol, and phenotolamine are at hand to correct dysrrhythmias or blood pressure changes.

Approaches
- *Anterior*: both glands
- *Lateral*: obese patients, tumour localized to one side

Position of patient
- *Anterior*: Patient supine, foot-down, 30°, lateral tilt
- *Lateral*: Patient in full lateral position, table broken to 30° over tenth and eleventh ribs

Strap the patient to the table for both approaches.

Incisions
- *Anterior*: Long left paramedian. Extend laterally through the transpyloric plane if access is difficult
- *Lateral*: Over eleventh rib from lateral border of sacrospinalis to abdominal wall

Procedure

Anterior approach

Left adrenal Pack small intestine downwards. Retract costal margin upwards. Place left hand over spleen and divide the posterior parietal peritoneum from splenic flexure to oesophageal hiatus. Apply several pairs of forceps to retract the peritoneum downwards. Identify the kidney with adrenal above it. Identify and divide between ligatures the adrenal vein which drains into the left renal vein. Mobilize and remove the gland by blunt dissection. Cauterize bleeders.

Right adrenal Retract liver upwards and hepatic flexure downwards. Note the vena cava and kidney. Divide the posterior parietal peritoneum along the upper pole of the kidney to the vena cava. Identify the gland. Divide the two or three adrenal veins which drain into the vena cava between ligatures. Remove the gland by blunt dissection. Secure haemostasis.

Lateral approach Deepen the incision in the rib bed. Beware of the pleura. Use a self-retaining retractor. Identify kidney and upper pole. Divide its fascia. Remove the gland carefully as in the anterior approach.

Closure

Layers. Suction drain. Nylon to skin.

The surgery of coronary artery disease

Until the prevention of atherosclerosis becomes a reality surgery will continue to play a major role in the treatment of coronary artery disease.

Indications

Chronic myocardial ischaemia left main stem stenosis, triple vessel disease (the 'widow-maker'), combined proximal and anterior branch disease. Patients who have severe and unstable angina are less likely to survive with medical treatment.

Acute myocardial ischaemia and infarction in selected patients

Coronary artery disease associated with valvular heart disease

Mechanical consequences of infarction ventricular septal defect, ventricular aneurysm, papillary muscle dysfunction

Investigations

- Ask questions to exclude GI disease which can mimic angina.
- Determine the family history. Test for diabetes mellitus.
- Examine the patient. Remember that aortic stenosis can present as angina in a patient with healthy coronary arteries.
- Is the long saphenous vein present? If not (due to stripping) use the short saphenous or an arm vein.
- Assess the patient's general suitability for surgery and the extent of his symptoms by ECG and exercise ECG. Do echocardiography to demonstrate valvular disease and assess left ventricular function. Assess the extent of disease by radionuclide scan and coronary angiography.

Procedures

- Coronary artery bypass grafting (CABG = cabbage). The conduits used may be the long saphenous vein, the internal mammary artery, or the arm veins.
- Combined methods include CABG + endarterectomy or percutaneous transluminal angioplasty and intracoronary thrombolysis.

Complications

Postoperative haemorrhage. Myocardial infarction. Cerebral complications—stroke and personality changes. Urethral strictures (catheterization). Wound infection.

Results

Mortality rate is 2.6 per cent. The risk for women is twice that for men. The prognosis is poorer in patients with impaired left ventricular function or urgent surgery. Angina is relieved but life may not be lengthened.

Cost

Each procedure costs about £4000 (UK, 1989), but patients can return to work and are no longer invalids.

Prospect

The need for CABG will probably increase until prevention becomes effective. However, the use of thrombolytic agents like streptokinase may reduce the need for surgery in selected patients.

Signs, syndromes, and rarities for reference

Acanthosis nigricans Pigmentation of the axillary skin associated with the presence of breast or gastric cancer.

Achondroplasia Familial dwarfism in which there is defective growth of the long bones and skull.

Acute torsion of the gall bladder This is rare but incidence increases with age. 4–6 per cent of people have a gall bladder mesentery. It is thought that loss of supportive tissue with age leads to elongation of the mesentery and abnormal movement of the gall bladder. Subsequent changes in the weight of the gall bladder or sudden peristalsis of adjacent organs leads to torsion. *Clinical features*: Consider the diagnosis in the elderly. Severe and sudden onset of right upper quadrant pain. Tender mass in right upper quadrant. Previous history of gall bladder disease is uncommon. Absence of toxaemia is common.
Treatment: Cholecystectomy to prevent gangrene, perforation and biliary peritonitis.

Adenolymphoma (Warthin's tumour). A benign tumour, usually affecting the lower pole of the parotid gland in middle aged or elderly women.

AIDS Autoimmune deficiency syndrome is an infection caused by retroviruses which affect the T-lymphocytes and monocytes leading to immune depression. Opportunistic infections and neoplasms are characteristic and may affect any body system but especially the respiratory system (Pneumocystitis carinii pneumonia) and the skin (Kaposi's sarcoma). The virus is spread by infected body fluids and almost always by sexual intercourse (especially anal intercourse) although it is also transmitted by drug abusers using infected needles and by blood and blood products. Infected mothers can transmit the virus to the foetus. Infected individuals are so for life. Although the virus may remain dormant, all eventually develop AIDS. Diagnosis is by detecting antibodies in the serum. Treatment is that of opportunistic infections. There is no specific cure.

Ainhum A disease of negroes in which a fissure develops at an interphalangeal joint of the fifth toe. The fissure becomes a fibrous band as it heals and leads to digital gangrene. Treatment is by division of the fissure and reconstruction by Z-plasty. Advanced cases need amputation.

Albright's hereditary osteodystrophy An X-linked dominant form of pseudo-hypoparathyroidism. There is hypoparathyroidism with a low serum calcium, mental retardation, cataracts, and tetany. Metastatic calcification of the basal ganglia is a feature. The patients are below average height with short first, fourth, or fifth metacarpals.

Albright's polyostotic fibrous dysplasia A condition of unknown aetiology in which fibrous dysplastic changes involve several bones. There is precocious puberty in girls with skin pigmentation. The affected bones (often the tibia) become softened and deformed.

Amaurosis fugax Transient blindness associated with embolization of carotid artery atheroma into a central retinal artery.

Amazia (amastia) Congenital absence of one or both breasts. The condition is more common in males.

Ameloblastoma A locally invasive neoplasm originating from the dental lamina which usually affects the posterior aspect of the mandible. Treatment is by excision with a 1 cm margin of normal tissue.

Angiodysplasia (colon) Vascular anomalies, usually a-v malformations can be a cause of bleeding from the large bowel which may be massive. The site of bleeding may be determined by angiography (small bowel and colon) or colonoscopy. The right side of the colon is the commonest. Treatment is partial or total colectomy with anastomosis or restorative proctocolectomy. Alternatively laser coagulation of angiodysplastic areas may be possible.

Angioplasty (Grunzig) A technique for increasing blood flow through areas of atheromatous narrowing of arteries. A catheter with an inflatable balloon is passed into the stricture, visualised radiographically and inflated to a pressure of around 10 atmospheres for 20–30 seconds. The atheromatous tissue is firmly compressed and the lumen of the artery increased.

Antibioma The long-term use of antibiotics instead of draining pus from an abscess can lead to the formation of a hard, oedematous swelling containing sterile pus. Treatment involves exploration and drainage if the lump mimics a carcinoma. Spontaneous resolution often takes several weeks.

Apert's syndrome An autosomal dominant condition with an incidence of 1/160 000 births. The skull is tower-shaped (brachycephaly) with premature fusion of the sutures. There is also syndactyly of the middle three fingers. Associations are oesophageal atresia, renal anomalies, and congenital heart disease. Mental retardation may be due to cerebral compression. Craniofacial surgery improves the child's appearance and may prevent mental retardation.

ARDS (Adult respiratory distress syndrome, shock lung). This is acute respiratory failure associated with major trauma, smoke inhalation, sepsis, haemorrhagic shock, embolism, metabolic disorders, etc. The common factor initiating the process is interference with surfactant or its precursors within the alveolar cells which extrude their contents into the alveoli. There is increased interstitial fluid, oedema, increased vascular congestion and fibrosis. Clinically, the patient becomes centrally cyanosed while breathing air with hyperventilation and respiratory alkalosis. This progresses to respiratory distress with pulmonary oedema on chest X-ray and later respiratory failure.

Treatment is that of the cause, e.g. restore blood volume, treat sepsis, etc. Ventilation with positive end expiratory pressure

(PEEP) is indicated if $PaO_2 < 10K$ Pa. Pulmonary hypertension responds to fluid restriction, diuretics e.g. frusemide (up to 180 mg/24 h IV) and inotropic drugs like dopamine (2–10 μg/kg/min IV). Steroids may be of value.

The prognosis is poor. More than 50 per cent of patients will die.

Auriculo-temporal syndrome   See 'Frey's syndrome'.

Baghdad sore   Indolent ulceration on exposed areas due to a protozoon *Leishmania tropica*. Treatment is with IV antimony tartrate or local cryosurgery.

Baker's cyst   This is a central swelling of the popliteal fossa most evident when the patient is standing. It represents a synovial membrane diverticulum and is almost always associated with knee joint pathology like meniscus tears or arthritis.

Ballance's sign   (ruptured spleen). On percussion of both flanks there is dullness which is shifting on the right side but static on the left due to coagulated blood around the splenic rupture.

Barrett's ulcer   Ulceration of the oesophagus proximal to the oesophagogastric junction. Thought to be caused by ectopic gastric mucosa or oesophageal reflux.

Bazin's disease   Affects adolescent girls. It is characterized by an indurated erythematous rash (erythema induratum) which breaks down to form indolent ulcers. Tuberculosis may be a cause.
Treatment: Antituberculous drugs if indicated, or sympathectomy.

Blind loop syndrome   Stasis in a gut loop (diverticulum, long afferent loop, intestinal stenosis or stricture) leads to colonization with abnormal bacterial flora which prevent proper digestion and absorption of food. Vitamin deficiencies (B_{12}), anaemia, and steatorrhea are features.

Treatment is surgical correction of the anomaly.

Brailsford's disease   Infarction of the navicular in an adult, usually a middle-aged woman, which may ultimately lead to degeneration of the mid-tarsal joint.

Bowens disease   Intraepithelial carcinoma *in situ* of the skin. Treatment is by excision.

Budd–Chiari syndrome   This is due to obstruction of the hepatic veins or their tributaries due to tumour growth, clotting diseases, and fertility hormones. In one congenital form there is a membrane in the suprahepatic inferior vena cava. An acute form of veno-occlusive disease is seen in Jamaica where drinking herbal teas leads to subintimal fibrosis and occlusion of the hepatic veins. The cause is alkaloids present in the plants *Senecio* and *Crotolaria* used for the tea.
Treatment:
• Transatrial meatotomy (for membranous IVC obstruction)
• Portacaval, mesocaval, mesoatrial shunt

Buschke–Loewenstein tumour   A rare type of penile carcinoma which is locally invasive. Treatment is surgical excision.

Cachexia ovarica   This is caused by a massive ovarian cyst which causes lower limb and perineal oedema with lordosis of the spine.

Calot's triangle   The triangle formed by the liver above, the common bile duct medially, and the cystic duct below. It is an important landmark in cholecystectomy.

Campbell de Morgan spots   Red spots on the skin which show no sign of emptying on pressure. Once thought to be associated with gastric cancer, they are of no clinical significance.

Caput medusae   Dilatation of the superficial veins radiating from the umbilicus due to portal hypertension in cirrhosis of the liver.

Carnett's test   Determines whether an abdominal lump is intraperitoneal or in the abdominal wall. Ask the patient to lie flat and raise the extended legs from the couch. An intraperitoneal lump disappears, whereas one in the abdominal wall persists.

Charcot's triad   Jaundice, pain, and fever associated with acute cholangitis. Rigors are also a feature.

Chvostek's sign   In latent tetany tapping the facial nerve as it emerges at the angle of the jaw leads to facial twitching, an exaggerated reflex seen in 10 per cent of normal people.

Chylothorax   A complication of trauma to the thoracic duct and its tributaries. Chyle leaks into the thoracic cavity and may look like pus when aspirated. Confirmation of a chylous effusion can be made by lipoprotein electophoresis. Treatment is conservative by repeated aspiration and a low-fat diet. In recalcitrant cases a thoracotomy and suture of the duct is indicated.

Codman's triangle   A radiological sign seen in osteosarcoma. As the tumour grows it elevates the periosteum and new bone is laid under it.

Collar-stud abscess   A tuberculous abscess which lies in two fascial planes. Pus originates (in the neck) deep to the deep cervical fascia but erodes through it to lie beneath the superficial fascia thus adopting a collar-stud or dumbbell appearance.

Conn's syndrome   Primary hyperaldosteronism due to an aldosterone secreting tumour of the adrenal cortex. There is sodium retention and potassium depletion leading to myasthenia, polyuria, and polydipsia.

Contre-coup injury   This relates to head injuries. Injury to the brain may occur at the site of trauma or be transmitted due to cerebral movement within the skull to the opposite side (contrecoup).

Corrigan's pulse   This is a collapsing pulse, found in the presence of an arteriovenous fistula.

Courvoisier's law   In the jaundiced patient if the gall bladder is palpable and distended this is more likely to be caused by malignant obstruction of the bile ducts than stone disease in which the gall bladder is usually contracted.

Coxa vara   In malunion of femoral neck fractures, osteomalacia, and slipped upper femoral epiphysis the angle between the neck of the femur and the shaft is reduced producing a positive Trendelenburg's sign.

Cozen's test   (tennis elbow). The patient clenches his fist and extends his wrist. Forced palmarflexion against the patient's resistance produces pain at the lateral epicondyle.

Cronkhite–Canada syndrome   Gastrointestinal polyps, alopecia, and atrophy of the finger nails. The changes are not neoplastic but due to an unidentified deficiency state.

Cruveilhier's sign   (saphena varix). This is positive if an impulse is imparted to the saphenofemoral junction when the erect patient coughs or blows his nose.

Cubitus valgus/varus   The carrying angle of the elbow may be increased (valgus) or decreased (varus) as a result of fracture, epiphyseal injury, or arthritis.

Cullen's sign   Discolouration around the umbilicus associated with intraperitoneal bleeding. Although rare it may occasionally be seen in ruptured ectopic pregnancy, pancreatitis, and abdominal trauma (especially liver and spleen).

Curling's ulcer   Acute gastroduodenal ulceration associated with extensive burns.

Cushing's ulcer   Acute gastroduodenal ulceration associated with stress, e.g. haemorrhage, toxicity, myocardial infarction.

Cystic hygroma   This is a cavernous lymphangioma usually of the lower third of the neck which first appears in childhood. It is transluscent and because it communicates with other compartments of the neck it is both compressible and inflatable by coughing, etc. The cysts contain lymph and are best treated by early excision and sclerosing injections.

Cystosarcoma phylloides   (serocystic disease of Brodie). This is a rapidly growing neoplasm of the breast which generally remains well encapsulated but can become enormous. 25 per cent will become frankly sarcomatous and metastasize. Treatment is by total mastectomy.

Dercum's disease   (adiposis dolorosa). Multiple lipomatosis. Usually a lipoma is painless. This condition is characterized by pain and tenderness in the lumps.

Dermatofibroma   This is a form of cutaneous angioma which is nodular and may be multiple. It may be resemble a melanoma due to staining with haemosiderin.

Desmoid tumour   This is a fibroma of the aponeurosis and muscle of the anterior abdominal wall below the umbilicus. It usually affects middle-aged multiparous women and has a tendency to recur unless widely excised. Such tumours may be encountered in patients with familial adenomatous polyposis.

De Quervain's disease
1. Stenosing teno-synovitis of the sheath and tendons of the abductor pollicus longus and extensor pollicus brevis at the wrist.
2. A viral form of thyroiditis.

Dietl's crisis The passage of large amounts of urine following acute intermittent hydronephrosis characterized by renal colic, a swelling in the loin (distended kidney) and the disappearance of the swelling after passing urine.

Dumping syndrome A postgastrectomy syndrome which occurs after eating and may be early or late.

Early dumping occurs immediately after meals. There is hypotension, tachycardia, sweating, and diarrhoea associated with gastric bloating. It is related to the osmotic effect of a carbohydrate load in the small intestine and is most common after Pólya gastrectomy and persists in around 5 per cent of patients. It is usually transient. Symptoms can be relieved by small frequent meals and lying down.

Late dumping is related to hypoglycaemia and occurs about two hours after eating. It is characterized by sensations of nausea and emptiness, and is relieved by more food. It is less common.

Enophthalmos The eye(s) appear sunken and smaller. This is a feature of Horner's syndrome (damage or division of the cervical sympathetic chain) or fracture of the zygomatic arch or maxilla.

Epiplocoele A hernial sac which contains omentum (an omentocoele).

ERCP Endoscopic retrograde choledocho pancreatography. This technique involves visualization and retrograde cannulation of the ampulla of Vater followed by injection of contrast medium to delineate the pancreatic and common bile ducts and carry out other procedures.

Erythroplasia of querat Is a premalignant condition of the penis associated with chronic balanitis. Treatment is by radiotherapy or excision with diathermy (synonym: Paget's disease of the penis).

Ewing's tumour A malignant tumour of uncertain origin occurring in childhood and adolescence which usually affects the long bones and presents as a hot tender swelling. Treatment is by chemo- and radiotherapy. The prognosis has improved recently with this regimen.

Exophthalmos Widening of the palpebral fissure which may progress to protruding eyeballs. It is a feature of primary thyrotoxicosis and may affect one or both eyes.

Extradural haematoma Most cases occur in association with fractures of the parietal or temporal bone. Branches of the middle meningeal artery are damaged as a result. The haematoma leads to extradural compression of the parietal region of the brain. A boggy swelling may also be present at the site of injury due to extracranial extravasation of blood through the fracture site. The patient (usually young) sustains a head injury which may produce unconsciousness. He then recovers consciousness (the lucid interval) only to suffer further deterioration in conscious level due to the expanding haematoma. The diagnosis is made by CAT scan followed by evacuation of the haematoma through burr holes or bone flap.

Fallot's tetralogy   This is the commonest form of congenital cyanotic heart disease. There is pulmonary stenosis, a ventricular septal defect, aortic overriding of the ventricles and right ventricular hypertrophy. Investigation is by cardiac angiography. The aim of treatment is complete surgical correction in the long term, but palliative bypass may be performed initially in severe deformities.

Fanconi's syndrome   An inherited, possible autosomal recessive, complex renal tubular disorder. There is rickets, dwarfing, amino-aciduria, glycosuria, high urinary phosphate but a low blood phosphate level.

Felty's syndrome   Leucopenia and moderate anaemia associated with splenomegaly in patients with chronic rheumatoid arthritis. Patients develop pyogenic infections and ulceration of the lower leg. Splenomegaly may improve the symptoms of fatigue and weight loss. The blood picture, however, shows only a transient return to normal.

Finkelstein's test   (De Quervain's disease. Stenosing tenosynovitis). The patient places his thumb in his palm then makes a fist over it. The examiner then pushes the hand into ulnar deviation. In a positive test pain is felt at the radial styloid which may shoot down the arm.

Fovea coccygea   Is a post-anal dimple at the level of the tip of the coccyx. It is of little clinical significance.

Foster–Kennedy syndrome   Optic atrophy in one eye and papilloedema of the other. The cause is a tumour in the frontal region which blocks the subarachnoid space on the side of the atrophy but leads to increased intracranial pressure with papilloedema on the other side.

Fournier's gangrene   Gangrene of the scrotum which may develop subsequent to injuries or operations to the scrotum or perineum. It is most commonly due to a mixed microaerophilic infection but can also be caused by haemolytic streptococci in association with *Staphylococcus*, *Cl. perfringens*, or *E. coli*. The result is sudden inflammation, cellulitis, and gangrene. Treatment involves wide excision of the dead tissue and IV gentamicin in combination with metronidazole (to deal with anaerobic organisms).

Freiberg's disease   Osteochondritis juvenilis of the head of the second or third metatarsal.

Frey's syndrome   Following surgery or injury to the parotid gland or temporomandibular joint a syndrome of flushing and sweating of the skin supplied by the auriculo-temporal nerve accompanies salivation. The cause is believed to be traumatic synthesis of postganglionic parasympathetic ganglion fibres and sympathetic nerves destined for the sweat glands. Severe cases may be treated by intratympanitic parasympathetic neurectomy.

Froment's sign (ulnar nerve paresis). The affected thumb becomes flexed when the patient pinches a piece of paper between thumb and index. This is due to weakness of adductor pollicis which permits uncontrolled contraction of flexor pollicis longus (innervated by the median nerve).

Galactocoele (rare) A postlactational solitary, subalveolar cyst which contains milk.

Galeazzi fracture–dislocation Dislocation of the inferior radio-ulnar joint combined with a fracture of the radius at the junction of its lower and middle thirds.

Gamekeeper's thumb A sprain of the metacarpophalangeal joint of the thumb may be associated with rupture of the ulnar collateral ligament. Non-healing may lead to chronic instability of the joint with weakness of the 'pinch' grip. Early recognition of the rupture with surgical repair, advancement of the insertion of adductor pollicis to the phalangeal base or buttressing the ligament with extensor pollicis brevis tendon prevents both arthritis and instability.

Gaucher's disease There is active storage of abnormal lipoids in the spleen and bone marrow in this disorder. This leads to massive splenomegaly in late childhood and early adolescence. The diagnosis is made on this finding plus cutaneous discolouration with thickening of the conjunctiva, anaemia, and bone marrow examination. Splenectomy is usually carried out to alleviate the discomfort caused by the splenic enlargement.

Glomus tumour A cutaneous glomus is a specialized skin thermoregulatory organ found especially around the nailbeds. When an angioma develops in the organ (glomus tumour) the patient usually complains of severe pain which is described as burning in nature. Treatment is by excision.

Glomus jugulare tumour A rare slowly growing tumour which resembles a carotid body tumour. Symptoms and signs include seventh nerve palsy, tinnitus synchronous with the pulse and bleeding. Invasion occurs locally. Treatment is by radiotherapy or cryosurgery.

Goodsall's rule (anal fistulas) Fistulas which open on the skin around the anterior half of the anus are usually direct. Those which open on the skin around the posterior half may have multiple external openings and may be curved or horseshoeshaped.

Gradenigo's syndrome Acute mastoiditis with homolateral paralysis of the sixth cranial nerve and retro-orbital pain. It is due to thrombophlebitis of the inferior petrosal sinus with compression of the nerve as it leaves the posterior cranial fossa.

Graves' disease Diffuse toxic goitre with primary thyrotoxicosis due to abnormal circulating thyroid stimulating antibodies. Young women are usually affected. Eye signs are common.

Grawitz tumour Adenocarcinoma of the kidney (syn. Hypernephroma). See Cancer of the Kidney.

Gynaecomastia Enlargement of the male breast; may be unilateral or bilateral. Causes include idiopathic, drug-related (cimetidine, spironolactone), abnormal hormone levels (due to stilboestrol therapy or associated with testicular or bronchial tumours), leprosy (due to testicular atrophy), hepatic failure, and sexual anomalies (e.g. Klinefelter's syndrome).

Hammer toe Hyperextension of the metatarsophalangeal joint and the distal interphalangeal joint with flexion of the proximal interphalangeal joint. Bursae and callosities subsequently develop. Associations are hallux valgus, overcrowding of the toes and diabetic neuropathy. Treatment is by arthrodesis of the p.i.p. joint with extensor tenotomy of the m.t. joint.

Hangman's fracture Fracture of the pars interarticularis of cervical vertebra 2 (C_2). It is caused by hyperextension of the neck (due to trauma). Treatment is by traction and immobilization of the patient in a Stryker frame. In judicial hanging the fracture led to traumatic spondylolisthesis and transection of the spinal cord.

Hashimoto's disease Auto-immune thyroiditis. The thyroid is diffusely enlarged and lobulated. It is more common in menopausal women with measurable serum antibodies. There is an association with lymphoma and papillary carcinoma.

Hidradenitis suppurativa Chronic infection of the axillary (usually) or perianal sweat glands (when multiple perianal sinuses may develop). Treatment is by excision of the infected tissue allowing healing to take place by granulation or skin grafting.

Hippocratic facies The anxious appearance and sunken eyes associated with terminal peritonitis.

Hoffa's disease (obese subjects) Tender fat pads on either side of the ligamenta patellae on palpation. Forced extension of the knee also causes pain.

Homans' sign Dorsiflexion of the foot causes pain in the calf in the presence of popliteal and posterior tibial vein thrombosis (deep vein thrombosis). Unreliable in over 70 per cent of cases.

Horner's syndrome Ptosis, myosis, ipsilateral anhidrosis, and apparent enophthalmos. The syndrome occurs secondary to invasion of or damage to the cervical sympathetic chain.

Horseshoe kidney The lower poles of the kidney are fused together at the level of the fourth lumbar vertebra. The adrenals are in normal position. The condition is important because the kidneys are prone to disease due to the abnormal disposition of the ureters as they pass over the fused lower poles. Urinary stasis with infection and calculus formation are common. More common in men. The incidence is approximately 1 in 10 000.

Housemaid's knee Chronic bursitis of the prepatellar bursa caused by the trauma of repeated kneeling (as in scrubbing floors). Less common today.

Howship–Romberg sign   When an obturator hernia strangulates the pain may be referred along the genicular branch of the obturator nerve to the knee.

Hunner's ulcer   Painful micturition caused by interstitial cystitis. The aetiology is unknown. Most often seen in women. The bladder becomes contracted with avascular necrosis of areas of the mucosa and ulceration characteristically in the fundus. The bladder capacity is markedly reduced. Urinary frequency, pain and distension of the bladder (with urine) and relief following micturition are characteristic. Cystocopy confirms the presence of a fissure in the fundus of the bladder which bleeds easily. Treatment is difficult. Local steroid infiltration or hydrostatic dilatation may be temporarily effective. Complete relief may necessitate urinary diversion.

Hutchinson's melanotic freckle   (lentigo maligna). Histologically this is a melanoma *in situ*. It begins as a small pigmented spot usually on the sun-exposed skin of the face of an elderly person. The lesion enlarges slowly. Malignant change may supervene at anytime. Spontaneous regression has also been reported. When recognized the affected area should be treated by wide local excision and skin grafting.

Hutchinson's teeth   (congenital syphilis). Peg-shaped upper central incisors (second dentition).

Hutchinson's triad   (congenital syphilis). Interstitial keratitis, Hutchison's teeth, eighth nerve deafness were described by Sir Jonathon Hutchison, (a surgeon to the London Hospital in the nineteenth century), as the classical stigmata of late congenital syphilis.

Hyperhidrosis   Excessive sweating of the axilla, hands, or feet can be socially unacceptable and extremely distressing. Sympathectomy (Cervical or lumbar) is of value in patients who do not respond to conservative measures.

Hydatid of morgagni   A degenerative cyst of the appendix of the epididymis (pedunculated hydatid) or the appendix of the testis (sessile hydatid). Both are tense transilluminable swellings of the epididymis or upper anterior surface of the testis. Surgical excision of those causing discomfort is the treatment of choice.

Hypersplenism   Thrombocytopaenia secondary to an enlarged hyperactive spleen. White cell destruction also occurs to varying degrees. The condition is most often associated with portal hypertension. Splenectomy and relief of portal hypertension are indicated.

Icterus gravis neonatorum   Physiological jaundice (icterus neonatorum) is a feature of 20–25 per cent normal new-born babies 2–5 days after birth. Icterus gravis neonatorum (erythroblastosis foetalis) is a feature of Rh incompatability (an Rh+ve foetus born to an Rh−ve mother). The child is born jaundiced and becomes progressively so. Untreated (by exchange transfusion) the condition leads to deposition of bile pigment in the basal ganglia (kernicterus) and death. Screening is effected in the UK by testing the maternal serum for antibodies.

Impetigo An intradermal infection caused by *Streptococcus* or *Staphylococcus*. Classically bullae develop which burst and become encrusted. This is a contagious infection. Treatment is to remove the crusts and apply antiseptic washes. Systemic antibiotics are only occasionally necessary.

Induratio penis plastica Peyronie's disease (see text).

Insulinoma A beta-cell tumour of the pancreatic islet cells. It usually affects young adults under 40 years. Attacks of hypoglycaemia which are unpredictable and progressively severe are characteristic. The diagnosis is based on fasting hypoglycaemic attacks (Blood sugar less than 2.5 mmol) completely relieved by oral or intravenous glucose (Whipple's triad). Treatment is surgical excision of the tumour.

Intertrigo In obese patients intertrigo (chafing) is common in the inframammary fold or between the scrotum and thigh in males. There is frequently an associated fungal infection.

Intraperitoneal rupture of the bladder (post-traumatic) The patient experiences severe pain and may faint. This soon passess off and abdominal distension develops due to intestinal ileus (caused by the irritation of intra-abdominal urine). Clinically there may be abdominal distension, reduced or absent bowel sounds, and shifting dullness. Frequently there is no wish to pass urine and the bladder dullness associated with retention is absent. Unless the abdomen is drained diffuse peritonitis will develop.

Intraperitoneal spaces These are spaces lying above and below the liver.

Suprahepatic spaces: The right suprahepatic space lies between the right dome of the diaphragm and the superior, right and anterior surfaces of the liver. It is limited medially by the falciform ligament. The left suprahepatic space separates the diaphragm from the left lobe of the liver, fundus of the stomach, and spleen.

Infrahepatic spaces: The right infrahepatic space (pouch of Morison) is bounded by the right lobe of the liver above and the gall bladder anteriorly. Inferiorly and posteriorly lie the upper pole of the right kidney, the right suprarenal gland, the head of the pancreas, second part of the duodenum, right colic flexure, part of transverse colon and mesocolon. There are two infrahepatic spaces on the left side—the left anterior and posterior infrahepatic spaces. The left anterior space lies below the left lobe of the liver with stomach and lesser omentum behind. The left posterior infrahepatic space is the lesser sac.

Jansen's disease A severe form of metaphyseal dysostosis inherited as an autosomal dominant. The characteristics are deafness and extreme dwarfism.

Jarisch–Herxheimer reaction Some hours after beginning treatment for syphilis with penicillin G 600 000 units daily more than 50 per cent of patients will develop pyrexia and rigors. This usually passes off in an hour or so but patients should be forewarned.

Jefferson's fracture A bursting fracture of the ring of the atlas. Most patients need stabilization in a collar and neurological damage is uncommon. However, if there is lateral shift suggesting rupture of the transverse ligament cervical fusion is indicated.

Jeffrey's sign (exophthalmos). In moderate exophthalmos there is absent wrinkling of the forehead when the patient, with the head bent downwards, looks upwards.

Kalokerinos' sign A filling defect in the gastric fundus which mimics a gastric neoplasm. It is produced by that portion of the fundus which is about to herniate and lies to the left of the cardio-oesophageal hiatus.

Kanavel's sign (infected ulnar bursa). Greatest tenderness is elicited between the transverse palmar creases on the ulnar side.

Kantor's string sign (Crohn's disease). Following involvement of the terminal ileum scarring may lead to constriction which can be demonstrated radiologically by barium follow-through. The terminal ileum appears as a thread-like structure—the 'string' sign of Kantor.

Kaposi's sarcoma Brown macules of the face, nose, eyelids and conjunctivae, are seen in male AIDS sufferers. They are often an early sign.

Kaposi's angiosarcoma (AIDS). Brown/bluish nodules of the limbs which may mimic melanoma.

Kehr's sign (ruptured spleen). Left shoulder-tip pain in association with splenic rupture. The pain is referred from irritation of the diaphragm by blood via afferent fibres of the phrenic nerve.

Kenaway's sign (bilharzial cirrhosis of liver). In the presence of splenomegaly associated with bilharzial cirrhosis of the liver a venous hum can be heard, louder on inspiration, on auscultation with the bell of the stethoscope held below the xiphisternum.

Keratoacanthoma (molluscum sebaceum). A lesion which occurs on the face or hands. It grows with crusting for 6–8 weeks, becoming a firm nodule which then spontaneously regresses. Correct treatment is by excision and microscopy to exclude the possibility of squamous cell carcinoma.

Kienbock's disease Avascular necrosis of the lunate. It may occur after trauma. The patient complains of a painful, stiff wrist. Treatment is by splinting to limit movement, excision, and replacement with a prosthesis or arthrodesis.

Klein's sign (mesenteric lymphadenitis). Pain in the right iliac fossa moves to the left of the original site when the patient lies on the left side.

N.B. This does not exclude inflammation in a Meckel's diverticulum.

'Kiss' cancer A cancer implanted in one area by local contact from another affected site, e.g. cancer of the lip, vulval labium, etc.

Kohler's disease Osteochrondritis of the navicular bone. It is a cause of a painful limp in children under 5 years. Treatment is by reducing activity and applying strapping for several weeks.

Krukenberg's tumour An ovarian tumour secondary to transcoelomic spread from a primary gastric cancer.

Ladd's band Failure of the caecum to rotate to its correct position in the right iliac fossa leads to it lying in the upper right quadrant. Its fixation to the abdominal wall leaves bands lying across the duodenum. They can lead to duodenal obstruction and volvulus of the midgut, due to its narrow superior mesenteric pedicle.

Leriche's syndrome Impotence secondary to disease of the aortic bifurcation or internal iliac arteries.

Linitis plastica (leather bottle stomach). Submucosal proliferation of fibrous tissue secondary to carcinoma of the stomach leads to thickening of the gastric wall. Biopsy often fails to reveal any evidence of malignancy. The condition may affect the whole stomach or the pyloric antrum only. The prognosis is poor.

Little's area A plexus of veins on the lower part of the nasal septum which is the commonest site of nose-bleeding and may require cauterization.

Littre's hernia A hernia, the sac of which contains a Meckel's diverticulum.

Lupus vulgaris Tuberculosis of the skin. The face is the commonest site. Treatment is by antituberculous drugs.

McBurney's point The point of maximum tenderness in acute appendicitis. It lies one third of the way along a line drawn from the anterior superior iliac spine to the umbilicus. It is also the point over which the grid-iron incision should be centred.

McMurray's test (medial mensicus tear). The patient lies supine with the knee flexed. The examiner grasps the knee with one hand and the foot with the other. The foot is rotated medially or laterally whilst the knee is brought to a right angle. A click is felt or heard in the knee when the test is positive. The patient may also experience discomfort in the knee similar to episodes of 'the knee giving way'.

Madelung's deformity There is dorsal subluxation of the lower end of the ulna. The condition may be congenital but is more likely to be traumatic. Patients are 10–18 years. Treatment is by excision of the lower end of the ulna or radial osteotomy.

Madura foot (mycetoma pedis). A chronic fungal infection of the foot which results in multiple sinuses and resembles actinomycosis but does not disseminate. The infecting agent is present in dust in tropical countries. Treatment with broad spectrum antibiotics eliminates secondary infection. Dapsone (100 mg twice daily) deals specifically with the fungus.

Maisonneuve's fracture Medial malleolar fracture, diastasis of the inferior tibiofibular joint, and spiral fracture of the neck of the fibula.

Malgaigne's bulgings Bulges seen above the inguinal ligaments in thin individuals on coughing or straining. They are a normal variant and are not inguinal hernias.

Mallory–Weiss syndrome A longitudinal tear in the gastric mucosa near the cardia which leads to haematemesis. The history is usually of violent vomiting following alcoholic intoxication. Most patients respond to conservative treatment but laparotomy and oversewing is occasionally necessary.

Marfan's syndrome An inherited disorder in which dissecting aneurysm of the thoracic aorta or aortic valve incompetence due to aortic root dilatation are found. These are associated with arachnodactyly (spider fingers), excessive height (Abraham Lincoln and Mary Queen of Scots were believed to be sufferers), subluxation of the lens of the eye, and anomalies of the ear and skeleton.

Marion's disease (infants). Bladder-neck obstruction analagous to hypertrophic pyloric stenosis. Treatment is by V-Y plasty.

Marjolin's ulcer Squamous carcinoma developing in a scar.

Meig's syndrome Ascites and left-sided pleural effusions related to the presence of an ovarian fibroma. Excision of the tumour leads to complete resolution.

Meleney's synergistic bacterial gangrene Postoperative gangrene, often idiopathic, which develops secondary to cellulitis. The causal organisms are aerobic and anaerobic bacteria. Treatment is with a cephalosporin, metronidazole, and complete excision of the gangrenous tissue.

Menetrier's disease Hypertrophy of the gastric fundal mucosa, of unknown cause, associated with protein loss and the development of gastric carcinoma. Treatment is by partial gastrectomy.

Meralgia paraesthetica A compression syndrome of the lateral cutaneous nerve of the thigh as it passes through the inguinal ligament. There is hyperaesthesia in the lateral aspect of the thigh, worse on sitting. The syndrome was common in gunslingers (caused by the position of the belt of their holsters) and more recently has been described in gymnasts (caused by flexion of the hips against parallel bars). It may also present as a complication of surgery in the region of the inguinal ligament. Treatment is by block or excision of part of the nerve.

Mikulicz disease A variant of Sjogren's syndrome. There is symmetrical enlargement of all the salivary glands, the lacrimal glands, and dryness of the mouth.

Mill's manoeuvre (tennis elbow). Pronation of the forearm with the elbow held straight produces pain over the common extensor origin (if positive).

Milroy's disease Congenital lymphoedema due to hypoplasia, aplasia, or varicosities of the lymphatic vessels.

Molluscum sebaceum See keratoacanthoma.

Mondor's disease Superficial thrombophlebitis of the veins of the anterior chest wall, the breast, and the axilla. The cause is unknown and the condition self-limiting.

Mucormycosis This is infection with fungi of the Mucoraceae group—*Rhizopus*, *Mucor*, or *Absidia*. They are a single multinucleate cell which forms the entire fungal mass.
Source: Ubiquitous in soil, decaying fruit, and vegetables.
Pathogenicity: Opportunistic pathogens which infect debilitated, immunocompromized, or uncontrolled diabetic patients and lead to localized sepsis and gangrene.
Features: the rhinocerebral form is the commonest, originating in the nose and nasal sinuses.
Treatment: Excision of gangrenous tissue. Amphotericin B.
Prognosis: Poor. Mortality rate is high.

Murphy's sign Gently palpate the right upper quadrant whilst asking the patient to breathe in. Inspiration pushes the liver (and gall bladder) downwards. The patient feels pain as the inflamed gall bladder comes into contact with the area of the abdominal wall under palpation and cannot complete inspiration.

Myositis ossificans Calcification with subsequent ectopic bone formation occurring in a haematoma (usually of skeletal muscle). It is common after injuries around the elbow and the quadriceps femoris.

Nail–Patella syndrome An inherited disorder associated with subluxation of the radial head. The patient has characteristically small patellae and deformed or absent nails.

Ochsner–Scherren regimen (appendix mass). In the presence of an appendix mass conservative management with IV fluids, antibiotics, and observation of vital functions is undertaken and surgery is postponed until the mass settles, when the appendix is more easily removed. If the patient's condition deteriorates, surgery is indicated.

Orf A self-limiting viral infection of the hands contracted from the saliva of sheep. An example of zoonosis.

Ortolani's test (congenital dislocation of the hip). The infant lies on a firm surface. The hips are abducted and flexed with the examining index fingers placed over the trochanters and the thumbs over the hip joints. Pressure is then applied over the trochanters. In a positive sign a 'click' is felt as the femoral head relocates in the acetabulum.

Osgood–Schlatter's disease Osteochondritis of the tibial tubercle.

Osteitis deformans Paget's disease of bone.

Osteitis fibrosia cystica The bone changes of hyperparathyroidism.

Osteogenesis imperfecta (brittle bone disease, fragilitas ossium). An inherited disease (autosomal dominant) caused by defective collagen throughout the body. The cortices of the long bones are thin and the trabeculae of cancellous bone are sparse. The condition predisposes to fractures which may occur soon after birth and with minimal trauma. Some children have more than 100 fractures but the rate slows down around puberty. Associated features are blue sclerae (not always present), inguinal hernia, deafness, and occasionally hypopituitarism.

Osteopetrosis (Albers–Schonberg disease, marble bone disease, osteosclerosis fragilis generalisata). An inherited disorder which occurs in severe (autosomal dominant) and benign forms (autosomal recessive) characterized by generalized increase in bone density, multiple fractures, and anaemia.

Osteopoikilosis (spotted bones). An incidental radiological finding of no significance. There are areas of sclerosis in the long bones of infants associated with yellowish skin lesions (dermatofibrosis lenticularis disseminata). Inheritance is by autosomal dominant.

Oxycephaly Premature fusion of the sutures may lead to an egg-shaped skull. Patients are prone to develop increased intracranial pressure.

Paget's disease of the nipple An eczematous-like condition of the nipple secondary to an intraductal carcinoma of breast. Treatment is by mastectomy. The prognosis is good if there is no palpable lump.

Painful arc (supraspinatus tendonitis). Patients experience pain during the middle third of abduction of the arm as the inflamed portion of the supraspinatus tendon passes between the acromion process and the head of the humerus.

Pancoast tumour (bronchial carcinoma). This is a peripheral carcinoma of the lung apex which invades the brachial plexus, sympathetic chain, ribs, and vertebrae. Such lesions lead to intractable pain, brachial plexus lesions, and Horner's syndrome.

Paterson–Brown Kelly syndrome (Plummer–Vinson syndrome). Iron-deficiency anaemia in a middle-aged female associated with dysphagia high in the oesophagus due to proximal web formation. Associations are achlorhydria, koilonychia, angular cheilosis, and a smooth, pale tongue. Treatment is by oesophageal dilatation and iron replacement therapy.

Paronychia A subcuticular infection of the nail bed. Treatment is with broad-spectrum antibiotics in the early stages. If there is pus it must be drained, usually through a wedge incision at the nail margin or base.

Peau d'orange (carcinoma of the breast). Dimpling of the skin due to involvement from an underlying scirrhous carcinoma. This is a sign of locally advanced disease.

Pendred's syndrome Thyroid goitre associated with congenital deafness—a form of genetically determined dyshormonogenesis.

Peritoneoscopy (laparoscopy). A method of examination of the intra-abdominal contents. The peritoneoscope consists of a trocar, cannula and telescope. The trocar and cannula are introduced into the peritoneal cavity under general anaesthetic. CO_2

gas (1–1.5 l) is then insufflated and the telescope introduced via a separate incision. The patient is then tilted to permit examination of the organs. Target biopsy of organs is an advantage of the technique.

Peutz–Jegher syndrome Melanosis of the lips and mucous membrane of the mouth associated with jejunal polyposis. Complications include haemorrhage and intussusception.

Plummer–Vinson syndrome The Paterson–Brown Kelly syndrome.

Pneumatosis cystoides intestinalis Asymptomatic (usually) gas-filled cysts of the intestinal wall or mesentery. The small intestine is most commonly affected. Gas (usually nitrogen) enters the bowel wall through a tear in the mucosa and is propelled by peristalsis. The cysts usually occur in clusters and are seen on plain radiography.

Pneumaturia The passage of flatus in the urine. A common cause is colovesical fistula secondary to diverticular disease of the colon.

Porphyria A hereditary disorder of haemoglobin breakdown which leads to porphyrinuria. Acute colicky abdominal pain which may be precipitated by barbiturates is a feature. A negative laparotomy should arouse suspicion. Diagnosis is by specific examination of the urine for porphyrinuria.

Pott's disease Tuberculosis of the spine. The vertebral bodies, usually one or two, are involved initially. Abscess formation, destruction of bone and vertebral collapse are characteristic. Pus may track along the psoas fascia to present as a lump or discharging abscess in the groin. Angulation of the spine follows collapse and paraplegia may supervene in advanced disease. Treatment is drainage of the abscess, spinal stabilization, and antituberculous drugs.

Proctalgia fugax Severe, recurrent rectal pain in the absence of organic disease. Attacks may occur at night, after evacuation of the bowels or following ejaculation. Anxiety is said to be an associated feature. There is no treatment but anxiolytics may help. The patient should be reassured. Fortunately attacks are short-lived.

Pseudomyxoma peritonei A rare condition, more common in women, which follows rupture of a pseudomucinous ovarian cystadenoma or a mucocoele of the appendix or gall bladder. Clinical features are abdominal discomfort and distension. At laparotomy the peritoneal cavity contains large quantities of jelly-like material. This and the underlying ovarian cyst or appendix should be removed. Although locally malignant and recurrent, metastases are not a feature.

Ranula An extravasation cyst of a sublingual gland containing saliva. Ranulas occur on the floor of the mouth presenting as soft bluish submucosal cysts. Treatment is by excision.

Raspberry tumour A tender granulomatous mass arising from the posterior urethral meatus which bleeds easily. Treatment is by excision.

Regional enteritis Crohn's disease.

Redcurrant jelly stool The description attributed to the blood-stained stool of intussusception.

Reiter's syndrome Venereally acquired urethritis, conjunctivitis, and arthritis, sometimes associated with changes in the skin of the foot (keratoderma blennorrhagicum). Most patients settle spontaneously in 4–6 weeks. Recurrent attacks may occur. Treatment is supportive with mydriatics and local steroids for iritis. Other causes of urethritis should always be excluded (e.g. gonococcus).

Reye's syndrome Affects children between 2–18 months usually after a minor illness. It produces encephalopathy and fatty degeneration of the viscera. The cause is unknown.
Treatment: IV glucose 10 per cent at 1–2 l/M 24 h. Positive pressure respiration. Mannitol 2 g/kg IV over 20 mins.
Prognosis: Mortality rate is 40 per cent.

Sabre-tibia (late syphilis). Bowing of the tibia caused by subperiosteal new bone formation, the result of syphilitic periostitis.

Saint's triad Cholelithiasis, colonic diverticular disease, and hiatus hernia often co-exist. If one condition is diagnosed the others should be considered and investigated if there are symptoms.

Scheuermann's disease (adolescent kyphosis). Osteochondritis affecting the bodies of thoracic vertebrae 6–10 (commonly) and the lumbar vertebral bodies. The bodies become narrowed anteriorly and a smooth rigid kyphosis develops.

Schmorl's node Extrusion of the nucleus pulposus of an intervertebral disc into a vertebral body may be seen radiologically as a lucent area surrounded by new bone formation.

Sever's disease Osteochrondritis of the posterior epiphysis of the os calcis near the site of insertion of the Achilles tendon.

Shenton's line (congenital dislocation of the hip). This line is drawn from the lesser trochanter of the femur upwards along the medial border of the femur and inferior border of the superior pubic ramus. The line is a continuous curve, broken in cases of congenital dislocation of the hip.

Shieie's syndrome An autosomal recessive inherited disorder of the skeleton. Clinical characteristics include clouding of the cornea, cardiac anomalies, and epiphyseal dysplasia. Life expectation is normal.

Shoveller's fracture (clay shoveller's fracture). A stable cervical injury in which direct trauma or indirect muscular contra action leads to a fracture of the spinous process of C7. Treatment is by providing pain relief and occasionally bed rest.

The 'sick-cell' syndrome (hypovolaemic shock). As a result of poor tissue perfusion due to low cardiac output the cells lose their ability to actively pump sodium out. Thus sodium remains within the cell and potassium leaks out and the serum potassium rises. Calcium also enters the cell leading to hypocalcaemia. Ultimately further cellular damage is mediated as enzymes are released from intracellular lysosomes.

Sinding–Larsen's disease Osteochrondritis of the distal pole of the patella.

Sister Joseph's nodule A sign of advanced intra-abdominal carcinoma (usually gastric). The nodule appears at the umbilicus. It may also result from colonic, ovarian, or even breast cancer.

Swan–Ganz catheter A balloon-tipped flotation catheter used to obtain the pulmonary capillary wedge pressure (PCWP) which is an indicator of left ventricular function and circulating blood volume. Cannulation is carried out under aseptic conditions as for insertion of a central venous line. The catheter is introduced into the pulmonary artery through the right atrium and ventricle. The position is checked radiographically and the readings recorded by a transducer connected to the catheter which should be removed within 72 hours. The technique is valuable in fluid therapy and can also be used to measure cardiac output by a thermodilution method.

Thalassaemia An hereditary disorder of haemoglobin synthesis affecting the alpha, beta (commoner), or gamma peptide chains. Patients suffer from a reduced haemoglobin level and premature red cell breakdown leading to anaemia, jaundice, and splenomegaly. Inheritance is by a dominant trait. Grades of the disease vary depending on whether the patient is hetero- or homozygous. The diagnosis is by red cell electrophoresis. Treatment involves blood transfusion and occasionally splenectomy.

Thrombophlebitis migrans Increased coagulability of the blood is a feature of some visceral cancers (especially pancreas). The patient may develop pulmonary emboli which may be both the presenting feature and the terminal event.

Tietze's disease Painful swelling of the second or third costal cartilage. The cause is unknown and X-rays are negative. Treatment involves reassurance and local heat treatment. Persistent symptoms may respond to steroid injections.

Treacher–Collins syndrome (Mandibular-facial dysostosis). The inheritance is sporadic. One variant is autosomal recessive. There are facial deformities—malar, mandibular hypoplasia with malformed ears, and conductive deafness. Characteristic limb abnormalities are absence of the thumbs and radius. Associated anomalies of the heart and vertebrae are common.

Trench foot Moist gangrene due to prolonged immersion of the foot in water associated with tight-fitting footwear and muscular inactivity. Treatment involves gradual rewarming, hyperbaric oxygen, control of infection, and excision of dead tissue.

Trichobezoar (rare) A hair ball of the stomach. Women are most often affected and most have a psychiatric history. Treatment is by laparotomy, gastrostomy, and removal.

Troisier's sign Enlargement of the left supraclavicular lymph nodes due to metastatic carcinoma of stomach. A sign of advanced disease.

Trousseau's sign Phlebothrombosis of the superficial leg veins associated with gastric cancer.

Umbolith An umbilical concretion of desquamated skin which is dark in colour. It can lead to infection. Treatment is by removing the calculus and advising the patient to keep the area scrupulously clean. Recurrence may need to be treated by excision of the umbilicus.

Ureterocoele Cystic dilatation of the ureter due to congenital atresia of the ureteric orifice. Treatment is indicated only if symptoms develop, e.g. infection, stone formation. Cystoscopic cauterization through the wall of the cyst will allow drainage into the bladder.

Urachus A patent urachus results from persistence of the allantois leading to a *congenital urinary fistula* (rare). *A urachal cyst* (rare) may develop when the umbilical end of the urachus persists. This becomes an *umbilical abscess* when infected. Treatment is by drainage of the abscess and excision of the persistent remnants when the infection has settled.

Vermooten's sign (intrapelvic rupture of the prostatic urethra). Digital examination of the rectum reveals a doughy, displaced, or absent prostate.

Vincent's angina Infection around the tonsillar crypts which spreads to cause gingival ulceration and bleeding. The causal organisms, *Borrelia vincentii* (a spirochaete) and *Fusiformis fusiformis* (rod-shaped) are Gram-negative anaerobes which are normal mouth commensals. In specific circumstances they become opportunistic pathogens infection being more frequent in those already ill or malnourished. Treatment is local dental intervention and antibiotics (penicillin, metronidazole).

Von Recklinghausen's disease (neurofibromatosis). An autosomal dominantly inherited disease of nerves which leads to nodular and diffuse thickening of nerve trunks. There is associated skin pigmentation (café-au-lait colouration). Complications may arise from local pressure by the neuromata in enclosed areas (e.g. chest) or from sarcomatous change. The name is also attached to the osteitis fibrosa cystica of hyperparathyroidism.

Von-Rosen's sign (congenital dislocation of the hip). The click elicited when the hip is flexed, adducted, then abducted. The head of the femur dislocates and reduces with a clunk as the procedure is performed and reversed.

Von-Willebrand's disease A haemorrhagic disease associated with low levels of Factor VIII.

Waldenstrom's disease (slipped upper femoral epiphysis). Necrosis of the articular cartilage of the head of the femur following a slipped upper femoral epiphysis. The condition is commoner in African races. Stiffness or complete loss of movement may result. Treatment is supportive, mainly directed at achieving a position of maximum function.

Waterhouse–Friderichsen syndrome   Bilateral adrenocortical necrosis due to septicaemia (meningococcal, pneumococcal, staphylococcal, streptococcal) haemorrhage or burns. Treatment is the correction of shock, IV broad-spectrum antibiotics, IV hydrocortisone (up to 400 mg/24 h), oxygen therapy, and measurement of urine output.

Whipple's triad   (insulinoma). Fasting hypoglycaemic attacks (1) with a blood glucose of less than 2.5 mmol/l (2) which is relieved by glucose (3). Laboratory confirmation is measurement of the fasting blood sugar (during the attack) and the insulin level (by radioimmunoassay).

Zollinger–Ellison syndrome   Recurrent peptic ulceration, sometimes multiple, associated with gastric mucosal hypertrophy and the presence of a non-B-islet-cell gastrin-secreting tumour of the pancreas (commoner). Treatment is by partial gastrectomy to remove the antrum and excision of the tumour if it can be located. Some patients respond well to H_2 receptor antagonists.

Useful addresses

Financial help

The National Society for Cancer Relief
Michael Sobell House
30 Dorset Square
London NW1 6QL   Tel: 01-402 8125

Patients' self-help groups

CARE (Cancer Aftercare and Rehabilitation Education)
Lodge Cottage
Church Lane
Tinsbury
Bath
Somerset   Tel: 0761-70731

CLAPA (Cleft Lip and Palate Association)
Tel: 01 405 9200 Ext. 256

Colostomy Welfare Group
38–9 Eccleston Square
2nd Floor
London SW1V 1PB   Tel: 01-828 5175

The National Association of Laryngectomee Clubs
Fourth Floor
Michael Sobell House
30 Dorset Square
London NW1 6QL   Tel: 01-402 6007

Mastectomy Association
25 Brighton Road
South Croydon
Surrey CR2 6EA   Tel: 01-654 8643

Organizations for the bereaved

CRUSE
126 Sheen Road
Richmond
Surrey TW9 1CR   Tel: 01-940 4818/9047

Gingerbread
35 Wellington Street
London WC2   Tel: 01-240 0953

The National Association of Widows
Stafford District Voluntary Service Centre
Chell Road
Stafford ST16 2QA   Tel: 0785-45465

The Society of Compassionate Friends
50 Woodwaye
Watford
Hertfordshire   Tel: 0923-24279

Other relevant organizations

The British Assocation of Cancer United Patients and their
   Families (BACUP)
121/121 Charterhouse Street
London EC1M 6AA

The British Organ Donor Society
Balsham
Cambridge CB1 6DL   Tel: (0223) 893636

British Red Cross
9 Grosvenor Crescent
London SW1X 7EJ   Tel: 01-235 5454

Hospice Information
St Christopher's Hospice
51–3 Lawrie Park Road
Sydenham
London SE26 6DZ   Tel: 01-778 1240

The Samaritans
17 Uxbridge Road
Slough SL1 1SN   Tel: 0753-32713/4
(Details of local branches in the telephone directory)

WRVS (Women's Royal Voluntary Service)
17 Old Park Lane
London W1Y 4AJ   Tel: 01-499 6040

Main sources

Blandy, J. (1988). *Lecture notes on urology* (3rd edn). Blackwell Scientific Publications, Oxford.

Decosse, J. J. and Todd, I. P. (1984). *Anorectal surgery*. Churchill Livingstone, Edinburgh.

Forrest, A. P. M., Carter, D. C., and McLeod, I. B. (1985). *Principles and practice of surgery*. Churchill Livingstone, Edinburgh.

Freidin, J. and Marshall, V. (1984). *Illustrated guide to surgical practice*. Churchill Livingstone, Edinburgh.

Harding Rains, A. J. and Mann, Charles V. (1988). *Bailey and Love's short practice of surgery*. H. K. Lewis, London.

Kester, R. and Leveson, S. H. (1981). *A practice of vascular surgery*. Pitman, London.

McGregor, I. A. (1983). *Fundamental techniques of plastic surgery* (7th edn). Churchill Livingstone, Edinburgh.

MacMahon, R. A. (1984). *An aid to paediatric surgery*. Churchill Livingstone, Edinburgh.

Moffat, D. B. (1987). *Lecture notes on anatomy*. Blackwell Scientific Publications, Oxford.

Russell, R. C. G. (1986, 1988). *Recent advances in surgery*, Volumes 12 and 13. Churchill Livingstone, Edinburgh.

Index

चे० टी० सू

नज़्मी